Photonic Neural Networks with Spatiotemporal Dynamics

Hideyuki Suzuki · Jun Tanida ·
Masanori Hashimoto

Editors

Photonic Neural Networks with Spatiotemporal Dynamics

Paradigms of Computing and Implementation

 Springer

Editors
Hideyuki Suzuki
Graduate School of Information Science
and Technology
Osaka University
Osaka, Japan

Jun Tanida
Graduate School of Information Science
and Technology
Osaka University
Osaka, Japan

Masanori Hashimoto
Department of Informatics
Kyoto University
Kyoto, Japan

ISBN 978-981-99-5071-3   ISBN 978-981-99-5072-0   (eBook)
https://doi.org/10.1007/978-981-99-5072-0

This Springer imprint is published by the registered company Springer Nature Singapore Pte Ltd.
The registered company address is: 152 Beach Road, #21-01/04 Gateway East, Singapore 189721, Singapore

Paper in this product is recyclable.

Preface

Light is an excellent carrier of information with the ability to transmit signals at extraordinarily fast speeds, as exemplified by the global network of optical fibers. Moreover, its spatiotemporal properties suggest the capability to perform a variety of information processing and computation. Optical computing technologies have been demonstrated to uniquely achieve high-speed, parallel, and massive computation, which is becoming increasingly important in the era of the smart society. Accordingly, optical computing is expected to serve as a crucial foundation for future information technologies.

Artificial neural network models, such as those employed in deep learning, have become a fundamental technology in information processing. These models were originally inspired by biological neural networks in the brain, which achieve robust and highly advanced information processing through the spatiotemporal dynamics arising from a large number of unreliable elements, all while consuming remarkably low amounts of energy. The increasing demand for highly efficient computing technologies highlights the importance of further development of brain-inspired computing.

Given the potential of optical computing and brain-inspired computing, the development of photonic neural networks is considered promising. A number of attempts have already been made to develop photonic neural networks, through which it has become clear that photonic neural networks should not merely be photonic implementations of existing neural network models. Instead, photonic neural networks need to be developed as a fusion of optical computing and brain-inspired computing, where the spatiotemporal aspects of light and the spatiotemporal dynamics of neural networks are expected to play crucial roles.

This book presents an overview of recent advances in photonic neural networks with spatiotemporal dynamics. It particularly focuses on the results obtained in the research project "Computing Technology Based on Spatiotemporal Dynamics of Photonic Neural Networks" (grant number JPMJCR18K2), which is conducted from October 2018 to March 2024 in CREST Research Area "Technology for Computing Revolution for Society 5.0" of Japan Science and Technology Agency (JST).

The computing and implementation paradigms presented here are outcomes of interdisciplinary studies by collaborative researchers from the three fields of nonlinear mathematical science, information photonics, and integrated systems engineering. This book offers novel multidisciplinary viewpoints on photonic neural networks, illustrating recent advances in three types of computing methodologies: fluorescence energy transfer computing, spatial-photonic spin system, and photonic reservoir computing.

The book consists of four parts: The first part introduces the backgrounds of optical computing and neural network dynamics; the second part presents fluorescence energy transfer computing, a novel computing technology based on nanoscale networks of fluorescent particles; the third and fourth parts review the models and implementation of spatial photonic spin systems and photonic reservoir computing, respectively.

These contents can be beneficial to researchers in a broad range of fields, including information science, mathematical science, applied physics, and engineering, to better understand the novel computing concepts of photonic neural networks with spatiotemporal dynamics.

This book would not have been possible without the invaluable contributions of the members of our project. We would like to thank the contributors for writing excellent chapters: Ángel López García-Arias, Yuichi Katori, Takuto Matsumoto, Masaki Nakagawa, Takahiro Nishimura, Yusuke Ogura, Jun Ohta, Kiyotaka Sasagawa, Suguru Shimomura, Ryo Shirai, Sho Shirasaka, Michihisa Takeuchi, Masafumi Tanaka, Naoya Tate, Takashi Tokuda, Hiroshi Yamashita, and Jaehoon Yu. We especially appreciate the considerable assistance from Hiroshi Yamashita in coordinating the manuscripts toward publication. We also thank Ken-ichi Okubo and Naoki Watamura for their contributions in reviewing manuscripts.

We would like to express our gratitude to Prof. Shuichi Sakai, the supervisor of JST CREST Research Area "Technology for Computing Revolution for Society 5.0," for his farsighted advice and encouragement to our project. We would also like to extend our appreciation to the area advisors for their insightful comments: Shigeru Chiba, Shinya Fushimi, Yoshihiko Horio, Michiko Inoue, Toru Shimizu, Shinji Sumimoto, Seiichiro Tani, Yaoko Nakagawa, Naoki Nishi, and Hayato Yamana. We also thank the editorial office of Springer for the opportunity to publish this book.

Osaka, Japan   Hideyuki Suzuki
Osaka, Japan   Jun Tanida
Kyoto, Japan   Masanori Hashimoto
June 2023

Contents

Introduction

Revival of Optical Computing

Jun Tanida

Abstract Optical computing is a general term for high-performance computing technologies that effectively use the physical properties of light. With the rapid development of electronics, its superiority as a high-performance computing technology has diminished; however, there is momentum for research on new optical computing. This study reviews the history of optical computing, clarifies its diversity, and provides suggestions for new developments. Among the methods proposed thus far, those considered useful for utilizing optical technology in information systems are introduced. Subsequently, the significance of optical computing in the modern context is considered and directions for future development is presented.

1 Introduction

Optical computing is a general term for high-performance computing technologies that make effective use of the physical properties of light; it is also used as the name of a research area that attracted attention from the 1980s to the 1990s. This was expected to be a solution for image processing and large-capacity information processing problems that could not be solved by electronics at that time, and a wide range of research was conducted, from computing principles to device development and architecture design. Unfortunately, with the rapid development of electronics, its superiority as a high-performance computing technology was lost, and the boom subsided. However, as seen in the recent boom in artificial intelligence (AI), technological development has repeated itself. In the case of AI, the success of deep learning has triggered another boom. It should be noted that technologies with comparable potential for further development have been developed in the field of optical computing. Consequently, the momentum for research on new optical computing is increasing.

J. Tanida (✉)
Osaka University, Graduate School of Information Science and Technology, 1-5 Yamadaoka, Osaka 565-0871, Japan
e-mail: tanida@ist.osaka-u.ac.jp

© The Author(s) 2024
H. Suzuki et al. (eds.), *Photonic Neural Networks with Spatiotemporal Dynamics*,
https://doi.org/10.1007/978-981-99-5072-0_1

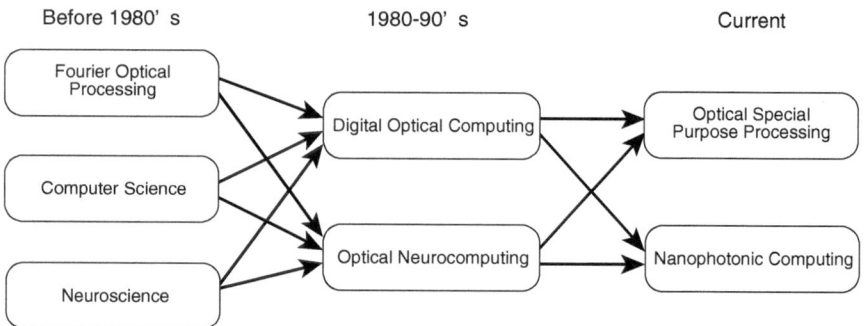

Fig. 1 Relationship between research areas related to optical computing. Fourier optical processing is a root of optical computing in the 1980s and the 1990s. Computer science and neuroscience accelerated the research fields of optical digital computing and optical neurocomputing. Currently, both research fields have been extended as optical special purpose processing and nanophotonic computing

Optical computing is not a technology that emerged suddenly. Figure 1 shows the relationship among the research areas related to optical computing. Its roots are *optical information processing*, represented by a Fourier transform using lenses, and technologies such as spatial filtering and linear transform processing. They are based on the physical properties of light waves propagating in free space, where the superposition of light waves is a fundamental principle. Light waves emitted from different sources in space propagate independently and reach different locations. In addition, light waves of different wavelengths propagate independently and can be separated. Thus, the light-wave propagation phenomenon has inherent parallelism along both the spatial and wavelength axes. Utilizing this property, large-capacity and high-speed signal processing can be realized. In particular, optical information processing has been researched as a suitable method for image processing. From this perspective, optical information processing can be considered an information technology specialized for image-related processing.

Optical computing in the 1980s and the 1990s was characterized by the realization of general-purpose processing represented by *digital optical computing*. Compared with previous computers, the development of the *Tse computer* [1], which processes two-dimensional signals such as images in parallel, was a turning point. In the Tse computer, the concept of parallelizing logical gates that compose a computer to process images at high speed was presented. Inspired by this study, various parallel logic gates based on nonlinear optical phenomena have been developed. The development of optical computers composed of parallel logic gates has become a major challenge. Several computational principles and system architectures were proposed. Two international conferences on *Optical Computing* (OC) and *Photonic Switching* (PS) were started and held simultaneously. Notably, the proposal of *optical neurocomputing* based on neural networks was one of the major achievements in optical computing research during this period.

At that time, optical computing showed considerable potential; however, the development boom eventually waned because of the immaturity of peripheral technologies that could support ideas, as well as the improvement of computational performance supported by the continuous development of electronics. Nevertheless, the importance of light in information technology has increased with the rapid development of *optical communication*. In addition, optical functional elements and *spatial light modulators*, which have been proposed as devices for optical computing, continue to be developed and applied in various fields. Consequently, the development of optical computing technology in a wider range of information fields has progressed, and a research area called *information photonics* has been formed. The first international conference on *Information Photonics* (IP) was held in Charlotte, North Carolina, in June 2005 [2] and has continued to date.

Currently, optical computing is in the spotlight again. Optical computing is expected to play a major role in *artificial intelligence* (AI) technology, which is rapidly gaining popularity owing to the success of deep learning. Interesting methods, such as AI processors using nonphotonic circuits [3] and optical deep networks using multilayer optical interconnections [4], have been proposed. In addition, as a variation of neural networks, a *reservoir model* suitable for physical implementation was proposed [5], which increased the feasibility of optical computing. Reservoir implementation by Förster resonant energy transfer (FRET) between quantum dots was developed as a low-energy information processor [6]. In addition to AI computation, optical technology is considered a promising implementation of quantum computation. Owing to the potential capabilities of the optics and photonics technologies, optical computing is a promising solution for a wide range of high-performance computations.

This study reviews the history of optical computing, clarifies its diversity, and provides suggestions for future developments. Many methods have been developed for optical computing; however, only a few have survived to the present day, for various reasons. Here, the methods proposed thus far that are considered useful for utilizing optical technology in information technology are introduced. In addition, the significance of optical computing in the modern context is clarified, and directions for future development are indicated. This study covers a part of works based on my personal opinion and is not a complete review of optical computing as a whole.

2 Fourier Optical Processing

Shortly after the invention of lasers in the 1960s, a series of optical information-processing technologies was proposed. Light, which is a type of electromagnetic wave, has wave properties. Interference and diffraction are theoretically explained as basic phenomena of light; however, they are based on the picture of light as an ideal sinusoidal wave. Lasers have achieved stable light-wave generation and made it possible to perform various types of signal processing operations by superposing the amplitudes of light waves. For example, the Fraunhofer diffraction, which describes the diffraction phenomenon observed at a distance from the aperture, is equivalent to a two-dimensional Fourier transform. Therefore, considering that an infinite point is equivalent to the focal position of the lens, a Fourier transform using a lens can be implemented with a simple optical setup.

Figure 2 shows a *spatial filtering* system as a typical example of optical information processing based on the optical Fourier transform. This optical system is known as a telecentric or double diffraction system. For the input function $f(x, y)$ given as the transmittance distribution, the result of the convolution with the point spread function $h(x, y)$, whose Fourier transform $H(\mu, \nu)$ is embodied as a filter with the transmittance distribution at the filter plane, is obtained as the amplitude distribution $g(x, y)$ as follows:

$$g(x, y) = \iint f(x', y')h(x - x', y - y')\mathrm{d}x'\mathrm{d}y'. \tag{1}$$

The following relationship is obtained using the Fourier transforms of Eq. (1), which is performed at the filter plane of the optical system:

$$G(\mu, \nu) = F(\mu, \nu)H(\mu, \nu) \tag{2}$$

where $F(\mu, \nu)$ and $G(\mu, \nu)$ are the spectra of the input and output signals, respectively. The filter function $H(\mu, \nu)$ is called the *optical transfer function*. The signal processing and filter operations can be flexibly designed and performed in parallel with the speed of light propagation. The Fourier transform of a lens is used effectively in the optical system.

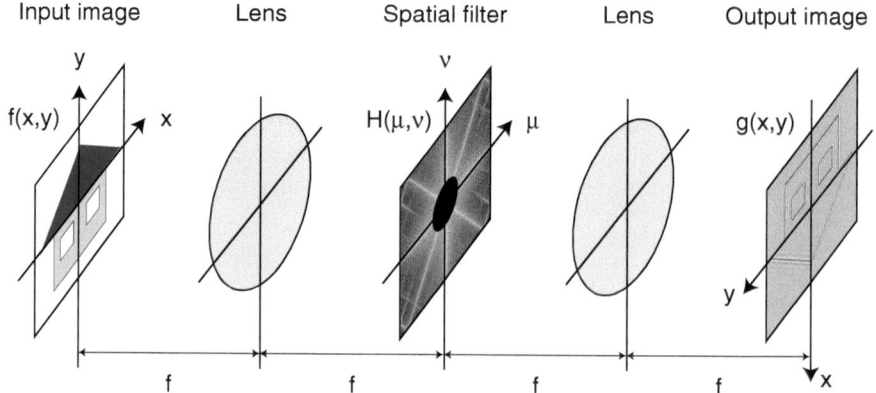

Fig. 2 Spatial filtering in optical systems. The input image $f(x, y)$ is Fourier transformed by the first lens, and the spectrum is obtained at the filter plane. The spatial filter $H(\mu, \nu)$ modulates the spectrum and the modulated signal is Fourier transformed by the second lens. The output image $g(x, y)$ is observed as an inverted image because the Fourier transform is performed instead of an inverse Fourier transform

In this optical system, because the input and filter functions are given as transmittance distributions, *spatial light modulators*, such as liquid crystal light bulbs (LCLV) [7] and liquid crystals on silicon devices [8] which can dynamically modulate light signals on a two-dimensional plane are introduced. In the spatial light modulator, the amplitude or phase of the incident light over a two-dimensional plane is modulated by various nonlinear optical phenomena. Spatial light modulators are important key devices not only in Fourier optical processing but also in general optical computing. Additionally, they are currently being researched and developed as universal devices for a wide range of optical applications.

Pulse shaping has been proposed as a promising optical information technology [9]. As shown in Fig. 3, the optical setup comprises symmetrically arranged gratings and Fourier transform lenses. Because the Fourier time spectrum $F(\nu)$ of the incident light $f(t)$ is obtained at the spatial filter plane, the filter function located therein determines the modulation properties. Subsequently, the shaped output pulse $g(t)$ is obtained after passing through the optical system. Note that the grating and cylindrical lens perform a one-dimensional Fourier transform by dispersion, and that the series of processes is executed at the propagation speed of light.

Fourier optical processing is also used in current information technologies such as *image encryption* [10] and *biometric authentication* [11]. In addition, owing to their capability to process ultrahigh-speed optical pulses, signal processing in *photonic networks* is a promising application. Furthermore, in combination with holography, which fixes the phase information of light waves through interference, high-precision measurements and light-wave control can be achieved [12].

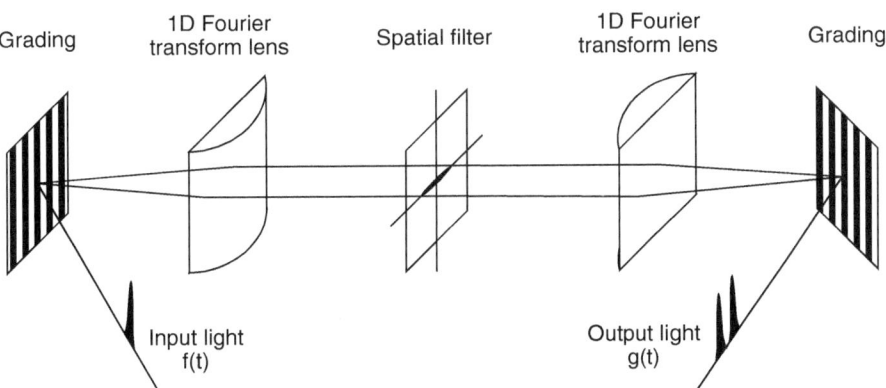

Fig. 3 Pulse shaping technique. The input light $f(t)$ is Fourier transformed in the time domain by the first grating and cylindrical lens. The spectrum is modulated by the filter and the modulated signal is Fourier transformed by the second cylindrical lens and grating to generate the output light $g(t)$

3 Digital Optical Computing

The significance of the Tse computer [1] is that it introduced the concept of digital arithmetic into optical computing. Optical information processing based on the optical Fourier transform is typical of optical analog arithmetic. Digital operations over analog operations are characterized by the elimination of noise accumulation, arbitrary computational precision, and moderate device accuracy requirements [13]. Although analog optical computing is suitable for specific processing, many researchers have proposed digital optical computing for general-purpose processing. The goal is to achieve computational processing of two-dimensional signals, such as figures and images, with high speed and efficiency.

A Tse computer consists of Tse devices that process two-dimensional information in parallel. Several prototype examples of the Tse devices are presented. In the prototype system, the parallel logic gate is composed of logic gate elements mounted on electronics, and the parallel logic gate is connected by fiber plates or optical fiber bundles. In contrast, using *free-space propagation* of light, two-dimensional information can be transmitted more easily. Various optical logic gate devices have been developed as free-space optical devices. ATT Bell Labs developed a series of optical logic gate devices called SEED [14] to implement an optical switching system. These developments directly contributed to the current optoelectronic devices and formed the foundation of photonic technologies for optical communication.

Digital operations are designed by a combination of logical operations. Logical operations on binary signals include logical AND, logical OR, exclusive logic, etc., and higher level functions such as flip-flops and adding circuits are realized by combining them. Logical operations are nonlinear, and any type of nonlinear processing is required for their implementation. Optoelectronic devices directly utilize nonlinear optical phenomena. In addition, by replacing nonlinear operations with a type of procedural process called *spatial coding*, a linear optical system can be used to perform logical operations.

The authors proposed a method for performing parallel logical operations using *optical shadow casting* [15]. As shown in Fig. 4, parallel logical operations between two-dimensional signals were demonstrated by the spatial coding of the input signals and multiple projections with an array of point light sources. Using two binary images as inputs, an encoded image was generated by replacing the combination of the corresponding pixel values with a spatial code that is set in an optical projection system. Subsequently, the optical signal provides the result of a logical operation through the decoding mask on the screen. The switching pattern of the array of point light sources enables all 16 types of logical operations, including logical AND and OR operations, as well as operations between adjacent pixels.

Parallel operations based on spatial codes have been extended to parallel-computing paradigms. *Symbolic substitution* [16, 17] is shown in Fig. 5 which illustrates the process in which a set of specific spatial codes is substituted with another set of spatial codes. Arbitrary parallel logical operations can be implemented using a substitution rule. In addition, a framework was proposed that extends parallel logic

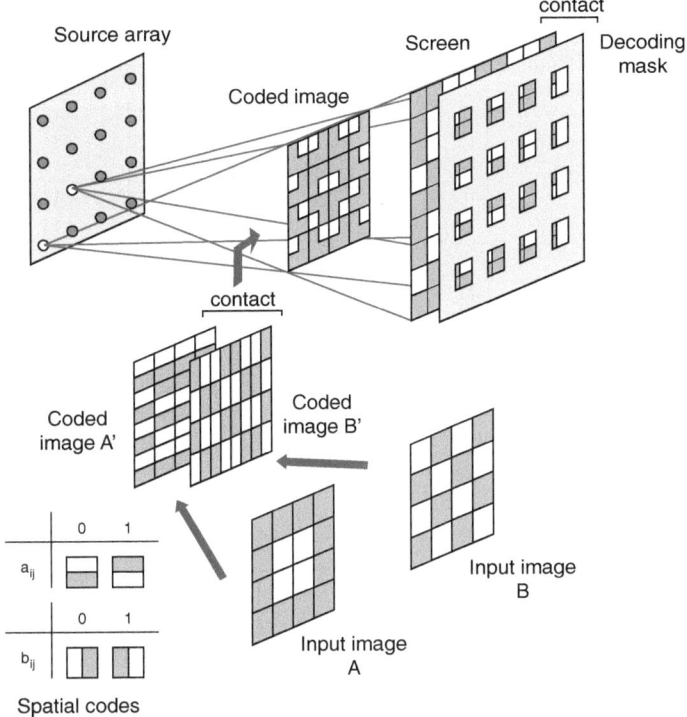

Fig. 4 Optical shadow casting for parallel logical operations. Two binary images are encoded pixel by pixel with spatial codes, and then, the coded image is placed in the optical projection system. Each point light source projects the coded image onto the screen and the optical signals provide the result of the logical operation through the decoding mask

operations using optical shadow casting to arbitrary parallel logic operations, called *optical array logic* [18, 19]. These methods are characterized by data fragments distributed on a two-dimensional image and a parallel operation on the image enables flexible processing of the distributed data in parallel. Similarly, *cellular automaton* [20] and *life games* [21] are mentioned as methods of processing data deployed on a two-dimensional plane. The parallel operations assumed in these paradigms can easily be implemented optically, resulting in the extension of gate-level parallelism to advanced data processing.

Construction of an optical computer is the ultimate goal of optical computing. The Tse computer is a clear target for the development of digital optical logical operations, and other interesting architectures have been proposed for optical computers. These architectures are categorized into *gate-oriented architectures*, in which parallel logic gates are connected by optical interconnections; *register-oriented architectures*, which combine processing on an image-by-image basis; and *interconnection-oriented architectures*, in which electronic processors are connected by optical interconnections. The gate-oriented architecture includes an optical sequence circuit pro-

Fig. 5 Symbolic
substitution. A specific
spatial pattern is substituted
with another spatial pattern
by the process of pattern
recognition and substitution.
A set of substitution rules
enables the system to
perform arbitrary parallel
logical operations such as
binary addition. Various
optical systems can be used
to implement the procedure

Jun Tanida

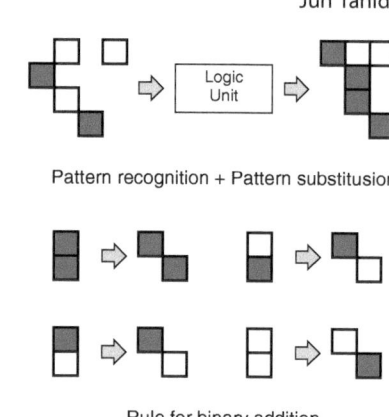

Pattern recognition + Pattern substitusion

Rule for binary addition

cessor [22] and a multistage interconnect processor [23]. The register-oriented architecture is exemplified by a Tse computer and parallel optical array logic system [24]. The interconnection-oriented architecture includes *optoelectronic computer using laser arrays with reconfiguration* (OCULAR) systems [25] and a three-dimensional optoelectronic stacked processor [26]. These systems were developed as small-scale systems with proof-of-principle, but unfortunately, few have been extended to practical applications.

4 Optical Neurocomputing

Digital optical logic operations were a distinctive research trend that distinguished optical computing in the 1980s from optical information processing. However, other important movements were underway during the same period. *Neurocomputing* was inspired by the neural networks of living organisms. As shown in Fig. 6, the neural network connects the processing nodes modeled on the neurons and realizes a computational task with signal processing over the network. Each node receives input signals from other processing units and sends an output signal to the successive processing node according to the response function. For example, when processing node i receives a signal from processing node j ($1 \le j \le N$), the following operations are performed at the node:

$$u_i = \sum_{j=1}^{N} w_{i,j} x_j \tag{3}$$

$$x_i' = f(u_i) \tag{4}$$

Fig. 6 Neural network.
Processing nodes modeled
on neurons of life organisms
are connected to form a
network structure. Each
processing node performs a
nonlinear operation for the
weighted sum of the input
signals and sends the output
signal to the next processing
node. Various forms of
networks can be made, and a
wide range of functionalities
can be realized

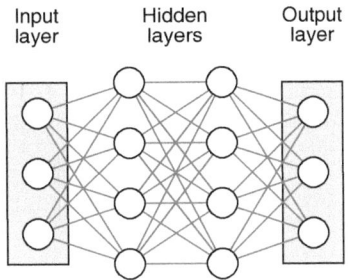

Multi-layer nerual network

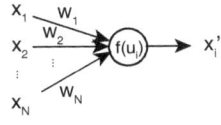

Processing node i

where u_i is the internal state signal of processing node i, $w_{i,j}$ is the connection weight from processing node j to node i, $f(\cdot)$ is the response function, and x_i' is the output signal of processing node i. All processing nodes perform the same operation, and the processing achieved by the entire network is determined by the connection topology of the processing node, connection weight of the links, and response function.

Therefore, neural networks are suitable for optical applications. Several same type of processing nodes must be operated in parallel and connected effectively, which can be achieved using parallel computing devices and free-space optical interconnection technology. Even if the characteristics of the individual processing nodes are diverse, this is not a serious problem because the connection weight compensates for this variation. In addition, if a free-space optical interconnection is adopted, physical wiring is not required, and the connection weight can be changed dynamically.

Anamorphic optics was proposed as an early optical neural network system to perform vector-matrix multiplication [27, 28]. The sum of the input signals for each processing node given by Eq. (3) was calculated using vector-matrix multiplication. As shown in Fig. 7, the input signals arranged in a straight line were stretched vertically and projected onto a two-dimensional image, expressing the connection weight as the transmittance. The transmitted signals were then collected horizontally and detected by a one-dimensional image sensor. Using this optical system, the weighted sum of the input signals to multiple processing nodes can be calculated simultaneously. For the obtained signal, a response function was applied to obtain the output signal sequence. If the output signals are fed back directly as the input signals of successive processing nodes, a Hopfield type neural network is constructed. In this system, the connection weight is assumed to be trained and is set as a two-dimensional image before optical computation.

Fig. 7 Anamorphic optical processor. Optical signals emitted from the LED array express vector **x**, and the transmittance of the two-dimensional filter corresponds to matrix H. The result of vector-matrix multiplication, **y** = H**x**, is obtained on the one-dimensional image censor. Cylindrical lenses are employed to stretch and collect the optical signals along one-dimensional direction

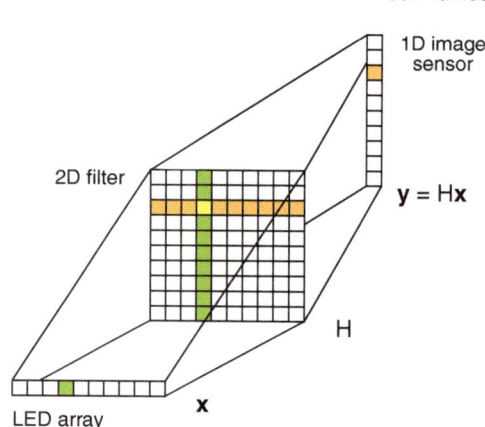

To implement a neural network, it is important to construct the required network effectively. An optical implementation based on volumetric holograms was proposed to address this problem. For example, a dynamic volumetric hologram using a non-linear optical crystal, such as lithium niobate, is an interesting method in which the properties of materials are effectively utilized for computation [29]. The application of nonlinear optical phenomena is a typical approach in optical computing, which is still an efficient method of natural computation.

Because of the recent success in deep neural networks, novel optical neuroprocessors have been proposed. One is the diffractive deep neural network (D^{2}NN) [4] which implements a multilayer neural network by stacking diffractive optical elements in multiple stages. The other is the optical integrated circuit processor [3] which is composed of beamsplitters and phase shifters by optical waveguides and efficiently performs vector-matrix multiplications. This is a dedicated processor specialized in AI calculations. These achievements are positioned as important movements in optical computing linked to the current AI boom.

Reservoir computing has been proposed as an approach opposite to multi-layered and complex networks [30]. This is a neural network model composed of a reservoir layer in which processing nodes are randomly connected and input and output nodes to the reservoir layer, as shown in Fig. 8. Computational performance equivalent to that of a recurrent neural network (RNN) can be achieved by learning only the connection weight from the reservoir layer to the output node. Owing to unnecessary changes in all connection weights, the neural network model is suitable for physical implementation. Optical fiber rings [31], lasers with optical feedback and injection [32], and optical iterative function systems [33] have also been proposed for the implementation.

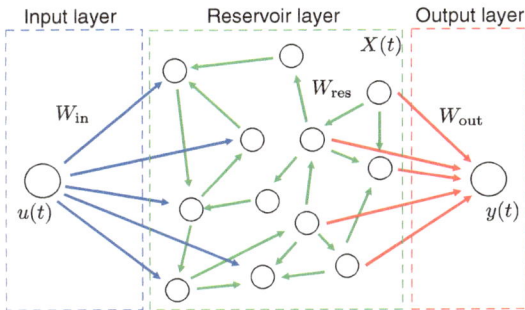

Fig. 8 A schematic diagram of reservoir computing. A neural network consisting of input, reservoir, and output layers is employed. The reservoir layer contains multiple processing nodes connected randomly, and the input and output layers are connected to the reservoir layer. Only by training the connection weight between the reservoir and the output layers, W_{out}, equivalent performance to a RNN can be achieved

5 Optical Special Purpose Processing

Electronics continued to develop far beyond what was expected when optical computing was proposed in the 1980s and the 1990s. Therefore, it is impractical to construct an information processing system using only optical technology; therefore, a system configuration that combines technologies with superiority is reasonable. Although electronic processors provide sufficient computational performance, the superiority of optical technology in signal transmission between processors is clear. Optical technology, also known as *optical interconnection* or *optical wiring*, has been proposed and developed along with optical computing. In particular, it has been shown that the higher the signal bandwidth and the longer the transmission distance, the more superior the optical interconnection becomes to that in electronics.

System architectures based on optical interconnection have encouraged the development of *smart pixels* with optical input and output functions for semiconductor processors. In addition, under the concept of a more generalized systemization technology called *VLSI photonics*, various kinds of systems-on-chips were considered in the US Department of Defense, DARPA research program, and the deployment of various systems has been considered [34]. In Japan, the element processor developed for the *optoelectronic computer using laser arrays with reconfiguration* (OCULAR) system [25] was extended to an image sensor with a processing function called VisionChip [35]. Until now, semiconductor materials other than silicon, such as GaAs, have been used as light-emitting devices. However, advances in *silicon photonics* have made it possible to realize integrated devices that include light-emitting elements [36]. This is expected to lead to an optical interconnection at a new stage.

Visual cryptography is an optical computing technology that exploits human visibility [37], which is an important feature of light. This method applies logical operations using spatial codes and effectively encrypts image information. As shown in Fig. 9, the method divides each pixel into 2×2 sub-pixels and assigns image

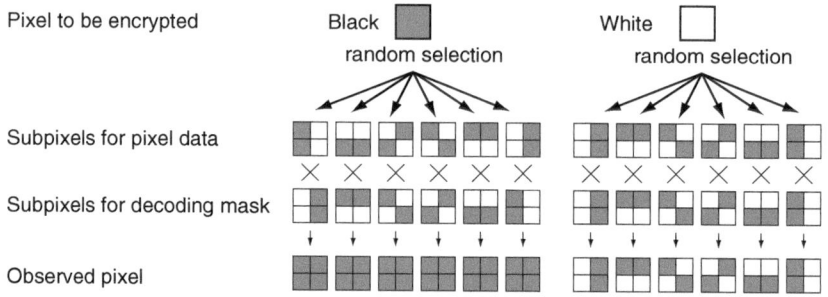

Fig. 9 Procedure of visual cryptography. Each pixel to be encrypted is divided into 2×2 sub-pixels selected randomly. To decode the pixel data, 2×2 sub-pixels of the corresponding decoding mask must be overlapped correctly. Exactly arranged decoding masks are required to retrieve the original image

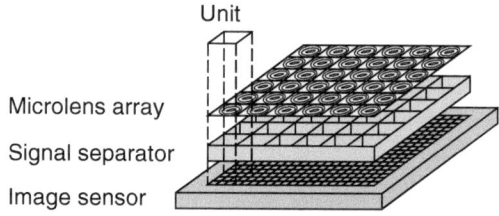

Fig. 10 The compound-eye imaging system TOMBO. The system is composed of a microlens array, a signal separator, and an image sensor. The individual microlens and the corresponding area of the image sensor form an imaging system called a *unit*. This system is a kind of multi-aperture optical system and provides various functionalities associated with the post processing

information and a decoding mask to a couple of the spatial codes in the sub-pixels. Owing to the arbitrariness of this combination, an efficient image encryption can be achieved. Another method for realizing secure optical communication was proposed by dividing logical operations based on spatial codes at the transmission and receiving ends of a communication system [38]. This method can be applied to information transmission from *Internet of Things* (IoT) edge devices.

The authors proposed a compound-eye imaging system called *thin observation module by bound optics* (TOMBO) which is inspired by a compound eye found in insects [39]. As shown in Fig. 10, it is an imaging system in which a microlens array is arranged in front of the image sensor, and a plurality of information is captured at once. This is a kind of *multi-aperture optical system* and provides various interesting properties. The system can be made very thin, and by appropriately setting the optical characteristics of each ommatidium and combining it with post-processing using a computer, the object parallax, light beam information, spectral information, etc. can be acquired. The presentation of TOMBO was one of the triggers for the organization of the OSA Topical Meeting on *Computational Optical Sensing and Imaging* (COSI) [40].

The imaging technique used in COSI is called *computational imaging*. In conventional imaging, the signal captured by the imaging device through an optical system faithfully reproduces the subject information. However, when considering signal acquisition for information systems such as machine vision, such requirements are not necessarily important. Instead, by optimizing optical and arithmetic systems, new imaging technologies beyond the functions and performance of conventional imaging can be developed. This is the fundamental concept of computational imaging, which combines optical encoding and computational decoding. This encoding style enables various imaging modalities, and several imaging techniques beyond conventional frameworks have been proposed [41, 42].

The task of optical coding in computational imaging is the same as that in optical computing, which converts object information into a form that is easy to handle for the processing system. This can be regarded as a modern form of optical computing dedicated to information visualization [43]. In imaging and object recognition using scattering media, new achievements have been made using deep learning [44, 45]. In addition, the concept of a *photonic accelerator* was proposed for universal arithmetic processing [46]. This concept is not limited to imaging and also incorporates optical computing as a processing engine in high-performance computing systems. This strategy is more realistic than that of the optical computers of the 1980s, which were aimed at general-purpose computation. Furthermore, when considering a modern computing system connected to a cloud environment, *IoT edge devices* as the interface to the real world will be the main battlefield of optical computing. If this happens, the boundary between optical computing and optical coding in computational imaging will become blurred, and a new optical computing image is expected to be built.

6   Nanophotonic Computing

A major limitation of optical computing is the spatial resolution limit due to the diffraction of light. In the case of visible light, the spatial resolution of sub-μm is much larger than the cell pitch of a semiconductor integrated circuit of several nanometers. Near-field probes can be used to overcome this limitation and an interesting concept has been proposed that uses the restriction imposed by the diffraction of light [47]. This method utilizes the hierarchy in optical near-field interaction to hide information under the concept of a *hierarchical nanophotonic system*, where multiple functions are associated with the physical scales involved. Unfortunately, a problem is associated with the near-field probe: the loss of parallelism caused by the free-space propagation of light. To solve this problem, a method that combines optical and molecular computing has been proposed [48]. The computational scheme is called *photonic DNA computing* which combines autonomous computation by molecules and arithmetic control by light.

DNA molecules have a structure in which four bases, adenine, cytosine, guanine, and thymine, are linked in chains. Because adenine specifically forms hydrogen bonds with thymine and cytosine with guanine, DNA molecules with this com-

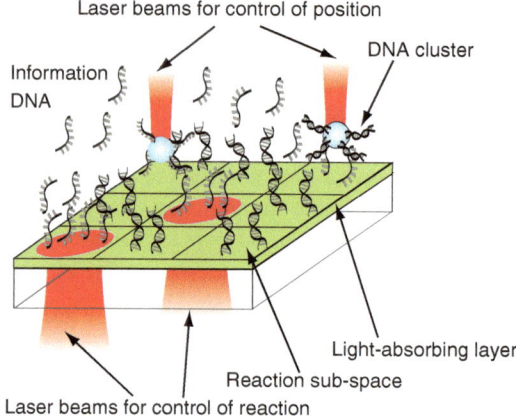

Fig. 11 A schematic diagram of photonic DNA computing. Similar to the conventional DNA computing, DNA molecules are used for information media as well as an autonomous computing engine. In this system, several reacting spaces are prepared to proceed multiple sets of DNA reactions concurrently in the separated reacting spaces. Laser trapped micro beads are employed to transfer the information-encoded DNA over the reacting spaces. DNA denaturation induced by light illumination is used to control the local reaction

plementary base sequence are synthesized in vivo, resulting in the formation of a stable state called a double-stranded structure. This is the mechanism by which DNA molecules store genetic information and can be referred to as the blueprint of living organisms. The principle of *DNA computing* is to apply this property to matching operations of information encoded as base sequences. Single-stranded DNA molecules with various base sequences are mixed, and the presence of specific information is detected by the presence or absence of double-strand formation. This series of processes proceeds autonomously; however, the processing content must be encoded in the DNA molecule as the base sequence. Therefore, it is difficult to perform large-scale complex calculations using naïve principles.

To solve this problem, the author's group conceived photonic DNA computing [48] which combines optical technology and DNA computing and conducted a series of studies. It is characterized by the microfineness, autonomous reactivity, and large-scale parallelism of DNA molecules, as shown in Fig. 11. Photonic DNA computing realizes flexible and efficient massive parallel processing by collectively manipulating the DNA molecules using optical signals. In this system, DNA molecules act as nanoscale computational elements and optical technology provides an interface. Because it is based on molecular reactions, it has the disadvantage of slow calculation speed; however, it is expected to be used as a nanoprocessor that can operate in vivo.

For accelerating computational speed restricted by molecular reactions, an arithmetic circuit using *Förster resonance energy transfer* (FRET) has been developed [49]. FRET is an energy-transfer phenomenon that occurs at particle intervals of several nanometers, and high-speed information transmission is possible. Using DNA molecules as scaffolds, fluorescent molecules can be arranged with nanoscale preci-

Fig. 12 Logic gates implemented by FRET. **a** DNA scaffold logic using FRET signal cascades. **b** A fluorescence dye corresponding to site i of a DNA scaffold switches between the ON and OFF states according to the presence of an input molecule (input a). **c** Configuration for the AND logic operation. Fluorescent molecules of a FRET pair are assigned to neighboring sites, i and j. **d** Configuration for the OR logic operation. Multiple input molecules can deliver a fluorescent molecule to a single site [49]

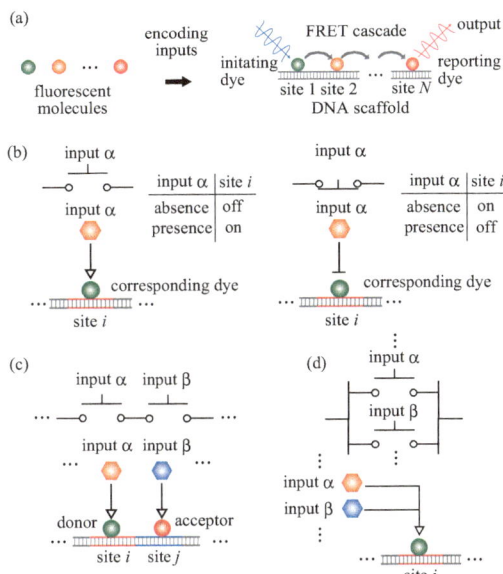

sion to generate FRET. As shown in Fig. 12, AND, OR, and NOT gates are implemented. By combining these gates, nanoscale arithmetic circuits can be constructed. This technology is characterized by its minuteness and low energy consumption. In contrast to a system configuration with precisely arranged fluorescent molecules, a random arrangement method is also promising. Based on this concept, our group is conducting research to construct a *FRET network* with *quantum dots* and use the output signal in the space, time, and spectrum domains as a physical reservoir [6]. For more details, please refer to the other chapters of this book.

7 Toward Further Extension

First, the significance of optical computing technology in the 1980s and the 1990s is considered. Despite the introduction of this study, explicit computing systems have not yet been developed for several technologies. However, some technologies have come to fruition as optical devices and technologies that support current information and communication technologies. The exploration of nonlinear optical phenomena has promoted the development of optoelectronics, and numerous results have been obtained, including those for quantum dot optical devices. They are used in current optical communications, and their importance is evident. Many researchers and engineers involved in optical computing have contributed to the development of general optical technologies. These fields have evolved into *digital holography* [50] and *information photonics* [51]. From this perspective, it can be said that the previous optical computing boom laid the foundation for the current optical information technology.

However, considering computing tasks, research at that time was not necessarily successful. One of the reasons for this is that they were too caught up in digital arithmetic methods, for better or worse. The diffraction limit of light is in the order of submicrometers, which is too large compared to the information density realized using semiconductor integration technology. Therefore, a simple imitation or partial application of digital arithmetic methods cannot demonstrate the uniqueness of computation using light. In addition, the peripheral technologies that could constitute the system are immature. Systematization requires not only elemental technologies but also design theories and implementation technologies to bring them together and applications that motivate their development. Unfortunately, the technical level at that time did not provide sufficient groundwork for development.

What about neural networks based on analog arithmetic other than digital arithmetic methods? As mentioned above, optical neural networks are architectures suitable for optical computing that can avoid these problems. However, the immaturity of the elemental devices is similar, and it has been difficult to develop it into a practical system that requires interconnections with a very large number of nodes. In addition, studies on neural networks themselves are insufficient, and we must wait for the current development of deep neural networks. Thus, the ingenious architecture at that time was difficult to develop owing to the immaturity of peripheral technology, and it did not lead to superiority over competitors. Furthermore, the problems that require high-performance computing are not clear, and it can be said that the appeal of the proposition of computing, including social demands, was insufficient.

Based on these circumstances, we will consider the current status of optical computing. Now that the practical applications of *quantum computing* are on the horizon, expectations for optical quantum computing are increasing [52]. Although not discussed in this article, quantum computing continues to be a promising application of optical computing. In addition, optical coding in computational imaging can be considered as a modern form of optical computing with limited use and purpose [43]. Computational imaging realizes highly functional and/or high-performance imaging combined with optical coding and arithmetic decoding. In this framework, the optical coding that proceeds through the imaging process determines the function of the entire process and enables high throughput. For optical coding, various methods, such as point spread function modulation, called *PSF engineering* [53], object modulation by projecting light illumination [54], and multiplexing imaging by multiple apertures [55], form the field of modern optical computing.

Optical computing, which is closer to practical applications, is *neuromorphic computing* [56]. The superior connection capability of optical technology makes it suitable for implementing neural networks that require many interconnections, and many studies are underway. However, in the human brain, approximately 100 billion neurons are connected by approximately 10 thousand synapses per neuron; therefore, a breakthrough is required to realize them optically. Deep learning models have advanced computational algorithms; however, their implementation relies on existing computers. The concept of a photonic accelerator is a realistic solution used specifically for processing in which light can exert superiority [46]. An optical processor that performs multiply-accumulate operations necessary for learning processing can be cited as a specific example [3].

Furthermore, *nanophotonic computing*, based on optical phenomena at the molecular level, has significant potential. Reservoir computing can alleviate connectivity problems and enable the configuration of more realistic optical computing systems [57]. Our group is developing a reservoir computing system that utilizes a FRET network, which is the energy transfer between adjacent quantum dots [6]. FRET networks are expected to perform computational processing with an extremely low energy consumption. The background of this research is the explosive increase in the amount of information and arithmetic processing, which require enormous computational power, as represented by deep learning. Computing technologies that consume less energy are expected to become increasingly important in the future. In addition, cooperation between real and virtual worlds, such as metaverse [58] and cyber-physical systems [59], is becoming increasingly important. As an interface, optical computing is expected to be applied in IoT edge devices that exploit the characteristics of optical technologies.

In conclusion, optical computing is by no means a technology of the past and is being revived as a new computing technology. In addition, optical information technology research areas that are not aimed at computing, such as digital holography and information photonics, are being developed. By combining the latest information technology with various optical technologies cultivated thus far, a new frontier in information science is being opened.

References

1. D. Schaefer, J. Strong, Tse computers. Proc. IEEE **65**(1), 129–138 (1977). https://doi.org/10.1109/PROC.1977.10437
2. Adaptive optics: Analysis and methods/computational optical sensing and imaging/information photonics/signal recovery and synthesis topical meetings on CD-ROM (2005)
3. Y. Shen, N.C. Harris, S. Skirlo, M. Prabhu, T. Baehr-Jones, M. Hochberg, X. Sun, S. Zhao, H. Larochelle, D. Englund, M. Soljačić, Deep learning with coherent nanophotonic circuits. Nat. Photonics **11**(7), 441–446 (2017). https://doi.org/10.1038/nphoton.2017.93
4. X. Lin, Y. Rivenson, N.T. Yardimci, M. Veli, Y. Luo, M. Jarrahi, A. Ozcan, All-optical machine learning using diffractive deep neural networks. Science **8084**, eaat8084 (2018). arXiv:1406.1078, https://doi.org/10.1126/science.aat8084, http://www.sciencemag.org/lookup/doi/10.1126/science.aat8084
5. G. Tanaka, T. Yamane, J.B. Héroux, R. Nakane, N. Kanazawa, S. Takeda, H. Numata, D. Nakano, A. Hirose, Recent advances in physical reservoir computing: a review. Neural Netw. **115**, 100–123 (2019). https://doi.org/10.1016/j.neunet.2019.03.005
6. N. Tate, Y. Miyata, S. ichi Sakai, A. Nakamura, S. Shimomura, T. Nishimura, J. Kozuka, Y. Ogura, J. Tanida, Quantitative analysis of nonlinear optical input/output of a quantum-dot network based on the echo state property. Opt. Express **30**(9), 14669–14676 (2022). https://doi.org/10.1364/OE.450132,https://opg.optica.org/oe/abstract.cfm?URI=oe-30-9-14669

7. W.P. Bleha, L.T. Lipton, E. Wiener-Avnear, J. Grinberg, P.G. Reif, D. Casasent, H.B. Brown, B.V. Markevitch, Application of the liquid crystal light valve to real-time optical data processing. Opt. Eng. **17**(4), 174371 (1978). https://doi.org/10.1117/12.7972245

8. G. Lazarev, A. Hermerschmidt, S. Krüger, S. Osten, *LCOS Spatial Light Modulators: Trends and Applications, Optical Imaging and Metrology: Advanced Technologies* (2012), pp.1–29

9. A.M. Weiner, J.P. Heritage, E.M. Kirschner, High-resolution femtosecond pulse shaping. J. Opt. Soc. Am. B **5**(8), 1563–1572 (1988). https://doi.org/10.1364/JOSAB.5.001563, https://opg.optica.org/josab/abstract.cfm?URI=josab-5-8-1563

10. P. Refregier, B. Javidi, Optical image encryption based on input plane and fourier plane random encoding. Opt. Lett. **20**(7), 767–769 (1995). https://doi.org/10.1364/OL.20.000767, https://opg.optica.org/ol/abstract.cfm?URI=ol-20-7-767

11. O. Matoba, T. Nomura, E. Perez-Cabre, M.S. Millan, B. Javidi, Optical techniques for information security. Proc. IEEE **97**(6), 1128–1148 (2009). https://doi.org/10.1109/JPROC.2009. 2018367

12. X. Quan, M. Kumar, O. Matoba, Y. Awatsuji, Y. Hayasaki, S. Hasegawa, H. Wake, Three-dimensional stimulation and imaging-based functional optical microscopy of biological cells. Opt. Lett. **43**(21), 5447–5450 (2018). https://doi.org/10.1364/OL.43.005447, https://opg. optica.org/ol/abstract.cfm?URI=ol-43-21-5447

13. B.K. Jenkins, A.A. Sawchuk, T.C. Strand, R. Forchheimer, B.H. Soffer, Sequential optical logic implementation. Appl. Opt. **23**(19), 3455–3464 (1984). https://doi.org/10.1364/AO.23. 003455 . URL https://opg.optica.org/ao/abstract.cfm?URI=ao-23-19-3455

14. D.A.B. Miller, D.S. Chemla, T.C. Damen, A.C. Gossard, W. Wiegmann, T.H. Wood, C.A. Burrus, Novel hybrid optically bistable switch: the quantum well self-electro-optic effect device. Appl. Phys. Lett. **45**(1), 13–15 (1984). arXiv: https://pubs.aip.org/aip/apl/article-pdf/45/1/13/7753868/13_1_online.pdf , https://doi.org/ 10.1063/1.94985

15. J. Tanida, Y. Ichioka, Optical logic array processor using shadowgrams. J. Opt. Soc. Am. **73**(6), 800–809 (1983). https://doi.org/10.1364/JOSA.73.000800, https://opg.optica.org/ abstract.cfm?URI=josa-73-6-800

16. A. Huang, Parallel algormiivls for optical digital computers, in *10th International Optical Computing Conference*, vol. 0422, ed. by S. Horvitz (International Society for Optics and Photonics, SPIE, 1983), pp. 13–17. https://doi.org/10.1117/12.936118

17. K.-H. Brenner, A. Huang, N. Streibl, Digital optical computing with symbolic substitution. Appl. Opt. **25**(18), 3054–3060 (1986). https://doi.org/10.1364/AO.25.003054, https://opg. optica.org/ao/abstract.cfm?URI=ao-25-18-3054

18. J. Tanida, Y. Ichioka, Programming of optical array logic. 1: Image data processing. Appl. Opt. **27**(14), 2926–2930 (1988). https://doi.org/10.1364/AO.27.002926, https://opg.optica.org/ao/ abstract.cfm?URI=ao-27-14-2926

19. J. Tanida, M. Fukui, Y. Ichioka, Programming of optical array logic. 2: Numerical data processing based on pattern logic. Appl. Opt. **27**(14), 2931–2939 (1988). https://doi.org/10.1364/ AO.27.002931, https://opg.optica.org/ao/abstract.cfm?URI=ao-27-14-2931

20. B. Chopard, M. Droz, *Cellular Automata Modeling of Physical Systems* (Cambridge University Press, Cambridge, 1998). https://doi.org/10.1007/978-0-387-30440-3_57

21. B. Durand, Z. Róka, *The Game of Life: Universality Revisited* (Springer Netherlands, Dordrecht, 1999), pp. 51–74. https://doi.org/10.1007/978-94-015-9153-9_2

22. K.-S. Huang, A.A. Sawchuk, B.K. Jenkins, P. Chavel, J.-M. Wang, A.G. Weber, C.-H. Wang, I. Glaser, Digital optical cellular image processor (docip): experimental implementation. Appl. Opt. **32**(2), 166–173 (1993). https://doi.org/10.1364/AO.32.000166, https://opg.optica.org/ao/ abstract.cfm?URI=ao-32-2-166

23. M.J. Murdocca, *A Digital Design Methodology for Optical Computing* (MIT Press, Cambridge, MA, USA, 1991)

24. J. Tanida, Y. Ichioka, Opals: optical parallel array logic system. Appl. Opt. **25**(10), 1565–1570 (1986). https://doi.org/10.1364/AO.25.001565, https://opg.optica.org/ao/abstract.cfm? URI=ao-25-10-1565

25. N. McArdle, M. Naruse, M. Ishikawa, Optoelectronic parallel computing using optically inter-connected pipelined processing arrays. IEEE J. Sel. Top. Quantum Electron. **5**(2), 250–260 (1999)

26. G. Li, D. Huang, E. Yuceturk, P.J. Marchand, S.C. Esener, V.H. Ozguz, Y. Liu, Three-dimensional optoelectronic stacked processor by use of free-space optical interconnection and three-dimensional vlsi chip stacks. Appl. Opt. **41**(2), 348–360 (2002). https://doi.org/10.1364/AO.41.000348, https://opg.optica.org/ao/abstract.cfm?URI=ao-41-2-348

27. J.W. Goodman, A.R. Dias, L.M. Woody, Fully parallel, high-speed incoherent optical method for performing discrete fourier transforms. Opt. Lett. **2**(1), 1–3 (1978). https://doi.org/10.1364/OL.2.000001, https://opg.optica.org/ol/abstract.cfm?URI=ol-2-1-1

28. D. Psaltis, N. Farhat, Optical information processing based on an associative-memory model of neural nets with thresholding and feedback. Opt. Lett. **10**(2), 98–100 (1985). https://doi.org/10.1364/OL.10.000098, https://opg.optica.org/ol/abstract.cfm?URI=ol-10-2-98

29. D. Psaltis, D. Brady, K. Wagner, Adaptive optical networks using photorefractive crystals. Appl. Opt. **27**(9), 1752–1759 (1988). https://doi.org/10.1364/AO.27.001752, https://opg.optica.org/ao/abstract.cfm?URI=ao-27-9-1752

30. M. Lukoševičius, H. Jaeger, Reservoir computing approaches to recurrent neural network training. Comput. Sci. Rev. **3**(3), 127–149 (2009). https://doi.org/10.1016/j.cosrev.2009.03.005, https://www.sciencedirect.com/science/article/pii/S1574013709000173

31. F. Duport, B. Schneider, A. Smerieri, M. Haelterman, S. Massar, All-optical reservoir computing. Opt. Express **20**(20), 22783–22795 (2012). https://doi.org/10.1364/OE.20.022783, https://opg.optica.org/oe/abstract.cfm?URI=oe-20-20-22783

32. J. Nakayama, K. Kanno, A. Uchida, Laser dynamical reservoir computing with consistency: an approach of a chaos mask signal. Opt. Express **24**(8), 8679–8692 (2016). https://doi.org/10.1364/OE.24.008679, https://opg.optica.org/oe/abstract.cfm?URI=oe-24-8-8679

33. N. Segawa, S. Shimomura, Y. Ogura, J. Tanida, Tunable reservoir computing based on iterative function systems. Opt. Express **29**(26), 43164–43173 (2021). https://doi.org/10.1364/OE.441236, https://opg.optica.org/oe/abstract.cfm?URI=oe-29-26-43164

34. E.-H. Lee, Vlsi photonics: a story from the early studies of optical microcavity microspheres and microrings to present day and its future outlook, in *Optical Processes in Microparticles and Nanostructures: A Festschrift Dedicated to Richard Kounai Chang on His Retirement from Yale University* (World Scientific, 2011), pp. 325–341

35. T. Komuro, I. Ishii, M. Ishikawa, A. Yoshida, A digital vision chip specialized for high-speed target tracking. IEEE Trans. Electron Devices **50**(1), 191–199 (2003). https://doi.org/10.1109/TED.2002.807255

36. B. Jalali, S. Fathpour, Silicon photonics. J. Light. Technol. **24**(12), 4600–4615 (2006). https://doi.org/10.1109/JLT.2006.885782

37. H. Yamamoto, Y. Hayasaki, N. Nishida, Securing information display by use of visual cryptography. Opt. Lett. **28**(17), 1564–1566 (2003). https://doi.org/10.1364/OL.28.001564, https://opg.optica.org/ol/abstract.cfm?URI=ol-28-17-1564

38. J. Tanida, K. Tsuchida, R. Watanabe, Digital-optical computational imaging capable of end-point logic operations. Opt. Express **30**(1), 210–221 (2022). https://doi.org/10.1364/OE.442985, https://opg.optica.org/oe/abstract.cfm?URI=oe-30-1-210

39. J. Tanida, T. Kumagai, K. Yamada, S. Miyatake, K. Ishida, T. Morimoto, N. Kondou, D. Miyazaki, Y. Ichioka, Thin observation module by bound optics (tombo): concept and experimental verification. Appl. Opt. **40**(11), 1806–1813 (2001). https://doi.org/10.1364/AO.40.001806, https://opg.optica.org/ao/abstract.cfm?URI=ao-40-11-1806

40. D.J. Brady, A. Dogariu, M.A. Fiddy, A. Mahalanobis, Computational optical sensing and imaging: introduction to the feature issue. Appl. Opt. **47**(10), COSI1–COSI2 (2008). https://doi.org/10.1364/AO.47.0COSI1, https://opg.optica.org/ao/abstract.cfm?URI=ao-47-10-COSI1

41. E.R. Dowski, W.T. Cathey, Extended depth of field through wave-front coding. Appl. Opt. **34**(11), 1859–1866 (1995). https://doi.org/10.1364/AO.34.001859, https://opg.optica.org/ao/abstract.cfm?URI=ao-34-11-1859

42. J.N. Mait, G.W. Euliss, R.A. Athale, Computational imaging. Adv. Opt. Photon. **10**(2), 409–483 (2018). https://doi.org/10.1364/AOP.10.000409, https://opg.optica.org/aop/abstract.cfm?URI=aop-10-2-409
43. J. Tanida, Computational imaging demands a redefinition of optical computing. Jpn. J. Appl. Phys. **57**(9S1), 09SA01 (2018). https://doi.org/10.7567/JJAP.57.09SA01
44. T. Ando, R. Horisaki, J. Tanida, Speckle-learning-based object recognition through scattering media, Opt. Express **23**(26), 33902–33910 (2015). https://doi.org/10.1364/OE.23.033902, https://opg.optica.org/oe/abstract.cfm?URI=oe-23-26-33902
45. R. Horisaki, R. Takagi, J. Tanida, Learning-based imaging through scattering media. Opt. Express **24**(13), 13738–13743 (2016). https://doi.org/10.1364/OE.24.013738, https://opg.optica.org/oe/abstract.cfm?URI=oe-24-13-13738
46. K. Kitayama, M. Notomi, M. Naruse, K. Inoue, S. Kawakami, A. Uchida, Novel frontier of photonics for data processing—Photonic accelerator. APL Photonics **4**(9), 090901 (2019). arXiv: https://pubs.aip.org/aip/app/article-pdf/doi/10.1063/1.5108912/14569493/090901_1_online.pdf , https://doi.org/10.1063/1.5108912
47. M. Naruse, N. Tate, M. Ohtsu, Optical security based on near-field processes at the nanoscale. J. Opt. **14**(9), 094002 (2012). https://doi.org/10.1088/2040-8978/14/9/094002
48. J. Tanida, Y. Ogura, S. Saito, Photonic DNA computing: concept and implementation, in ICO20: Optical Information Processing, vol. 6027, eds. by Y. Sheng, S. Zhuang, Y. Zhang (International Society for Optics and Photonics, SPIE, 2006), p. 602724. https://doi.org/10.1117/12.668196
49. T. Nishimura, Y. Ogura, J. Tanida, Fluorescence resonance energy transfer-based molecular logic circuit using a DNA scaffold. Appl. Phys. Lett. **101**(23), 233703 (2012). arXiv: https://pubs.aip.org/aip/apl/article-pdf/doi/10.1063/1.4769812/13176369/233703_1_online.pdf , https://doi.org/10.1063/1.4769812
50. B. Javidi, A. Carnicer, A. Anand, G. Barbastathis, W. Chen, P. Ferraro, J.W. Goodman, R. Horisaki, K. Khare, M. Kujawinska, R.A. Leitgeb, P. Marquet, T. Nomura, A. Ozcan, Y. Park, G. Pedrini, P. Picart, J. Rosen, G. Saavedra, N.T. Shaked, A. Stern, E. Tajahuerce, L. Tian, G. Wetzstein, M. Yamaguchi, Roadmap on digital holography [invited]. Opt. Express **29**(22), 35078–35118 (2021). https://doi.org/10.1364/OE.435915, https://opg.optica.org/oe/abstract.cfm?URI=oe-29-22-35078
51. G. Barbastathis, A. Krishnamoorthy, S.C. Esener, Information photonics: introduction. Appl. Opt. **45**(25), 6315–6317 (2006). https://doi.org/10.1364/AO.45.006315, https://opg.optica.org/ao/abstract.cfm?URI=ao-45-25-6315
52. P. Kok, W.J. Munro, K. Nemoto, T.C. Ralph, J.P. Dowling, G.J. Milburn, Linear optical quantum computing with photonic qubits. Rev. Mod. Phys. **79**, 135–174 (2007). https://doi.org/10.1103/RevModPhys.79.135, https://link.aps.org/doi/10.1103/RevModPhys.79.135
53. P.T.C. So, H.-S. Kwon, C.Y. Dong, Resolution enhancement in standing-wave total internal reflection microscopy: a point-spread-function engineering approach. J. Opt. Soc. Am. A **18**(11), 2833–2845 (2001). https://doi.org/10.1364/JOSAA.18.002833, https://opg.optica.org/josaa/abstract.cfm?URI=josaa-18-11-2833
54. T. Ando, R. Horisaki, J. Tanida, Three-dimensional imaging through scattering media using three-dimensionally coded pattern projection. Appl. Opt. **54**(24), 7316–7322 (2015). https://doi.org/10.1364/AO.54.007316, https://opg.optica.org/ao/abstract.cfm?URI=ao-54-24-7316
55. J. Tanida, Multi-aperture optics as a universal platform for computational imaging. Opt. Rev. **23**(5), 859–864 (2016). https://doi.org/10.1007/s10043-016-0256-0
56. N. Zheng, P. Mazumder, *Learning in Energy-Efficient Neuromorphic Computing: Algorithm and Architecture Co-Design* (Wiley, New York, 2019)
57. D. Brunner, M.C. Soriano, G.V. der Sande (eds.), *Optical Recurrent Neural Networks* (De Gruyter, Berlin, Boston, 2019) [cited 2023-06-08]. https://doi.org/10.1515/9783110583496
58. S. Mystakidis, Metaverse. Encyclopedia **2**(1), 486–497 (2022). https://doi.org/10.3390/encyclopedia2010031, https://www.mdpi.com/2673-8392/2/1/31
59. E.A. Lee, Cyber physical systems: design challenges, in *2008 11th IEEE International Symposium on Object and Component-Oriented Real-Time Distributed Computing (ISORC)* (2008), pp. 363–369. https://doi.org/10.1109/ISORC.2008.25

Nonlinear Dynamics and Computing in Recurrent Neural Networks

Hideyuki Suzuki

Abstract Nonlinearity is a key concept in the design and implementation of photonic neural networks for computing. This chapter introduces the fundamental models and concepts of recurrent neural networks, with a particular focus on their nonlinear dynamics. We review several types of nonlinear dynamics that emerge in symmetrically connected recurrent neural networks, in which the energy function plays a crucial role. In addition, we introduce the concepts of reservoir computing, covering fundamental models and physical reservoir computing. Overall, this chapter provides a foundation for the theoretical aspects in the subsequent chapters of this book, which explore a variety of photonic neural networks with nonlinear spatiotemporal dynamics.

1 Introduction

For more than half a century, various artificial neural network models have been developed and studied as abstractions of thought processes in the brain and as constructive approaches to thinking machines [1, 2]. Artificial neural networks are currently a fundamental technology in artificial intelligence, applied across various fields and playing crucial roles in our daily lives.

Recurrent neural networks (RNNs) are a type of neural network that can be contrasted with feedforward neural networks, such as multilayer perceptrons (MLPs). Unlike feedforward neural networks, which perform unidirectional information processing from the input layer to the output layer, RNNs allow mutual interactions among the constituent neuron models. These interactions typically induce spatiotemporal dynamics as the network state evolves over time, which performs various computational tasks, such as processing time-sequence data, solving combinatorial optimization problems, and generative statistical modeling.

H. Suzuki (✉)
Graduate School of Information Science and Technology, Osaka University,
Osaka 565–0871, Japan
e-mail: hideyuki@ist.osaka-u.ac.jp

© The Author(s) 2024
H. Suzuki et al. (eds.), *Photonic Neural Networks with Spatiotemporal Dynamics*,
https://doi.org/10.1007/978-981-99-5072-0_2

Nonlinearity is an indispensable property of neuron models, which naturally leads to the emergence of nonlinear dynamics in RNNs. Thus, understanding and utilizing their nonlinear dynamics is especially important for realizing energy-efficient, high-speed, and large-scale implementations of RNNs using optical technologies.

Based on this motivation, this chapter introduces the fundamental models and concepts of RNNs for computing, with a particular focus on their nonlinear dynamics. In Sect. 2, we introduce the notion of the energy function in two fundamental models of symmetrically connected RNNs: the Amari–Hopfield network and the Boltzmann machine. We explore how these models exhibit computational functions such as associative memory, combinatorial optimization, and statistical learning. Section 3 presents an overview of various types of nonlinear dynamics that arise in RNNs. We discuss how chaotic dynamics contributes to computation in models such as the chaotic neural network and the chaotic Boltzmann machine. We also observe the important roles of nonlinear dynamics in several types of Ising machines, which serve as hardware solvers for combinatorial optimization problems. Moreover, we introduce a sampling algorithm, known as the herding system, which exhibits complex nonlinear dynamics related to the learning process of RNNs. Section 4 provides a brief overview of reservoir computing, which is a lightweight approach that leverages the rich nonlinear dynamics of RNNs for information processing. We introduce the basic models and concepts of reservoir computing, such as the echo state network and echo state property, and further discuss physical reservoir computing. Finally, in Sect. 5, we explain how the concepts introduced in this chapter underlie the studies covered in the subsequent chapters of this book, which explore various aspects of photonic neural networks with nonlinear spatiotemporal dynamics.

2 Fundamental RNN Models and Energy Function

In this section, we introduce the two fundamental models of symmetrically connected RNNs: the Amari–Hopfield network and the Boltzmann machine. We provide a brief overview of their definitions and behavior, highlighting the role of the energy function. These models have computational functions such as associative memory, combinatorial optimization, and statistical learning. Although their dynamics is fundamentally governed by the energy function, they lay the foundation for RNN models with rich nonlinear dynamics as discussed in the following sections.

2.1 Amari–Hopfield Network with Binary States

The Amari–Hopfield network [3, 4] is an RNN model composed of binary McCulloch–Pitts neurons [1] with symmetrical connections. The state of each ith neuron at time t is represented by $s_i(t) \in \{0, 1\}$, with values corresponding to the resting and firing states, respectively. Note that a formulation that employs -1, instead of 0, as the resting state is also widely used. The input to the ith neuron from the jth neuron is assumed to be $w_{ij}s_j(t)$, where $w_{ij} \in \mathbb{R}$ denotes the synaptic weight. The weights are assumed to be symmetric, $w_{ij} = w_{ji}$, and have no self-connections, $w_{ii} = 0$. The state of the ith neuron takes $s_i = 1$ if the total input, including constant input (or bias) $b_i \in \mathbb{R}$, exceeds zero and $s_i = 0$ if otherwise. Hence, the update rule for the ith neuron can be written as follows:

$$s_i(t+1) = \theta \left(\sum_{j=1}^{N} w_{ij}s_j(t) + b_i \right), \tag{1}$$

where N denotes the total number of neurons and $\theta(\cdot)$ is the Heaviside unit step function; i.e., $\theta(z) = 1$ for $z \geq 0$ and $\theta(z) = 0$ for $z < 0$. According to this equation, the state of the network, $\mathbf{s}(t) = (s_1(t), s_2(t), \ldots, s_N(t))^\top$, evolves over time in the state space $\{0, 1\}^N$.

The key notion for understanding the behavior of the Amari–Hopfield network is the energy function

$$H(\mathbf{s}) = -\frac{1}{2} \sum_{i,j=1}^{N} w_{ij}s_i s_j - \sum_{i=1}^{N} b_i s_i, \tag{2}$$

which is guaranteed to decrease over time. Specifically, assume that only the ith neuron is updated according to (1), while the states of the other neurons are kept unchanged, i.e., $s_j(t+1) = s_j(t)$ for $j \neq i$. Then, it holds that $H(\mathbf{s}(t+1)) \leq H(\mathbf{s}(t))$ because

$$H(\mathbf{s}(t+1)) - H(\mathbf{s}(t)) = -(s_i(t+1) - s_i(t)) \left(\sum_{j=1}^{N} w_{ij}s_j(t) + b_i \right) \leq 0. \tag{3}$$

Consequently, the network state $\mathbf{s}(t)$ evolves until it reaches a local minimum of the energy function, which has no neighboring states with lower energy. These local minima are considered attractors of the network because the network state eventually converges to one of the local minima.

This behavior of Amari–Hopfield network can be interpreted as the process of recalling memory stored within the network, which is referred to as associative memory. Memory patterns can be stored as local minima of the network by designing the weight parameters as follows:

$$w_{ij} = \sum_{k=1}^{K} (2\xi_i^{(k)} - 1)(2\xi_j^{(k)} - 1), \tag{4}$$

where $\boldsymbol{\xi}^{(k)} = (\xi_1^{(k)}, \ldots, \xi_N^{(k)})^\top \in \{0, 1\}^N$ is the kth memory pattern. This learning rule, known as the Hebbian learning, strengthens the connections between neurons that are simultaneously activated in the memory patterns.

The Amari–Hopfield network is a simple but important foundation of RNN models with energy functions. Furthermore, recent studies have revealed that the generalization of the Amari–Hopfield network, such as the modern Hopfield network [5] and the dense associative memory model [6], share similar attention mechanisms present in modern neural network models, such as transformers and BERT. These generalized models employ continuous state variables, as described below.

2.2 Amari–Hopfield Network with Continuous States

The Amari–Hopfield network with continuous variables [7, 8], proposed almost simultaneously as the binary version, is an RNN model composed of symmetrically connected leaky integrators. The continuous-time nonlinear dynamics of state $x_i(t)$ of each ith neuron is expressed by the following ordinary differential equation (ODE):

$$\frac{dx_i}{dt} = -\frac{1}{\tau_{\text{leak}}} x_i + \sum_{j=1}^{N} w_{ij} \phi(x_j) + b_i , \tag{5}$$

where $\tau_{\text{leak}} > 0$ is a time constant and $\phi(x)$ is the sigmoid function

$$\phi(x) = \frac{1}{1 + \exp(-x/\varepsilon)} . \tag{6}$$

The output from the neurons $\phi(\mathbf{x}(t))$ evolves in the hypercube $[0, 1]^N$, where ϕ operates on each component of vector $\mathbf{x}(t) = (x_1(t), \ldots, x_N(t))^\top$. In the limit $\varepsilon \to 0$, where $\phi(x)$ is the Heaviside unit step function, the energy function (2) for the discrete Amari–Hopfield network also applies to the continuous version with $\mathbf{s} = \phi(\mathbf{x})$. That is, $H(\phi(\mathbf{x}(t)))$ decreases as the system evolves. Therefore, the network state converges to a local minimum of the energy function, which is analogous to that of the discrete model. Note that an energy function exists for $\varepsilon > 0$, while it introduces an extra term to (2).

In the continuous Amari–Hopfield network, state $x_i(t)$ of each neuron takes a continuous value that attenuates over time, according to (5). This model, known as a leaky integrator, is the simplest neuron model that describes the behavior of the membrane potentials of real neurons.

The Hopfield–Tank model [9] utilizes the dynamics of the Amari–Hopfield network for finding approximate solutions to combinatorial optimization problems such as the traveling salesman problem. Specifically, it is applicable to combinatorial optimization problems that are formulated as the minimization of the energy function (2), which is often referred to as quadratic unconstrained binary optimization (QUBO).

As the network state evolves, it converges to one of the local minima of the target energy function. This provides an approximate solution to the optimization problem. Once the network state is trapped in a local minimum, the search process terminates as it can no longer escape to explore other solutions. The Hopfield–Tank model can be considered the origin of recent studies on Ising machines (Sect. 3.3).

2.3 Boltzmann Machine

The Boltzmann machine [10] is an RNN model composed of binary stochastic neurons with symmetrical connections. The construction of the model is essentially the same as that of the Amari–Hopfield network with binary neurons, except that the neurons behave stochastically. The update rule is given by the probability that the ith neuron takes $s_i = 1$ in an update as follows:

$$\text{Prob}[s_i(t+1) = 1] = \frac{1}{1 + \exp(-z_i(t)/T)} , \qquad z_i(t) = \sum_{j=1}^{N} w_{ij} s_j(t) + b_i , \quad (7)$$

where $T > 0$ denotes the model temperature and $z_i(t)$ is the total input to the neuron at time t. At the limit of $T \to 0$, the update rule is equivalent to the McCulloch–Pitts model; that is, the network dynamics is equivalent to that of the Amari–Hopfield network. In the limit $T \to \infty$, each neuron takes the states 0 and 1 with the same probability $1/2$, irrespective of the network configuration.

The state of the network $\mathbf{s}(t) = (s_1(t), s_2(t), \ldots, s_N(t))^\top$, evolves over time in the state space $\{0, 1\}^N$. The sequence of states $\{\mathbf{s}(t)\}_t$ eventually follows the Gibbs distribution

$$P(\mathbf{s}) = \frac{1}{Z} \exp\left(-\frac{1}{T} H(\mathbf{s})\right) , \qquad Z = \sum_{\mathbf{s}} \exp\left(-\frac{1}{T} H(\mathbf{s})\right) \qquad (8)$$

with respect to the energy function $H(\mathbf{s})$ in (2), where Z is the normalizing constant called the partition function. The Boltzmann machine is more likely to adopt lower-energy states, and this tendency is more intense at lower temperatures. This probabilistic model is essentially equivalent to the Ising model, which is an abstract model of ferromagnetism in statistical mechanics.

Conversely, the Boltzmann machine can be considered to perform sampling from the Gibbs distribution $P(\mathbf{s})$. The Gibbs sampler, one of the Markov chain Monte Carlo (MCMC) methods, yields a sample sequence by updating each variable s_i in each step according to the conditional probability $P(s_i \mid \mathbf{s}_{\backslash i})$ given the values $\mathbf{s}_{\backslash i}$ of all the other variables. If applied to the Gibbs distribution $P(\mathbf{s})$ in (8), the conditional probability of $s_i = 1$ given $\mathbf{s}_{\backslash i}$ is as follows:

$$P(s_i = 1 \mid \mathbf{s}_{\backslash i}) = \frac{P(\mathbf{s}|_{s_i=1})}{P(\mathbf{s}|_{s_i=0}) + P(\mathbf{s}|_{s_i=1})} = \frac{1}{1 + \exp(-(H(\mathbf{s}|_{s_i=0}) - H(\mathbf{s}|_{s_i=1}))/T)} \,,$$
(9)

where $\mathbf{s}|_{s_i=\{0,1\}}$ denotes the vector $\mathbf{s}$ whose ith variable is set to $s_i = \{0, 1\}$. This probability is consistent with the update rule (7). Therefore, the Boltzmann machine is equivalent to the Gibbs sampler applied to the Gibbs distribution $P(\mathbf{s})$.

The Boltzmann machine can be utilized to solve combinatorial optimization problems, following the same approach as the Hopfield–Tank model to minimize the energy function. The stochasticity can help the network state escape the local minima, which is a remarkable difference from the Hopfield–Tank model. This effect is stronger at higher temperatures, whereas low-energy states are preferred at lower temperatures. Therefore, we typically employ simulated annealing to solve combinatorial optimization problems [11, 12], which controls the stochasticity by starting from a high temperature and gradually decreasing it to $T = 0$.

Another remarkable feature of the Boltzmann machine is its learning ability [10]. The learning is performed by tuning the parameters w_{ij} and b_i, such that the model distribution $P(\mathbf{s}) \propto \exp(-H(\mathbf{s}))$ is close to the given data distribution. Here, we omit temperature T by setting $T = 1$ without loss of generality.

The distance from the data distribution is quantified using the log-likelihood as follows:

$$\log L = \langle \log P(\mathbf{s}) \rangle_{\text{data}} = -\langle H(\mathbf{s}) \rangle_{\text{data}} - \log Z \,,$$
(10)

where $\langle \cdot \rangle_{\text{data}}$ denotes the average over the data distribution. We can then derive the learning rule as a gradient ascent on the log-likelihood as follows:

$$w_{ij}(k + 1) = w_{ij}(k) + \alpha \frac{\partial}{\partial w_{ij}} \log L \,,$$
(11)

$$b_i(k + 1) = b_i(k) + \alpha \frac{\partial}{\partial b_i} \log L \,,$$
(12)

where $\alpha > 0$ is the learning rate. The gradients are given by:

$$\frac{\partial}{\partial w_{ij}} \log L = \langle s_i s_j \rangle_{\text{data}} - \langle s_i s_j \rangle_{\text{model}} \,,$$
(13)

$$\frac{\partial}{\partial b_i} \log L = \langle s_i \rangle_{\text{data}} - \langle s_i \rangle_{\text{model}} \,,$$
(14)

where $\langle \cdot \rangle_{\text{model}}$ denotes the average over the model distribution $P(\mathbf{s})$.

The expressive power of the model distribution $P(\mathbf{s})$ can be improved by introducing hidden units into the state variable $\mathbf{s}$ of the Boltzmann machine. Accordingly, the state $\mathbf{s} = (\mathbf{v}, \mathbf{h})$ is composed of the visible part $\mathbf{v}$ and hidden part $\mathbf{h}$. The learning here aims to minimize the distance between the data distribution and the marginal distribution $P(\mathbf{v})$ of the visible part of the Gibbs distribution $P(\mathbf{v}, \mathbf{h}) \propto \exp(-H(\mathbf{v}, \mathbf{h}))$. The hidden units serve as additional latent variables that do not directly correspond to the data, and describe the indirect interactions among the visible units. The introduction of hidden units does not alter the learning rule (11)–(14), whereas only the visible units are clamped to the data distribution in the averaging $\langle \cdot \rangle_{data}$.

In practice, learning a large-scale Boltzmann machine is challenging. Rigorous computation of the average over $P(\mathbf{s})$ is intractable as the size of the state space $\{0, 1\}^N$ increases exponentially. This expectation can be approximated by averaging over the sample sequence from the Boltzmann machine. However, obtaining an accurate approximation requires massive computation to generate a sufficiently long sample sequence, as the sampling process often gets stuck in local modes of the Gibbs distribution.

The restricted Boltzmann machine (RBM) [13, 14] is an important model of a Boltzmann machine with restricted connections. Specifically, an RBM is a two-layer neural network comprising visible and hidden units, with no connections within each layer. Because of its restricted structure, the RBM can be efficiently trained using the contrastive divergence algorithm to obtain the gradient of log-likelihood. The restricted structure accelerates the Gibbs sampling procedure, because it allows for alternate block sampling of the visible units, given the hidden units, and vice versa. The deep Boltzmann machine (DBM) [15, 16] is a Boltzmann machine with a multilayer structure. It is a type of deep neural network consisting of multiple layers of RBMs. Thus, RBMs and DBMs are important classes of the Boltzmann machine that have led to recent developments in deep learning.

3 Nonlinear Dynamics in Symmetrically Connected RNNs

This section presents several RNN models with symmetrical connections that exhibit various types of nonlinear dynamics effective for computing. First, we introduce the chaotic neural network model and the chaotic Boltzmann machine, which are variants of the Amari–Hopfield network and Boltzmann machine, respectively, involving nonlinear chaotic dynamics. Then, we review several types of Ising machines that employ more advanced approaches than the Hopfield–Tank model in utilizing their nonlinear dynamics to solve combinatorial optimization problems. We also explore the nonlinear dynamics that arises in the learning process of RNNs. As an example, we introduce the herding system, which is a sampling algorithm with complex nonlinear dynamics that can also be regarded as an extreme case of Boltzmann machine learning.

3.1 Chaotic Neural Network

The chaotic neural network [17] is a variation of the Amari–Hopfield network, which incorporates relative refractoriness and a continuous activation function in the constituent neurons. It exhibits spatiotemporal chaotic dynamics with the ability to perform parallel-distributed processing.

First, we introduce the refractoriness, which is a temporary reduction in excitability after firing, into the Amari–Hopfield network (1). We update the state of the ith neuron in the network as follows:

$$s_i(t+1) = \theta \left(\sum_{j=1}^{N} w_{ij} S_j^{\text{fb}}(t) - \alpha S_i^{\text{ref}}(t) + b_i \right), \tag{15}$$

where $S_i^{\{\text{fb,ref}\}}(t) = \sum_{r=0}^{t} k_{\{\text{fb,ref}\}}^r s_i(t-r)$ represents the accumulated past output of the ith neuron with an exponential decay parameterized by $k_{\text{fb}}, k_{\text{ref}} \in (0, 1)$ for the feedback connections and refractoriness, respectively. This model can be considered as a restricted form of Caianiello's neuronic equation [2]. The network dynamics is described by a hybrid dynamical system [18], involving the continuous variables $S_i^{\{\text{fb,ref}\}}(t)$ and a discontinuous function $\theta(\cdot)$. The constituent neuron model with refractoriness is called the Nagumo–Sato model. Its single-neuron dynamics has been investigated by assuming the first term, which represents the input from other neurons, is constant in time, and has been shown to exhibit a complex response with a devil's staircase [18–20].

Next, we introduce a continuous activation function to obtain the chaotic neural network model as follows:

$$s_i(t+1) = \phi \left(\sum_{j=1}^{N} w_{ij} S_j^{\text{fb}}(t) - \alpha S_i^{\text{ref}}(t) + b_i \right), \tag{16}$$

where the Heaviside unit step function $\theta(\cdot)$ is replaced by the sigmoid function $\phi(\cdot)$ in (6).

The chaotic neural network exhibits spatiotemporal chaotic dynamics. Although the energy function in (2) does not necessarily decrease, it helps us to understand its dynamics. Unlike the Amari–Hopfield network, the state of the chaotic neural network continues to move around in the phase space without becoming stuck at a local minimum of the energy function. This is because if the network state remains at a local minimum for a while, the accumulated effects of refractoriness destabilize the local minimum, helping the state escape. Thus, the spatiotemporal chaotic dynamics emerge from a combination of the stabilizing effect, resulting from the mutual interactions in the network, and the destabilizing effect due to the refractoriness.

When applied to associative memory constructed by the Hebbian rule (4), the chaotic neural network continues to visit stored patterns itinerantly [21]. Such associative dynamics, which is characterized by chaotic itinerancy [22], has been demonstrated for a large-scale network in [23].

The itinerant behavior of the chaotic neural network is useful for solving combinatorial optimization problems [24], because the destabilizing effect helps the network state escape from local minima, and the state continues to explore possible solutions.

For hardware implementation of the chaotic neural network, it is crucial to utilize analog computation to simulate chaotic dynamics described by continuous variables.

Large-scale analog IC implementations of the chaotic neural network demonstrate high-dimensional physical chaotic neuro-dynamics and offer efficient applications in parallel-distributed computing, such as solving combinatorial optimization problems [25–27].

3.2 Chaotic Boltzmann Machine

The chaotic Boltzmann machine [28, 29] is a continuous-time deterministic system that utilizes nonlinear chaotic dynamics to function as a Boltzmann machine without requiring randomness for the time evolution. This contrasts with the original Boltzmann machine, comprising stochastic neurons updated at discrete-time steps.

Each neuron in the chaotic Boltzmann machine is associated with an internal state $x_i(t) \in [0, 1]$ besides the binary state $s_i(t) \in \{0, 1\}$ of the Boltzmann machine. The internal state x_i evolves according to the differential equation

$$\frac{dx_i}{dt} = (1 - 2s_i)\left(1 + \exp\frac{(1 - 2s_i)z_i}{T}\right) , \tag{17}$$

where z_i is the total input as defined in (7). State s_i of the ith neuron flips when x_i reaches 0 or 1 as follows:

$$s_i(t + 0) = 0 \text{ when } x_i(t) = 0 \quad \text{and} \quad s_i(t + 0) = 1 \text{ when } x_i(t) = 1 . \tag{18}$$

The right-hand side of (17) is positive when $s_i = 0$ and negative when $s_i = 1$. Therefore, the internal state x_i continues to oscillate between 0 and 1. If the states of the other neurons are fixed, the total input z_i becomes constant, and the oscillation continues periodically. Specifically, s_i takes the value 0 for $(1 + \exp(z_i/T))^{-1}$ unit time as x_i increases from 0 to 1, and s_i takes the value 1 for $(1 + \exp(-z_i/T))^{-1}$ unit time as x_i decreases from 1 to 0. Accordingly, the probability of finding $s_i = 1$ at a random instant is $(1 + \exp(-z_i/T))^{-1}$, which is consistent with the update rule (7) of the Boltzmann machine. Note that while this explanation provides intuitive validity to the equation, it does not necessarily imply that the network state $s(t)$ follows the Gibbs distribution $P(s) \propto \exp(-H(s)/T)$.

Although the chaotic Boltzmann machine is completely deterministic, it exhibits apparently stochastic behavior because of the chaotic dynamics that emerges from equations (17) and (18), which can be considered a hybrid dynamical system [18] with continuous variables x_i and discrete variables s_i. The entire system can be regarded as a coupled oscillator system because each constituent unit oscillates between $x_i = 0$ and 1, interacting with each other through the binary state s_i. This can also be viewed as a pseudo-billiard [30] in the hypercube $[0, 1]^N$, because the internal state $\mathbf{x}(t) = (x_1(t), \ldots, x_N(t))^\top$ moves linearly inside the hypercube, as shown in (17), and changes its direction only at the boundary, as shown in (18).

It has been numerically demonstrated that the chaotic Boltzmann machine serves as a deterministic alternative to the MCMC sampling from the Gibbs distribution, $P(\mathbf{s}) \propto \exp(-H(\mathbf{s})/T)$. It can be used in simulated annealing to solve combinatorial optimization problems and exhibits computing abilities comparable to those of the conventional Boltzmann machine.

The chaotic Boltzmann machine enables an efficient hardware implementation of the Boltzmann machine, primarily because it eliminates the need for a pseudorandom number generator and also because its mutual interactions are achieved digitally via the binary states. These advantages contribute to large-scale, energy-efficient hardware implementations of the chaotic Boltzmann machine, as demonstrated in analog CMOS VLSI and digital FPGA implementations [31, 32].

3.3 Ising Machines

Ising machines [33] are a class of specialized hardware designed to solve combinatorial optimization problems by finding the (approximate) ground state of the Ising model, which is an abstract model of ferromagnetism in statistical mechanics. They have attracted considerable attention in recent years because of their potential to efficiently solve complex optimization problems.

The energy function of the Ising model is given by:

$$H(\boldsymbol{\sigma}) = -\frac{1}{2} \sum_{i,j=1}^{N} J_{ij} \sigma_i \sigma_j \, , \tag{19}$$

where $\sigma_i \in \{-1, +1\}$ denotes the ith Ising spin. As is evident from the energy function, the Ising model is almost equivalent to the Boltzmann machine. For simplicity, we omit the linear bias term, which can be represented by introducing an additional spin fixed at $+1$. Coefficient J_{ij} represents the coupling strength between spins i and j, which is assumed to be symmetric $J_{ij} = J_{ji}$.

Ising machines are designed to find a spin configuration $\boldsymbol{\sigma} = (\sigma_1, \dots, \sigma_N)^\top$ that approximately minimizes $H(\boldsymbol{\sigma})$. To solve a combinatorial optimization problem using an Ising machine, we need to formulate it as an Ising problem, in a way analogous to the Hopfield–Tank model and the Boltzmann machine.

We provide a brief overview of three types of Ising machines: the coherent Ising machine (CIM) [34], the simulated bifurcation machine (SBM) [35], and the oscillator Ising machine (OIM) [36]. We also introduce a continuous-time solver for boolean satisfiability (SAT) problems [37].

3.3.1 CIM: Coherent Ising Machine

The coherent Ising machine (CIM) [34] is a network of optical parametric oscillators (OPOs) designed to solve Ising problems.

The fundamental, noiseless dynamics of CIM can be described by the following ordinary differential equation:

$$\frac{dx_i}{dt} = (-1 + p - x_i^2)x_i + \sum_{j=1}^{N} J_{ij}x_j \, , \tag{20}$$

where x_i is the amplitude of the ith OPO mode and p represents the pump rate. Intuitively, the dynamics of CIM can be viewed as a variant of the Hopfield–Tank model with bistability introduced into each neuron. Basic dynamics of each OPO, without the coupling term, undergoes a pitchfork bifurcation at $p = 1$. That is, for $p < 1$, the equilibrium at $x_i = 0$ is stable; however, for $p > 1$, $x_i = 0$ becomes unstable and two symmetric stable equilibria $x_i = \pm\sqrt{p - 1}$ emerge. These two equilibria in each OPO correspond to the binary state of spin $\sigma_i \in \{-1, +1\}$. Therefore, by gradually increasing the pump rate p, we expect the state of the OPO network to converge to a low-energy spin state, as each OPO is forced to choose one of the binary states by the inherent bistability. Thus, the obtained low-energy state corresponds to an approximate solution to the Ising problem.

3.3.2 SBM: Simulated Bifurcation Machine

The simulated bifurcation machine (SBM) [35] is an Ising machine described as an Hamiltonian system, which is given by the following ordinary differential equations:

$$\frac{dx_i}{dt} = \Delta y_i \, , \tag{21}$$

$$\frac{dy_i}{dt} = -(Kx_i^2 - p + \Delta)x_i + \xi_0 \sum_{j=1}^{N} J_{ij}x_j \, , \tag{22}$$

where x_i and y_i denote the position and momentum of the ith unit, respectively, and Δ, K, and ξ_0 are constants. Parameter p controls the bistability of each unit. This Hamiltonian system conserves the Hamiltonian

$$H_{SB}(\mathbf{x}, \mathbf{y}) = \frac{\Delta}{2} \sum_{i=1}^{N} y_i^2 + V(\mathbf{x}) \, , \tag{23}$$

where the potential function $V(\mathbf{x})$ is given by

$$V(\mathbf{x}) = \sum_i \left(\frac{\Delta - p}{2} x_i^2 + \frac{K}{4} x_i^4 \right) - \frac{\xi_0}{2} \sum_{i,j=1}^{N} J_{ij} x_i x_j \ . \tag{24}$$

The SBM employs the symplectic Euler method, a structure-preserving time-discretization method, to conserve the Hamiltonian in simulating the Hamiltonian dynamics. Unlike the CIM, the SBM wanders the state space according to the complex Hamiltonian dynamics to explore low-energy states.

The SBM has been implemented on FPGA and GPUs, making it an efficient hardware solver of the Ising problems.

3.3.3 OIM: Oscillator Ising Machine

The oscillator Ising machine (OIM) [36] is a coupled nonlinear oscillator system that utilizes subharmonic injection locking (SHIL) to solve Ising problems. The dynamics of the OIM is given by:

$$\frac{d\phi_i}{dt} = - \sum_{j=1}^{N} J_{ij} \sin(\phi_i - \phi_j) - K \sin(2\phi_i) \ , \tag{25}$$

where ϕ_i denotes the phase of the ith oscillator. It has a global Lyapunov function,

$$E(\boldsymbol{\phi}) = - \sum_{i,j=1}^{N} J_{ij} \cos(\phi_i - \phi_j) - K \sum_i \cos(2\phi_i) \ , \tag{26}$$

which is guaranteed to never increase in time. The first term of the Lyapunov function corresponds to the Ising Hamiltonian, where the phase $\phi_i \in \{0, \pi\}$ modulo 2π represents the Ising spin $\sigma_i \in \{+1, -1\}$. The second term enforces the phase ϕ_i to be either 0 or π, where $\cos(2\phi_i) = 1$. As a result, the oscillator phases evolve to minimize the Ising Hamiltonian, converging toward a low-energy state that represents an approximate solution to the Ising problem. Further details regarding the OIM can be found in Chap. 9.

3.3.4 Continuous-time Boolean Satisfiability Solver

A continuous-time dynamical system (CTDS) for solving Boolean satisfiability (SAT) problems was proposed in [37]. This aims to find an assignment that satisfies the given Boolean formula. Although the system is not an Ising machine designed specifically for solving (quadratic) Ising problems, it seeks a set of binary states that minimizes a given objective function, which can be understood as an Ising Hamiltonian with high-order terms.

The CTDS solver explores the assignment of Boolean variables $X_1, \ldots, X_N$ satisfying the Boolean formula given in the conjunctive normal form (CNF). CNF is a conjunction (AND) of clauses, where each clause is a disjunction (OR) of literals, which can be a Boolean variable X_i or its negation $\neg X_i$.

Essentially, the CTDS solver is a gradient system of an objective function $V(\mathbf{x})$, defined on the search space $\mathbf{x} = (x_1, \ldots, x_N)^\top \in [-1, +1]^N$, where the ith component $x_i \in \{-1, +1\}$ corresponds to the Boolean variable $X_i \in \{\text{False}, \text{True}\}$. To define the objective function $V(\mathbf{x})$, the CNF is represented as a matrix $[c_{mi}]$; each component c_{mi} takes $+1$ or -1 if the mth clause includes X_i or $\neg X_i$, respectively, and $c_{mi} = 0$ if neither is included. The objective function is defined as

$$V(\mathbf{x}) = \sum_{m=1}^{M} a_m K_m(\mathbf{x})^2 , \qquad K_m(\mathbf{x}) = \prod_{i=1}^{N} \frac{1 - c_{mi} x_i}{2} , \qquad (27)$$

where $a_m > 0$ is the weight coefficient of the unsatisfiedness $K_m(\mathbf{x})$ to the current assignment $\mathbf{x}$ in the mth clause of the CNF. The objective function takes $V(\mathbf{x}) = 0$ if the CNF is satisfied, and takes a positive value otherwise. Therefore, the states with $V(\mathbf{x}) = 0$ constitute global minima of the objective function, regardless of the weight values $a_m > 0$. The CTDS solver is a gradient system of $V(\mathbf{x})$ with time-varying coefficients a_m defined as follows:

$$\frac{d\mathbf{x}}{dt} = -\nabla V(\mathbf{x}) , \qquad \frac{da_m}{dt} = a_m K_m(\mathbf{x}) . \qquad (28)$$

If the mth clause is not satisfied, the weight a_m increases because of the positive unsatisfiedness $K_m(\mathbf{x}) > 0$, which modifies the objective function $V(\mathbf{x})$. This effect helps the dynamics to escape from local minima, which is similar to the refractoriness of chaotic neural networks. The CTDS solver exhibits a transient chaotic behavior until it converges to a global minimum. The interaction dynamics of $\mathbf{x}$ and a_m was investigated in [38].

Although the original CTDS solver is described as a gradient system of a time-varying objective function, its variant is represented by a recurrent neural network [39]. For efficient numerical simulation of the CTDS solver, structure-preserving time discretization using the discrete gradient is effective in the gradient part of the solver [40].

3.4 Herding System

The RNNs discussed in this section thus far have fixed connection weights and exhibit nonlinear dynamics in their network states. In contrast, the learning process of neural networks, which modifies the connection weights through a learning rule, as in (11) and (12), introduces nonlinear dynamics into the parameter space.

Herding is a deterministic sampling algorithm that can be viewed as an extreme case of parameter learning in statistical models [41, 42]. It exhibits complex dynamics, yielding sample sequences guaranteed to satisfy the predefined statistics asymptotically, which are useful for estimating other statistics of interest. Thus, statistical learning and inference are combined in the single algorithm of herding.

In this section, we introduce the herding algorithm as the zero-temperature limit of the Boltzmann machine learning. A more general and detailed description of the herding algorithm is provided in Chap. 10. The learning rule of the Boltzmann machine $P(\mathbf{s}) \propto \exp(-H(\mathbf{s})/T)$ including the temperature parameter T is given as follows:

$$w_{ij}(t+1) = w_{ij}(t) + \frac{\alpha}{T} \left(\langle s_i s_j \rangle_{\text{data}} - \langle s_i s_j \rangle_{\text{model}} \right) , \tag{29}$$

$$b_i(t+1) = b_i(t) + \frac{\alpha}{T} \left(\langle s_i \rangle_{\text{data}} - \langle s_i \rangle_{\text{model}} \right) . \tag{30}$$

Let us consider the low-temperature limit, $T \to 0$, which corresponds to the Amari–Hopfield network. That is, the model distribution $P(\mathbf{s})$ reduces to a point distribution on the minimizer of the Hamiltonian $\arg\min_{\mathbf{s}} H(\mathbf{s})$. Because the minimizer is invariant under the positive scalar multiplication of parameters, we can omit the scaling factor α/T, without loss of generality, to obtain the following update rule:

$$w_{ij}(t+1) = w_{ij}(t) + \langle s_i s_j \rangle_{\text{data}} - s_i(t)s_j(t) , \tag{31}$$

$$b_i(t+1) = b_i(t) + \langle s_i \rangle_{\text{data}} - s_i(t) , \tag{32}$$

where $\mathbf{s}(t) = (s_1(t), \ldots, s_N(t))^\top$ is the minimizer of the energy function with parameters at the tth iteration, that is,

$$\mathbf{s}(t) = \arg\min_{\mathbf{s}} \left(-\frac{1}{2} \sum_{i,j=1}^{N} w_{ij}(t)s_i s_j - \sum_{i=1}^{N} b_i(t)s_i \right) . \tag{33}$$

Equations (31)–(33) describe the herding system applied to the Boltzmann machine, which is a nonlinear dynamical system on the parameter space of w_{ij}'s and b_i's. In each update of the parameter values, we obtain a sample $\mathbf{s}(t)$. The sequence of network states, $\{\mathbf{s}(t)\}_t$, can be considered as a sample sequence from the neural network.

Interestingly, the idea of updating the parameters of the Amari–Hopfield network away from the equilibrium state $\mathbf{s}(t)$ was proposed as "unlearning" by Hopfield et al. [43]. Weakly updating the parameters suppresses spurious memories, which are undesirable local minima that do not correspond to any of the memory patterns stored through the Hebbian rule. Thus, the herding algorithm can be viewed as performing strong unlearning within the Amari–Hopfield network.

The herding algorithm is described as a discrete-time nonlinear dynamical system that belongs to the class of piecewise isometries. As with many piecewise isometries

[18, 44–46], the herding system typically exhibits complex dynamics with a fractal attracting set [41, 42]. As the Lyapunov exponents of the dynamics are strictly zero, the complexity originates only from the discontinuities of the piecewise isometry. This non-chaotic dynamics of the herding system is closely related to chaotic billiard dynamics [47].

As a sampling method, the herding algorithm exhibits a prominent convergence rate $O(1/\tau)$, which is significantly faster than $O(1/\sqrt{\tau})$ of random sampling algorithms, such as MCMC. Specifically, for a sample sequence of length τ, the deviation of the sample average of $s_i(t)s_j(t)$ from the target $\langle s_i s_j \rangle_{\text{data}}$ is given by:

$$\frac{1}{\tau} \sum_{t=1}^{\tau} s_i(t)s_j(t) - \langle s_i s_j \rangle_{\text{data}} = -\frac{1}{\tau}(w_{ij}(\tau) - w_{ij}(0)) , \qquad (34)$$

which converges to zero as τ goes to infinity, at rate $O(1/\tau)$, because $w_{ij}(t)$ is assured to be bounded if the minimizer is obtained in each step (33).

The herded Gibbs sampling [48] is a deterministic sampling algorithm that incorporates the herding algorithm into the Gibbs sampling. It can be used as an alternative to the Gibbs sampling to promote efficient sampling from probabilistic models in general situations. The convergence behavior of the herded Gibbs sampling has been analyzed in detail [49].

4 Reservoir Computing

In the previous sections, we focused on the role of the energy function, which is crucial for understanding the dynamics of symmetrically connected RNNs. However, this approach is not applicable to RNNs with asymmetrical connections, which are more likely to exhibit complex nonlinear dynamics, making training more challenging. Reservoir computing is a lightweight approach that leverages the rich dynamics of an RNN for information processing without training the RNN itself, which is referred to as a reservoir. In this section, we present an overview of the fundamental models and concepts of reservoir computing, illustrating how the nonlinear dynamics of RNNs can be utilized in various computing applications.

4.1 Training Input–Output Relation of RNNs

In the RNNs presented in the previous sections, neurons do not receive explicit input from outside the networks, whereas in some cases, inputs are implicitly provided as the initial state for the autonomous dynamics of neural networks.

In this section, time-varying inputs are explicitly incorporated into neurons of an RNN, and we consider the input–output relation of the network. While feedforward neural networks, such as multilayer perceptrons (MLPs), learn input–output relations for static information, RNNs can handle sequential information because the effects of past inputs remain within the network and subsequently influence its current state and output.

Let us consider an RNN model with time-varying inputs. The state $x_i(t + 1)$ of the ith neuron at time $t + 1$ is given by

$$x_i(t + 1) = f \left(\sum_{j=1}^{N} w_{ij} x_j(t) + \sum_{k=1}^{K} w_{ik}^{\text{in}} u_k(t + 1) \right) , \qquad (35)$$

where $u_k(t + 1)$ denotes the kth input at time $t + 1$, and w_{ij} and w_{ik}^{in} are the synaptic weights of recurrent and input connections, respectively. The activation function f is assumed to be tanh throughout this section. For the sake of simplicity, the bias term is omitted. The output from the lth output neuron is determined by

$$y_l(t + 1) = f \left(\sum_{i=1}^{N} w_{li}^{\text{out}} x_i(t + 1) \right) , \qquad (36)$$

where w_{li}^{out} is the weight of the output (readout) connection from the ith neuron.

Next, we consider the training of the neural network using input–output relation data. Specifically, given a pair of an input sequence

$$\mathbf{u}(t) = (u_1(t), \dots, u_K(t))^\top , \qquad t = 1, \dots, \tau, \qquad (37)$$

and the corresponding output sequence

$$\mathbf{d}(t) = (d_1(t), \dots, d_L(t))^\top , \qquad t = 1, \dots, \tau, \qquad (38)$$

we adjust the connection weights in the network to minimize or decrease the output error

$$E = \sum_{t=1}^{\tau} \| \mathbf{d}(t) - \mathbf{y}(t) \|^2 , \qquad (39)$$

where $\{ \mathbf{y}(t) \}_t$ is the output sequence from the RNN given the input sequence $\{ \mathbf{u}(t) \}_t$.

The backpropagation through time (BPTT) [50] and real-time recurrent learning (RTRL) [51] are well-known gradient-based algorithms for training RNNs. In computing the gradients of the output error E in (39), the gradients are recursively multiplied at each time step due to the influence of past inputs on the outputs. This often leads to the gradients either vanishing or exploding, which makes the gradient-based learning of connection weights in RNNs challenging. Various techniques have been proposed to address this problem. Consequently, recent RNN models, such as long short-term memory (LSTM) [52], effectively handle sequential data, whereas the computation of the gradients for training remains computationally expensive.

4.2 Echo State Network

The echo state network (ESN) [53, 54] is a type of reservoir computing that employs a different approach to training the input–output relations of RNNs. In ESNs, input weights and recurrent weights are typically generated randomly and remain fixed, while only the output connections are trained using a simple linear regression algorithm. This makes ESNs computationally efficient and easy to train. Instead of the nonlinear activation function in (36), the lth output from the ESN is obtained linearly as

$$y_l(t + 1) = \sum_{i=1}^{N} w_{li}^{\text{out}} x_i(t + 1) \,. \tag{40}$$

The underlying principle is that the state of a random RNN, or a reservoir, reflects the input sequence through nonlinear transformations. If the nonlinear dynamics of the reservoir is sufficiently rich, inferences can be performed effectively using only linear regression methods, such as ridge regression and FORCE (first-order reduced and controlled error) learning.

Ridge regression is a batch algorithm that can be used to train the readout connection weights, which introduces an L_2 regularization term, parameterized by $\alpha > 0$, into the error function as follows:

$$E_{\text{ridge}} = \sum_{t=1}^{\tau} \|\mathbf{d}(t) - W^{\text{out}}\mathbf{x}(t)\|^2 + \frac{\alpha}{2} \sum_{l=1}^{L} \sum_{i=1}^{N} |w_{li}^{\text{out}}|^2 \,. \tag{41}$$

The readout connection weight $W^{\text{out}} = [w_{li}^{\text{out}}]$ that minimizes the error function is given by:

$$W^{\text{out}} = DX^{\top}(XX^{\top} + \alpha I)^{-1} \,, \tag{42}$$

where I is the identity matrix, and $D = [\mathbf{d}(1), \ldots, \mathbf{d}(\tau)]$ and $X = [\mathbf{x}(1), \ldots, \mathbf{x}(\tau)]$.

The FORCE learning [55] is an online regression algorithm that updates W^{out} iteratively as follows:

$$P(t + 1) = P(t) - \frac{P(t)\mathbf{x}(t)\mathbf{x}(t)^{\top} P(t)}{1 + \mathbf{x}(t)^{\top} P(t)\mathbf{x}(t)} \,, \tag{43}$$

$$W^{\text{out}}(t + 1) = W^{\text{out}}(t) - \frac{\mathbf{e}(t)\mathbf{x}(t)^{\top} P(t)}{1 + \mathbf{x}(t)^{\top} P(t)\mathbf{x}(t)} \,, \tag{44}$$

$$\mathbf{e}(t) = W^{\text{out}}(t)\mathbf{x}(t) - \mathbf{d}(t) \,, \tag{45}$$

where $P(0) = I/\alpha$.

These linear regression algorithms require significantly less computation time compared to conventional gradient-based learning methods for RNNs. However, as these algorithms still involve the manipulation of large matrices, more lightweight and biologically plausible learning algorithms [56, 57] can be employed for efficient reservoir computing.

4.3 Echo State Property and Reservoir Design

The primary principle of reservoir computing is that only the readout weights are trained. This implies that the performance of reservoir computing is largely dependent on the design of the reservoir.

The echo state property (ESP) is a concept that ensures a reservoir adequately reflects the input sequence, which is crucial for further information processing. When the inputs are transformed into reservoir states nonlinearly, it is important that the reservoir state becomes independent of its initial state after a sufficient amount of time has passed. This is critical because, without this property, the same input could yield different outputs, which is undesirable for the reproducibility of information processing. To prevent such inconsistencies, reservoirs are typically designed to satisfy the ESP.

Let $\mathbf{x}(t)$ and $\mathbf{x}'(t)$ represent the reservoir states with different initial states $\mathbf{x}(0)$ and $\mathbf{x}'(0)$, after receiving the same input sequence $\{\mathbf{u}(t)\}_t$. The ESP of a reservoir is defined as satisfying $\lim_{t\to\infty} \|\mathbf{x}(t) - \mathbf{x}'(t)\| = 0$ for any pair of different initial states $\mathbf{x}(0)$ and $\mathbf{x}'(0)$, and any input $\{\mathbf{u}(t)\}_t$.

Assuming tanh as the activation function, a sufficient condition for the ESP is that the largest singular value of $W = [w_{ij}]$ is less than 1. However, this condition is known to be empirically overly restrictive. Instead, we often require that W has the spectral radius of less than 1, which is expected to satisfy the ESP, though not necessarily in all cases [53].

However, using connection weights W with an excessively small spectral radius is undesirable, even though it indeed satisfies the ESP. If the spectral radius is small, the past input information is rapidly lost from the reservoir, making it difficult to effectively utilize a long input sequence for information processing. The memory to retain past input information can be measured using the memory capacity [58]. Empirically, the spectral radius is set slightly below 1 to promote both richer memory capacity and reservoir dynamics, expectedly without violating the ESP. Memory effects can also be enhanced by introducing leaky integrators as the constituent neurons.

The diversity of neuronal behavior is essential for the performance of reservoir computing, as the output is generated through a linear combination of neuronal activities in the reservoir. The diversity can be enhanced by introducing sparsity into the connection weight matrix W, because a dense W makes the inputs to the neurons become more similar. Increasing the number of neurons is also effective; however, this in turn can lead to an increased computation time and a higher risk of overfitting.

4.4 Neural Network Reservoirs

We have described basic concepts of reservoir computing using ESNs as a representative model for implementing a reservoir. However, it is important to note that reservoirs are not limited to this specific model.

Liquid state machines (LSMs) [59] are another important model of reservoir computing, proposed almost simultaneously and independently of ESNs. In contrast to ESNs, LSMs employ spiking neural network models, which are more biologically plausible and adequate for the modeling of information processing in the brain. LSMs also facilitate energy-efficient hardware implementations, such as FPGAs [60].

More generally, various RNN models, ranging from artificial neural network models to biologically plausible models, can serve as reservoirs. As described in Sect. 3, RNNs often exhibit chaotic dynamics. However, chaotic behavior is considered undesirable in reservoir computing, as its sensitive dependence on initial conditions contradicts the ESP. Therefore, attenuation mechanisms need to be introduced to ensure consistency. For instance, in a study on reservoir computing based on the chaotic neural networks [61], attenuation is achieved through connection weights W with a small spectral radius. Similarly, the chaotic Boltzmann machine serves as a reservoir by incorporating a reference clock to which each component is attenuated [62]. Moreover, an analog CMOS VLSI implementation of the chaotic Boltzmann machine has been utilized for energy-efficient reservoir computing [31].

As demonstrated in these examples, the fundamental concept of reservoir computing, which does not require manipulation of the reservoir itself, increases flexibility and enables efficient hardware implementations.

4.5 Physical Reservoir Computing

Furthermore, the concept of reservoir computing is not limited to utilizing RNNs as reservoirs. As there is no need to train the reservoir itself, any physical system exhibiting rich nonlinear dynamics can potentially be utilized as reservoirs.

The approach that leverages physical devices and phenomena as reservoirs, instead of using simulated models, is referred to as physical reservoir computing [63]. The nonlinear dynamics of the reservoir implemented physically is expected to be used for high-speed and energy-efficient computation.

A physical reservoir transforms the input to the reservoir state using its nonlinear dynamics. The output is obtained through a readout linear transformation of the measurements from the reservoir. As only the readout weights are trained, we do not have to manipulate the reservoir itself, which enables us to utilize various physical devices for computing. However, as in the case of ESNs, the performance of physical reservoir computing largely depends on the characteristics of the reservoir such as nonlinearity and memory effects. Building and tuning the physical reservoir, which may be sensitive to various environmental factors such as noise, can be challenging.

Therefore, it is crucial to select appropriate physical phenomena depending on the tasks performed by the reservoir.

Various types of physical devices and phenomena have been applied to reservoir computing [63]. Even if limited to photonic devices [64], there have been various studies utilizing optical node arrays [65, 66], optoelectric oscillators with delayed feedback [67–69], etc., as well as quantum dot networks [70] and optoelectric iterative-function systems [71] as presented in Chaps. 4 and 11 of this book.

5   Towards Photonic Neural Network Computing

We have seen how the nonlinear dynamics of RNNs can be utilized for computing. These basic models and concepts should serve as an important foundation for the implementation of neural networks using optical computing technologies.

Chapters in Part II of this book discuss fluorescence energy transfer (FRET) computing based on nanoscale networks of fluorescent particles, referred to as FRET networks. This can be considered as a type of physical reservoir computing in which FRET networks are employed as reservoirs.

Part III is devoted to spatial-photonic spin systems and is primarily related to Sect. 3 of this chapter. The spatial-photonic Ising machine (SPIM) introduced in Chap. 8 is an optical system capable of efficiently computing the energy function of the Ising model (19) using spatial light modulation. Recently, a new computing model for the SPIM has been proposed that improves its applicability to a variety of Ising problems and enables statistical learning as a Boltzmann machine [72]. Chapters 9 and 10 discuss the details of the herding system and OIM, which have been briefly introduced in this chapter.

Part IV consists of chapters discussing recent topics related to reservoir computing and its photonic implementation. In Chap. 11, a reservoir-computing system utilizing an electronic-optical implementation of iterated function systems (IFSs) as a reservoir is introduced. Chapter 12 introduces the hidden-fold network, which achieves high parameter efficiency by, in a sense, introducing the idea of reservoir computing into deep MLPs. Chapter 13 discusses brain-inspired reservoir computing, in which multiple reservoirs are hierarchically structured to model predictive coding for multimodal information processing.

Acknowledgements The author appreciates valuable comments from Professor Yuichi Katori and Dr. Hiroshi Yamashita.

References

1. W.S. McCulloch, W. Pitts, A logical calculus of the ideas immanent in nervous activity. Bull. Math. Biophys. **5**(4), 115–133 (1943). https://doi.org/10.1007/bf02478259
2. E.R. Caianiello, Outline of a theory of thought-processes and thinking machines. J. Theor. Biol. **1**, 204–235 (1961). https://doi.org/10.1016/0022-5193(61)90046-7
3. S. Amari, Learning patterns and pattern sequences by self-organizing nets of threshold elements. IEEE Trans. Comput. **C-21**, 1197–1206 (1972). https://doi.org/10.1109/T-C.1972.223477
4. J.J. Hopfield, Neural networks and physical systems with emergent collective computational abilities. Proc. Natl. Acad. Sci. **79**, 2554–2558 (1982). https://doi.org/10.1073/pnas.79.8.2554
5. H. Ramsauer, B. Schäfl, J. Lehner, P. Seidl, M. Widrich, L. Gruber, M. Holzleitner, T. Adler, D. Kreil, M.K. Kopp, G. Klambauer, J. Brandstetter, S. Hochreiter, Hopfield networks is all you need, in *International Conference on Learning Representations* (2021). https://openreview.net/forum?id=tL89RnzIiCd
6. D. Krotov, J.J. Hopfield, Large associative memory problem in neurobiology and machine learning, in *International Conference on Learning Representations* (2021). https://openreview.net/forum?id=X4y_10OX-hX
7. S. Amari, Characteristics of random nets of analog neuron-like elements. IEEE Trans. Syst. Man Cybern. **SMC-2**, 643–657 (1972). https://doi.org/10.1109/tsmc.1972.4309193
8. J.J. Hopfield, Neurons with graded response have collective computational properties like those of two-state neurons. Proc. Natl. Acad. Sci. **81**, 3088–3092 (1984). https://doi.org/10.1073/pnas.81.10.3088
9. J.J. Hopfield, D.W. Tank, "Neural" computation of decisions in optimization problems. Biol. Cybern. **52**, 141–152 (1985). https://doi.org/10.1007/bf00339943
10. D.H. Ackley, G.E. Hinton, T.J. Sejnowski, A learning algorithm for Boltzmann machines. Cogn. Sci. **9**, 147–169 (1985). https://doi.org/10.1016/S0364-0213(85)80012-4
11. S. Kirkpatrick, C.D. Gelatt, M.P. Vecchi, Optimization by simulated annealing. Science **220**, 671–680 (1983). https://doi.org/10.1126/science.220.4598.671
12. J.H.M. Korst, E.H.L. Aarts, Combinatorial optimization on a Boltzmann machine. J. Parallel Distrib. Comput. **6**, 331–357 (1989). https://doi.org/10.1016/0743-7315(89)90064-6
13. P. Smolensky, Information processing in dynamical systems: foundations of harmony theory, in *Parallel Distributed Processing: Explorations in the Microstructure of Cognition*, ed. by D.E. Rumelhart, J.L. McClelland (MIT Press, Cambridge, 1986)
14. G.E. Hinton, Training products of experts by minimizing contrastive divergence. Neural Comput. **14**, 1771–1800 (2002). https://doi.org/10.1162/089976602760128018
15. R. Salakhutdinov, G. Hinton, Deep Boltzmann machines, in *The 12th International Conference on Artificial Intelligence and Statistics (AISTATS 2009), vol. 5 of Proceedings of Machine Learning Research*, eds. by D. van Dyk, M. Welling (2009), pp. 448–455. https://proceedings.mlr.press/v5/salakhutdinov09a.html
16. R. Salakhutdinov, G. Hinton, An efficient learning procedure for deep Boltzmann machines. Neural Comput. **24**, 1967–2006 (2012). https://doi.org/10.1162/NECO_a_00311
17. K. Aihara, T. Takabe, M. Toyoda, Chaotic neural networks. Phys. Lett. A **144**, 333–340 (1990). https://doi.org/10.1016/0375-9601(90)90136-C
18. K. Aihara, H. Suzuki, Theory of hybrid dynamical systems and its applications to biological and medical systems. Philos. Trans. R. Soc. A **368**(1930), 4893–4914 (2010). https://doi.org/10.1098/rsta.2010.0237
19. J. Nagumo, S. Sato, On a response characteristic of a mathematical neuron model. Kybernetik **10**, 155–164 (1972). https://doi.org/10.1007/BF00290514
20. M. Hata, Dynamics of Caianiello's equation. J. Math. Kyoto Univ. **22**, 155–173 (1982). https://doi.org/10.1215/kjm/1250521865
21. M. Adachi, K. Aihara, Associative dynamics in a chaotic neural network. Neural Netw. **10**, 83–98 (1997). https://doi.org/10.1016/s0893-6080(96)00061-5

22. K. Kaneko, I. Tsuda, Chaotic itinerancy. Chaos **13**, 926–936 (2003). https://doi.org/10.1063/1.1607783
23. M. Oku, K. Aihara, Associative dynamics of color images in a large-scale chaotic neural network. Nonlinear Theory Appl. IEICE **2**, 508–521 (2011). https://doi.org/10.1587/nolta.2.508
24. M. Hasegawa, T. Ikeguchi, K. Aihara, Combination of chaotic neurodynamics with the 2-opt algorithm to solve traveling salesman problems. Phys. Rev. Lett. **79**, 2344–2347 (1997). https://doi.org/10.1103/PhysRevLett.79.2344
25. Y. Horio, K. Aihara, O. Yamamoto, Neuron-synapse IC chip-set for large-scale chaotic neural networks. IEEE Trans. Neural Netw. **14**, 1393–1404 (2003). https://doi.org/10.1109/tnn.2003.816349
26. Y. Horio, T. Ikeguchi, K. Aihara, A mixed analog/digital chaotic neuro-computer system for quadratic assignment problems. Neural Netw. **18**, 505–513 (2005). https://doi.org/10.1016/j.neunet.2005.06.022
27. Y. Horio, K. Aihara, Analog computation through high-dimensional physical chaotic neuro-dynamics. Physica D **237**, 1215–1225 (2008). https://doi.org/10.1016/j.physd.2008.01.030
28. H. Suzuki, J. Imura, Y. Horio, K. Aihara, Chaotic Boltzmann machines. Sci. Rep. **3**, 1610 (2013). https://doi.org/10.1038/srep01610
29. H. Suzuki, Monte Carlo simulation of classical spin models with chaotic billiards. Phys. Rev. E **88**, 052144 (2013). https://doi.org/10.1103/PhysRevE.88.052144
30. M. Blank, L. Bunimovich, Switched flow systems: pseudo billiard dynamics. Dyn. Syst. **19**, 359–370 (2004). https://doi.org/10.1080/14689360412331304309
31. M. Yamaguchi, Y. Katori, D. Kamimura, H. Tamukoh, T. Morie, A chaotic Boltzmann machine working as a reservoir and its analog VLSI implementation, in *2019 International Joint Conference on Neural Networks (IJCNN)* (2019). https://doi.org/10.1109/ijcnn.2019.8852325
32. I. Kawashima, T. Morie, H. Tamukoh, FPGA implementation of hardware-oriented chaotic Boltzmann machines. IEEE Access **8**, 204360–204377 (2020). https://doi.org/10.1109/access.2020.3036882
33. N. Mohseni, P.L. McMahon, T. Byrnes, Ising machines as hardware solvers of combinatorial optimization problems. Nat. Rev. Phys. **4**, 363–379 (2022). https://doi.org/10.1038/s42254-022-00440-8
34. T. Inagaki, Y. Haribara, K. Igarashi, T. Sonobe, S. Tamate, T. Honjo, A. Marandi, P.L. McMahon, T. Umeki, K. Enbutsu, O. Tadanaga, H. Takenouchi, K. Aihara, K.-I. Kawarabayashi, K. Inoue, S. Utsunomiya, H. Takesue, A coherent Ising machine for 2000-node optimization problems. Science **354**, 603–606 (2016). https://doi.org/10.1126/science.aah4243
35. H. Goto, K. Tatsumura, A.R. Dixon, Combinatorial optimization by simulating adiabatic bifurcations in nonlinear Hamiltonian systems. Sci. Adv. **5**, eaav2372 (2019). https://doi.org/10.1126/sciadv.aav2372
36. T. Wang, J. Roychowdhury, OIM: Oscillator-based Ising machines for solving combinatorial optimisation problems, in *Unconventional Computation and Natural Computation (UCNC 2019)*, eds. by I. McQuillan, S. Seki (2019), pp. 232–256. https://doi.org/10.1007/978-3-030-19311-9_19
37. M. Ercsey-Ravasz, Z. Toroczkai, Optimization hardness as transient chaos in an analog approach to constraint satisfaction. Nat. Phys. **7**, 966–970 (2011). https://doi.org/10.1038/nphys2105
38. H. Yamashita, K. Aihara, H. Suzuki, Timescales of Boolean satisfiability solver using continuous-time dynamical system. Commun. Nonlinear Sci. Numer. Simul. **84**, 105183 (2020). https://doi.org/10.1016/j.cnsns.2020.105183
39. B. Molnar, Z. Toroczkai, M. Ercsey-Ravasz, Continuous-time neural networks without local traps for solving boolean satisfiability, in *13th International Workshop on Cellular Nanoscale Networks and their Applications*. (IEEE, 2012). https://doi.org/10.1109/cnna.2012.6331411
40. H. Yamashita, K. Aihara, H. Suzuki, Accelerating numerical simulation of continuous-time Boolean satisfiability solver using discrete gradient. Commun. Nonlinear Sci. Numer. Simul. **102**, 105908 (2021). https://doi.org/10.1016/j.cnsns.2021.105908

41. M. Welling, Herding dynamical weights to learn, in *Proceedings of the 26th Annual International Conference on Machine Learning* (ACM, 2009). https://doi.org/10.1145/1553374.1553517
42. M. Welling, Y. Chen, Statistical inference using weak chaos and infinite memory. J. Phys. Conf. Ser. **233**, 012005 (2010). https://doi.org/10.1088/1742-6596/233/1/012005
43. J.J. Hopfield, D.I. Feinstein, R.G. Palmer, 'Unlearning' has a stabilizing effect in collective memories. Nature **304**(5922), 158–159 (1983). https://doi.org/10.1038/304158a0
44. A. Goetz, Dynamics of piecewise isometries. Illinois J. Math. **44**, 465–478 (2000). https://doi.org/10.1215/ijm/1256060408
45. H. Suzuki, K. Aihara, T. Okamoto, Complex behaviour of a simple partial-discharge model. Europhys. Lett. **66**, 28–34 (2004). https://doi.org/10.1209/epl/i2003-10151-x
46. H. Suzuki, S. Ito, K. Aihara, Double rotations. Discrete Contin. Dyn. Syst. A **13**, 515–532 (2005). https://doi.org/10.3934/dcds.2005.13.515
47. H. Suzuki, Chaotic billiard dynamics for herding. Nonlinear Theory Appl. IEICE **6**, 466–474 (2015). https://doi.org/10.1587/nolta.6.466
48. Y. Chen, L. Bornn, N. de Freitas, M. Eskelin, J. Fang, M. Welling, Herded Gibbs sampling. J. Mach. Learn. Res. **17**(10), 1–29 (2016). http://jmlr.org/papers/v17/chen16a.html
49. H. Yamashita, H. Suzuki, Convergence analysis of herded-Gibbs-type sampling algorithms: effects of weight sharing. Stat. Comput. **29**, 1035–1053 (2019). https://doi.org/10.1007/s11222-019-09852-6
50. P.J. Werbos, Backpropagation through time: what it does and how to do it. Proc. IEEE **78**, 1550–1560 (1990). https://doi.org/10.1109/5.58337
51. R.J. Williams, D. Zipser, A learning algorithm for continually running fully recurrent neural networks. Neural Comput. **1**, 270–280 (1989). https://doi.org/10.1162/neco.1989.1.2.270
52. S. Hochreiter, J. Schmidhuber, Long short-term memory. Neural Comput. **9**, 1735–1780 (1997). https://doi.org/10.1162/neco.1997.9.8.1735
53. H. Jaeger, The "echo state" approach to analysing and training recurrent neural networks, GMD Report 148, GMD Forschungszentrum Informationstechnik (2001). https://doi.org/10.24406/publica-fhg-291111
54. H. Jaeger, H. Haas, Harnessing nonlinearity: predicting chaotic systems and saving energy in wireless communication. Science **304**(5667), 78–80 (2004). https://doi.org/10.1126/science.1091277
55. D. Sussillo, L.F. Abbott, Generating coherent patterns of activity from chaotic neural networks. Neuron **63**, 544–557 (2009). https://doi.org/10.1016/j.neuron.2009.07.018
56. G.M. Hoerzer, R. Legenstein, W. Maass, Emergence of complex computational structures from chaotic neural networks through reward-modulated hebbian learning. Cereb. Cortex **24**, 677–690 (2012). https://doi.org/10.1093/cercor/bhs348
57. M. Nakajima, K. Inoue, K. Tanaka, Y. Kuniyoshi, T. Hashimoto, K. Nakajima, Physical deep learning with biologically inspired training method: gradient-free approach for physical hardware. Nat. Commun. **13**, 7847 (2022). https://doi.org/10.1038/s41467-022-35216-2
58. H. Jaeger, Short term memory in echo state networks, GMD Report 152, GMD Forschungszentrum Informationstechnik (2001). https://doi.org/10.24406/publica-fhg-291107
59. W. Maass, T. Natschläger, H. Markram, Real-time computing without stable states: a new framework for neural computation based on perturbations. Neural Comput. **14**, 2531–2560 (2002). https://doi.org/10.1162/089976602760407955
60. B. Schrauwen, M. D'Haene, D. Verstraeten, J.V. Campenhout, Compact hardware liquid state machines on FPGA for real-time speech recognition. Neural Netw. **21**, 511–523 (2008). https://doi.org/10.1016/j.neunet.2007.12.009
61. Y. Horio, Chaotic neural network reservoir, in *2019 International Joint Conference on Neural Networks (IJCNN)* (2019). https://doi.org/10.1109/IJCNN.2019.8852265
62. Y. Katori, H. Tamukoh, T. Morie, Reservoir computing based on dynamics of pseudo-billiard system in hypercube, in *2019 International Joint Conference on Neural Networks (IJCNN)* (2019). https://doi.org/10.1109/IJCNN.2019.8852329

63. G. Tanaka, T. Yamane, J.B. Héroux, R. Nakane, N. Kanazawa, S. Takeda, H. Numata, D. Nakano, A. Hirose, Recent advances in physical reservoir computing: a review. Neural Netw. **115**, 100–123 (2019). https://doi.org/10.1016/j.neunet.2019.03.005
64. G. Van der Sande, D. Brunner, M.C. Soriano, Advances in photonic reservoir computing. Nanophotonics **6**, 561–576 (2017). https://doi.org/10.1515/nanoph-2016-0132
65. K. Vandoorne, W. Dierckx, B. Schrauwen, D. Verstraeten, R. Baets, P. Bienstman, J.V. Campenhout, Toward optical signal processing using photonic reservoir computing. Opt. Express **16**, 11182 (2008). https://doi.org/10.1364/oe.16.011182
66. K. Vandoorne, P. Mechet, T.V. Vaerenbergh, M. Fiers, G. Morthier, D. Verstraeten, B. Schrauwen, J. Dambre, P. Bienstman, Experimental demonstration of reservoir computing on a silicon photonics chip. Nat. Commun. **5**, 3541 (2014). https://doi.org/10.1038/ncomms4541
67. L. Appeltant, M.C. Soriano, G. Van der Sande, J. Danckaert, S. Massar, J. Dambre, B. Schrauwen, C.R. Mirasso, I. Fischer, Information processing using a single dynamical node as complex system. Nat. Commun. **2**, 468 (2011). https://doi.org/10.1038/ncomms1476
68. L. Larger, M.C. Soriano, D. Brunner, L. Appeltant, J.M. Gutierrez, L. Pesquera, C.R. Mirasso, I. Fischer, Photonic information processing beyond Turing: an optoelectronic implementation of reservoir computing. Opt. Express **20**, 3241 (2012). https://doi.org/10.1364/oe.20.003241
69. Y. Paquot, F. Duport, A. Smerieri, J. Dambre, B. Schrauwen, M. Haelterman, S. Massar, Optoelectronic reservoir computing. Sci. Rep. **2**, 287 (2012). https://doi.org/10.1038/srep00287
70. N. Tate, Y. Miyata, S. Sakai, A. Nakamura, S. Shimomura, T. Nishimura, J. Kozuka, Y. Ogura, J. Tanida, Quantitative analysis of nonlinear optical input/output of a quantum-dot network based on the echo state property. Opt. Express **30**, 14669–14676 (2022). https://doi.org/10.1364/OE.450132
71. N. Segawa, S. Shimomura, Y. Ogura, J. Tanida, Tunable reservoir computing based on iterative function systems. Opt. Express **29**, 43164 (2021). https://doi.org/10.1364/oe.441236
72. H. Yamashita, K. Okubo, S. Shimomura, Y. Ogura, J. Tanida, H. Suzuki, Low-rank combinatorial optimization and statistical learning by spatial photonic Ising machine. Phys. Rev. Lett. **131**, 063801 (2023). https://doi.org/10.1103/PhysRevLett.131.063801

Fluorescence Energy Transfer Computing

Fluorescence Energy Transfer Computing

Takahiro Nishimura

Abstract This chapter presents the concept and implementation of fluorescence energy transfer computing, specifically utilizing Förster resonance energy transfer (FRET) between molecular fluorophores and quantum dots. FRET is a non-radiative form of excitation energy transfer that depends on the configuration and optical properties of molecular fluorophores and quantum dots. By designing energy flows through FRET, signal processing can be implemented to perform desired operations. Because the phenomenon occurs at the nanometer scale, miniaturization of information devices can be expected. This chapter reviews the concepts of FRET computing and the implementation of FRET computing devices. Then, a framework of DNA scaffold logic, which systematically handles FRET-based logic operations, is described. Finally, the idea of a FRET network is discussed as a method for enhancing FRET computing performance.

1 Fluorescence Energy Transfer

Nanoscale fluorophores such as molecular fluorophores and quantum dots serve as interfaces between the nano- and macro-worlds using optical excitation and emission [1]. Fluorescence imaging provides information with nanoscale spatial resolution [2]. Molecular beacons, which modulate the fluorescence properties in response to input stimuli in a molecular environment, have been widely used in biomedical measurements [3]. Designing combinations of input stimuli and fluorescence modulation also enables the fabrication of nanoscale devices including photonic wires [4] logic gates [5], and memory [6]. The integration of these diverse nanoscale devices is expected to enable sophisticated information exchange between the nanoscale and macroscale [7]. In systems that use the propagation of light, the spatial resolution and processing density are often limited by the diffraction of light [8]. Nanoscale fluorophores are advantageous for optical systems with resolutions beyond the diffraction limit [9].

T. Nishimura (✉)
Graduate School of Engineering, Osaka University, 2-1, Yamadaoka, Suita, Osaka, Japan
e-mail: nishimura-t@see.eng.osaka-u.ac.jp

© The Author(s) 2024
H. Suzuki et al. (eds.), *Photonic Neural Networks with Spatiotemporal Dynamics*,
https://doi.org/10.1007/978-981-99-5072-0_3

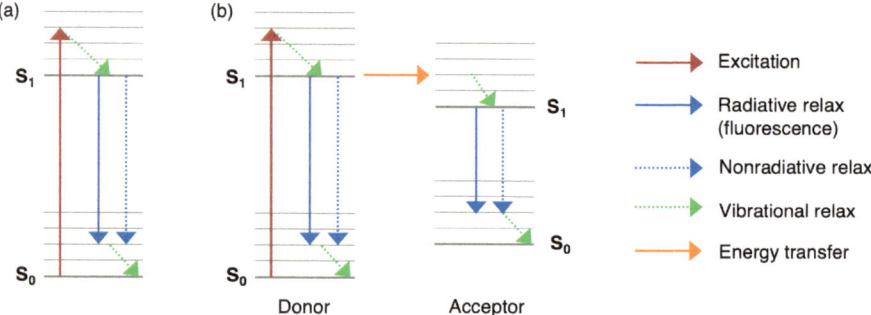

Fig. 1 Schematic illustration of energy diagram of **a** excitation of single fluorescent molecule and **b** FRET

The excited state energy of nanoscale fluorophores can be relaxed through various processes (Fig. 1a) [1]. In addition to radiative processes such as fluorescence, non-radiative processes such as thermal dissipation can also be involved in relaxation. Another feature of the relaxation process is fluorescence energy transfer, which allows the excitation energy to move non-radiatively between nearby fluorophores (Fig. 1b). Förster resonance energy transfer (FRET) is a dipole-dipole interaction-based fluorescence energy transfer technique that has been widely applied in various bio-applications [10, 11].

FRET is a phenomenon in which excitation energy is transferred from one molecular fluorophore or quantum dot to a nearby molecular fluorophore [12, 13] or quantum dot [14]. The energy transfer rate constant k is expressed as follows [1]:

$$k = \frac{9000(\ln 10)}{128\pi^5 \eta^4 N_A \tau_a} \cdot \frac{\phi_a \kappa^2}{R^6} \int f_a(\lambda)\epsilon_b(\lambda)\lambda^4 d\lambda, \qquad (1)$$

where η is the refractive index of the medium, N_A is Avogadro's number, τ_a is the donor fluorescence lifetime, ϕ_a is the donor fluorescence quantum yield, κ is the orientation factor, $f_a(\lambda)$ is the normalized fluorescence spectrum of the donor, $\epsilon_b(\lambda)$ is the absorption spectrum of the acceptor, R is the distance between the donor and acceptor, and ν is the frequency. The distance R at which k becomes 0.5 is known as the Förster radius and can be expressed as [1]

$$R_0 = 0.211[\kappa^2 \phi_a \eta^{-4} \int f_a(\lambda)\epsilon_b(\lambda)\lambda^4 d\lambda]^{1/6}. \qquad (2)$$

R_0 is typically between 5 and 10 nm [12, 15].

From Eq. (1), FRET occurs when the donor is in close proximity to the ground-state acceptor and the acceptor's absorption spectrum significantly overlaps with the donor's emission spectrum. Figure 2 shows an example of FRET. FRET serves as one of the relaxation pathways for the excited donor. The excitation energy transferred to the acceptor undergoes a relaxation process, including acceptor fluorescence, result-

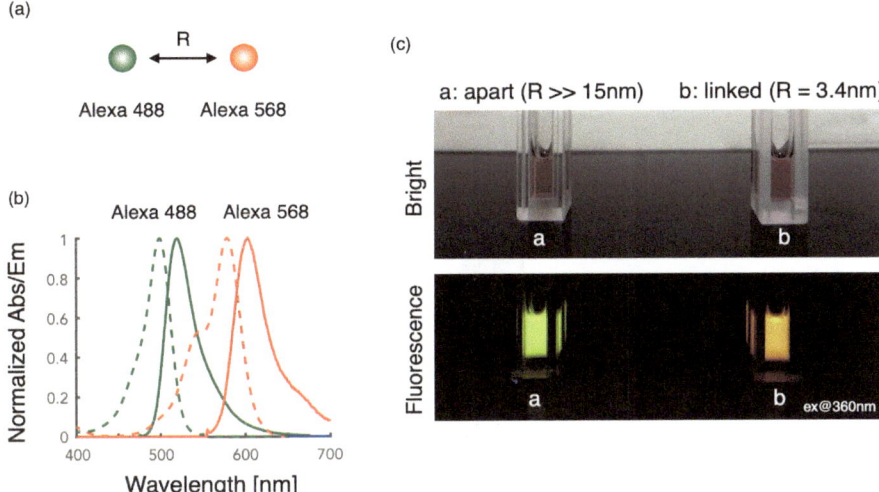

Fig. 2 FRET between Alexa 488 and Alexa 568. **a** Molecular fluorophores. **b** Absorption and fluorescence spectra of Alexa 488 and Alexa 568. **c** Fluorescence with and without FRET

ing in a signal with a wavelength different from that of the donor fluorescence. The FRET pathway competes with the natural fluorescence process, reducing the fluorophore's excitation state lifetime and fluorescence probability. FRET modulates the fluorescence intensity and lifetime signals. The probability of FRET depends on the absorption and emission properties and the relative positions of the fluorescent molecules, indicating that fluorescence signals can be modulated by FRET.

Fluorescence signal modulation through FRET using the relative positions of fluorescent molecules is widely used in biomolecular sensing and measurement [16, 17] (Fig. 3a). The FRET technique has been applied to molecular beacons by linking binding reactions in molecular sensing to changes in the positions of donor acceptors [18]. Molecular rulers have been proposed based on the fact that the donor–acceptor distance affects the FRET efficiency [19]. This allows nanometer-scale information to be read as a fluorescence signal. In addition, by using fast FRET measurements, molecular conformational changes can be monitored [20].

Another technique, fluorescence modulation based on FRET, can be achieved by changing the absorption of an acceptor molecule through light irradiation [21, 22] (Fig. 3b). According to Eq. (1), the efficiency of energy transfer in FRET is also determined by the degree of overlap between the fluorescence spectrum of the donor molecule and the absorption spectrum of the acceptor molecule. Photoresponsive proteins and photoswitchable fluorescent molecules can change their absorption spectra upon light irradiation through induced changes in their molecular structures or binding to functional groups. These molecules allow on/off switching of signal transmission through FRET in response to optical signals. Another method involves inhibiting the excitation transition from the donor by optically exciting the acceptor

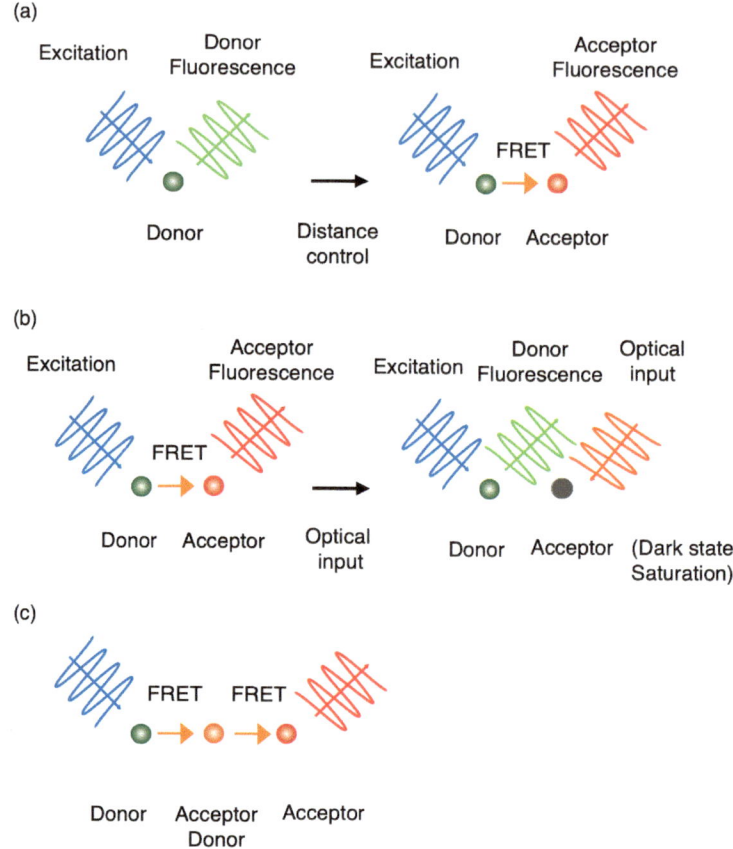

Fig. 3 Fluorescence signal modulation. FRET control using **a** relative positions of fluorescence molecules, **b** photoswitchable fluorescent molecules, and **c** multi-step transfers

to an excited state [23]. This technique usually requires high-power irradiation to saturate the acceptor.

FRET occurs in multiple steps between multiple phosphors [1, 24] (Fig.3c). Multi-step FRET (cascade of FRET) can also be understood in principle using Förster theory. Multi-step FRET has been experimentally demonstrated between fluorescent molecules [4] arranged in one dimension and between quantum dots [25]. Compared to simple single-step FRET, energy can be transferred across longer ranges and more diverse fluorescence signal generation is possible [26]. Furthermore, FRET can also occur when quantum dots of the same or multiple species are arranged in two dimensions [27]. Theoretical analyses are underway, and diverse fluorescence signal modulations have been experimentally verified [28, 29] and theoretically [30]. Applications of computing utilizing this fluorescence signal modulation are also being studied [31].

2 FRET-Based Device

Controlling energy flow through FRET enables the implementation of nanoscale signal processing devices. In FRET control, the distance between the fluorescent molecules is a key parameter. DNA self-assembly is a useful method for positioning fluorescent molecules to achieve FRET control. This method has been used to align multiple fluorescent molecules and quantum dots and to design FRET pathways between them. Dynamic FRET pathways can be established based on the binding/dissociation switches of fluorescent molecules using DNA reactions. Other important factors that may influence FRET-based devices include overlap and excitation states. This section reviews FRET-based devices that use DNA self-assembly.

2.1 DNA for FRET Devices

DNA is a polymer molecule composed of four types of nucleotides: adenine (A), guanine (G), cytosine (C), and thymine (T), which are linked together (Fig. 4a). Based on the Watson–Crick complementarity, as shown in Fig. 4b, A pairs specifically with T and G pairs specifically with C through hydrogen bonding. Single-stranded DNA with complementary sequences can selectively bind together to form a double-helix structure, as shown in Fig. 4c. Typically, double-stranded DNA has a diameter of 2 nm and contains 10.5 base pairs (bp) per helical turn, corresponding to a length of 3.4 nm [32]. By using appropriate sequence design, double-stranded DNA structures can be programmed. Various two- and three-dimensional nanostructures have been created using this approach [33]. Additionally, other functional molecules can be conjugated

Fig. 4 **a** Adenine (A), Thymine (T), Guanine (G), Cytosine (C). **b** Complementary binding. **c** Structure and size. **d** DNA hybridization reaction

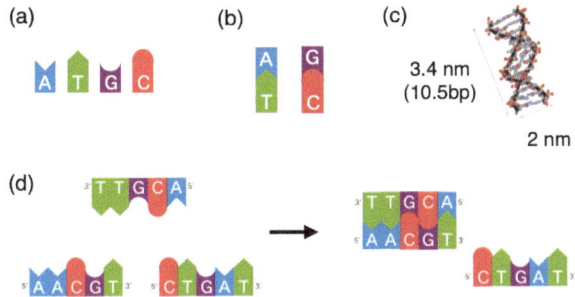

to DNA to introduce new functions. Fluorescent molecules and quantum dots can be attached to the ends of DNA to control the nanoscale arrangement, enabling the implementation of fluorescent configurations to control FRET efficiency [20].

2.2 FRET-Based Photonic Wire

Nanoscale photonic wires are of particular interest for nanoscale information devices because of their ability to control energy transfer via FRET and deliver light energy. The simplest and most common photonic wire design is multi-step FRET with a linear array of phosphors [26]. Such wires exhibit a linear array of dye absorption and emission that changes from blue to red, creating an energy landscape of downward energy transfer within the structure [34]. Photonic wires have been constructed using up to seven dye molecules sequentially arranged on DNA scaffolds [34]. Nanoscale photonic wires have also been realized using FRET between quantum dots and between QD-dye [35].

2.3 FRET-Based Photonic Switch

The FRET optical switch enables on/off control of the energy transfer along the FRET pathway and selection of the pathway. Control of the FRET pathway via light irradiation can be utilized for light access from macro- to nanoscale information devices. An optical switching technique using energy transfer between quantum dots and the saturation of the excitation level has been proposed [36]. Similar approaches have been employed for FRET pathway manipulation using fluorescent molecules [23]. In these techniques, the excitation levels of the acceptors are exclusively occupied to inhibit FRET and adopt a different FRET pathway from that in the non-excited state. Other techniques use the optical control of molecular separation or cleavage to regulate the distance between fluorescent molecules [37]. Although most of these methods are based on permanent changes, their application to high-density optical storage has been proposed [38].

2.4 FRET-Based Logic Gate

Logic gates based on FRET are interesting not only in terms of expanding the applications of information processing technology, but also for achieving on-site information processing in the nano-world. Several approaches have been proposed for logic gates based on FRET [39–41]. The inputs are represented by pH, molecules, and light stimulation, whereas the outputs are represented by the responsible fluorescent signals [42].

3   Scaffold DNA Logic

Scaffold DNA logic has been proposed as a method for systematically constructing information systems using FRET [43]. This is a molecular computing system that accepts molecules, such as nucleic acids, as inputs and processes them using signal processing based on FRET in an intra-single DNA structure. Molecular circuits design and implement complex molecular reaction systems to construct molecular signal systems, so that the desired processing is executed against the inputs. This section presents the scaffold DNA logic.

3.1   Basic Concept

The energy transfer efficiency of FRET depends on the type and spatial arrangement of the fluorescent molecules, and FRET transfer can be controlled by selecting and manipulating them. As shown in Fig. 5, by arranging or removing fluorescent molecules on scaffold DNA in response to the input, on/off switching of signal transmission via FRET can be achieved. When fluorescent molecules of FRET pairs are arranged on the scaffold DNA within the FRET-allowable distance, FRET occurs, and the signal transmission is turned on. On the other hand, when they are removed from the scaffold DNA, FRET does not occur, and the signal is not transmitted.

Because input signals are assumed to be molecular inputs, it is necessary to control the arrangement of fluorescent molecules on the scaffold DNA in response to the input molecule and to obtain the calculation results through signal processing based on FRET-based signal processing constructed on the scaffold DNA. An overview of this process is shown in Fig. 6. When the fluorescent molecule is excited, the output is represented by 1 or 0, depending on whether the output fluorescent molecule is excited via FRET. If the input signal meets these conditions, the FRET circuit is

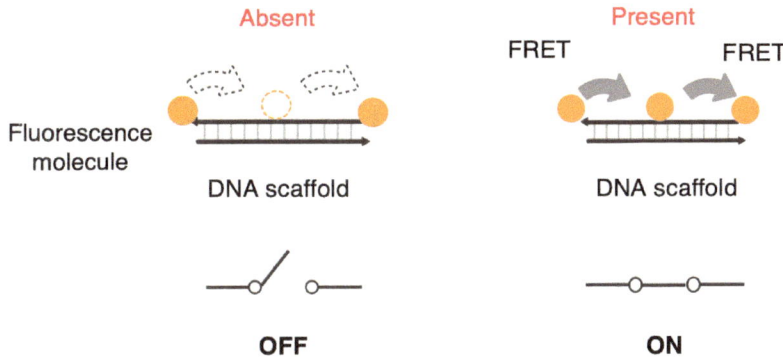

Fig. 5   FRET signal control through positional control of fluorophores

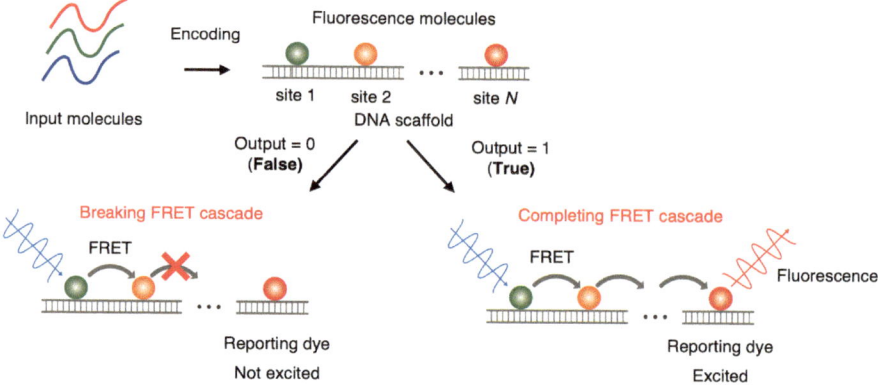

Fig. 6 FRET cascade control through positional control of fluorescent molecules

complete, the output fluorescent molecule is excited, and the fluorescent signal is the output. This state is defined as an output of 1. If the given logical conditions are not satisfied, the FRET pathway is not completely constructed, and the output fluorescent molecule is not excited. This state is defined as an output of 0. Local FRET phenomena on the scaffold DNA allow the necessary signal processing for the execution of logical operations.

3.2 Arrangement Control of Fluorescent Molecules

The positioning of the fluorescent molecules on the scaffold DNA is controlled using connecting DNA (Fig. 7a) and disconnecting DNA (Fig. 7b). The connecting and disconnecting DNA are modified with fluorescent molecules and are composed of recognition and address regions. The recognition region recognizes the input molecule, and the address region specifies binding to the prepared sites on the scaffold DNA. In the absence of an input molecule, the connecting DNA forms a closed hairpin structure. In this state, the address region is covered and the connecting DNA maintains a dissociated state from the scaffold DNA. Therefore, fluorescent molecules were not positioned on the scaffold DNA. In the presence of an input molecule, the connecting DNA (Fig. 7a) binds to the input molecule, and the hairpin structure opens. In this state, the address region is in a single-stranded state and binds to the scaffold DNA. As a result, the fluorescent molecules are positioned at the site specified on the scaffold DNA. The disconnecting DNA (Fig. 7b) binds to the specified site within the scaffold DNA when the input molecule is absent. In the presence of the input molecule, it forms a linear structure by binding to the recognition region, resulting in its dissociation from the scaffold DNA. Using these configurations, the positioning/removal of fluorescent molecules can be controlled depending on the presence or absence of the input molecule.

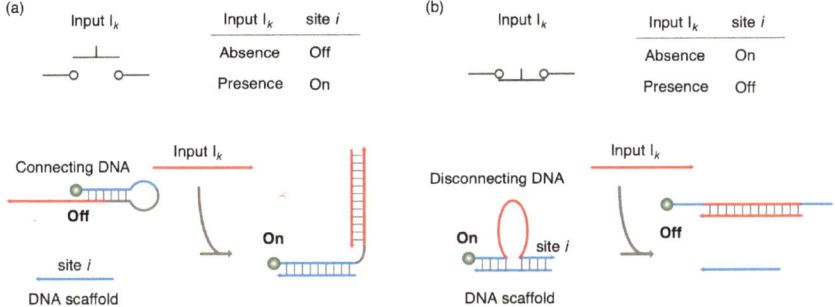

Fig. 7 Fluorescent molecule control on DNA scaffold using **a** connecting DNA and **b** disconnecting DNA. Reproduced with permission from Appl. Phys. Lett. 101, 233703 (2012). Copyright 2012, AIP Publishing LLC

3.3 Design of Logic Operations

By constructing connecting or disconnecting DNA, it is possible to pre-specify the input molecule species, placement of fluorescent molecules, and placement sites on the scaffold DNA. Through these combinations, it is possible to establish FRET pathways on the scaffold DNA for the input molecules. Placing FRET pair molecules on adjacent sites corresponds to connecting the FRET pathways in series, enabling the implementation of the AND operation (Fig. 8a). In contrast, placing the same type of fluorescent molecule at the same site for multiple inputs represented by different molecular species corresponds to connecting the FRET transmission pathways in parallel. This enables the implementation of the OR operation (Fig. 8b). In addition, the NOT operation can be implemented by utilizing the relationship between the input and signal transmission through the disconnecting DNA (Fig. 7b). Because any logical expression represented in a conjunctive normal form can be implemented, any logical operation can theoretically be performed. The following sections provide an overview of each operational experiment. For details on the experiments, including the DNA sequences, please refer to [43].

3.4 AND Operation

To confirm the AND operation, $I1 \wedge I2$ operation was performed using two types of single-stranded DNA, strands I1 and I2, as input molecules. Figure 9 shows the reaction scheme. FAM and Alexa 546 were used as FRET pairs. Strand C1 detects strand I1 and positions FAM at site 1. Strand C2 recognizes strand I2 and positions Alexa 546 at site 2. When the input (1, 1) is present, FAM and Alexa 546 bind to the DNA scaffold. The distance between the modification positions of FAM and Alexa 546 is 13 bp, which is calculated to be 4.6 nm. Because the Förster radius of FAM

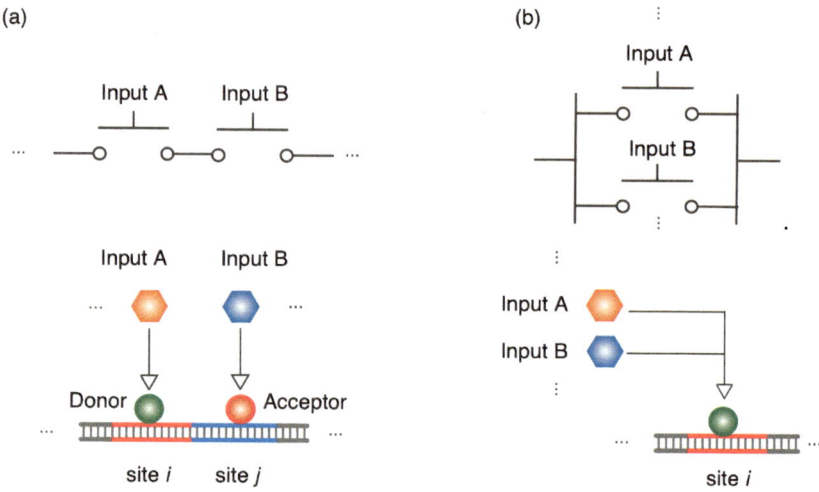

Fig. 8 Implementation of **a** AND and **b** OR operations. Reproduced with permission from Appl. Phys. Lett. 101, 233703 (2012). Copyright 2012, AIP Publishing LLC

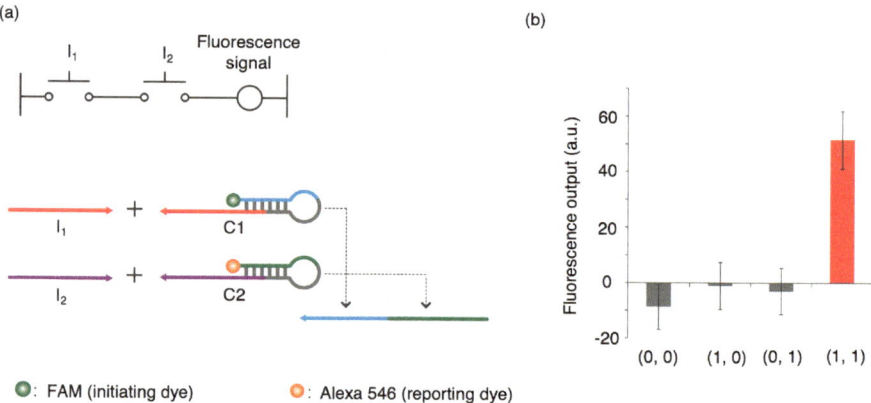

Fig. 9 **a** Design and **b** results for $I1 \wedge I2$ operation. Reproduced with permission from Appl. Phys. Lett. 101, 233703 (2012). Copyright 2012, AIP Publishing LLC

and Alexa 546 is 6.4 nm, they are sufficiently close for FRET to occur. At this point, the excitation energy is transferred from FAM to Alexa 546 via FRET, resulting in an output of 1. In other cases, at least one of FAM and Alexa 546 is not bound to the scaffold DNA, and FRET does not occur, resulting in Alexa 546 not being excited. This state results in an output of 0.

Figure 9a shows the reaction scheme for $I1 \wedge I2$ operation. The fluorescence intensity of FAM decreased only in the case of (1, 1), and that of Alexa 546 increased. The decrease in FAM fluorescence intensity in the case of (1, 0) may be attributed to the quenching effect of the DNA bases. The fluorescence output was evaluated as

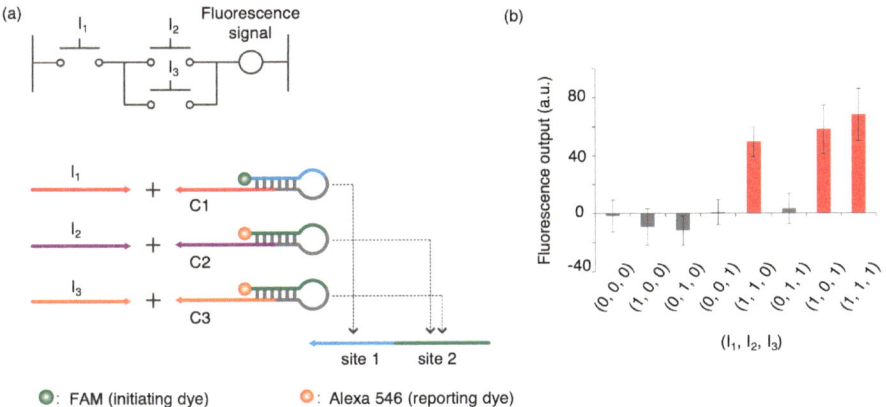

Fig. 10 **a** Design and **b** results for $I1 \land (I2 \lor I3)$ operation. Reproduced with permission from Appl. Phys. Lett. 101, 233703 (2012). Copyright 2012, AIP Publishing LLC

the change in intensity before and after the input at the fluorescence peak wavelength of the output molecule. The results are shown in Fig. 9b. The fluorescence output significantly increased in the case of (1, 1). The reason for the negative values could be the decrease in concentration due to the sample volume increase caused by the input. These results demonstrate that an appropriate AND operation can be achieved by positioning the FRET pair of fluorescent molecules at adjacent sites using different input molecules.

3.5 OR Operation

To confirm the OR operation, three types of strands, I1, I2, and I3, were used as inputs to perform $I1 \land (I2 \lor I3)$. The reaction scheme is shown in Fig. 10a. Strand C3, which recognizes strand I3 and places Alexa 546 at site 2, was added to the reaction system used for $I1 \land I2$. When strands I2 or I3 are present, Alexa 546 is placed at site 2. This allows OR operations on $I2$ and $I3$. When (1, 1, 0), (1, 0, 1), or (1, 1, 1) are input, FAM and Alexa 546 bind to the scaffold DNA. Only in this case does the excited energy transfer from FAM to Alexa 546 through FRET, resulting in the excitation of Alexa 546, which is defined as an output of 1. In other cases, because neither FAM nor Alexa 546 binds to the scaffold DNA, FRET does not occur, and Alexa 546 is not excited, which is defined as an output of 0.

The fluorescence outputs of each input are shown in Fig. 10b. The fluorescence output increases significantly for the (1, 1, 0), (1, 0, 1), and (1, 1, 1) cases. However, in the other cases, the fluorescence output is almost zero, indicating that $I1 \land (I2 \lor I3)$ is accurately performed.

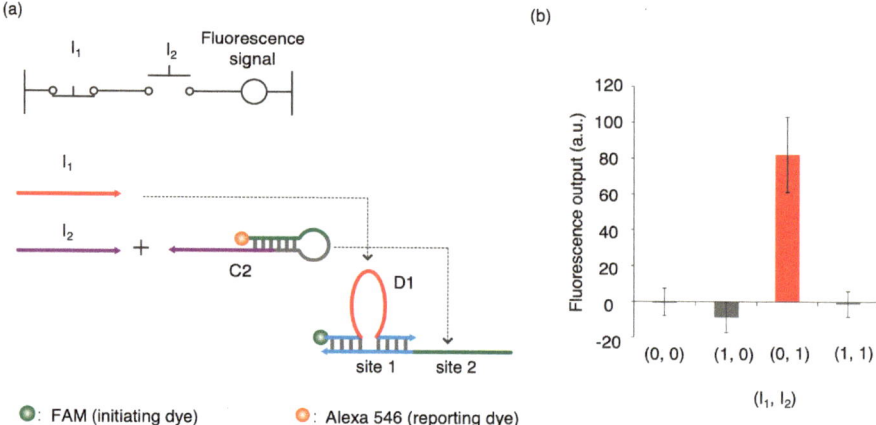

Fig. 11 **a** Design and **b** results for $\neg I1 \wedge I2$. Reproduced with permission from Appl. Phys. Lett. 101, 233703 (2012). Copyright 2012, AIP Publishing LLC

3.6 NOT Operation

The NOT operation was implemented using disconnecting DNA. The system for $\neg I1 \wedge I2$ is illustrated in Fig. 11a. Strand C1 used in the reaction system for $I1 \wedge I2$ is replaced with strand D1. Strand D1 releases FAM from site 1 when it recognizes strand I1, enabling the execution of the NOT operation. In this case, the fluorescence output increases significantly via FRET for the (0, 1) case. The fluorescence outputs for each input are shown in Fig. 11b. The fluorescence output increases significantly in the (0, 1) case, whereas it is almost zero in the other cases. Thus, logic operations, including NOT operations, can be performed accurately using disconnecting DNA.

3.7 Extended FRET Connection

To confirm that the FRET connections could be extended, $I1 \wedge I2 \wedge I3$ was implemented using multi-stage FRET. The reaction system used for this operation is shown in Fig. 12a. The fluorescent molecules used were FAM, Alexa 546, and Alexa 594. For the reaction system used in $I1 \wedge I2$, we changed strand S1 to strand S2, which has three sites. We also introduced strand C4, which places Alexa 594 at site 3 when recognizing strand I1. As shown in Fig. 12b, the output intensity increases for the input (1, 1, 1). It was demonstrated that circuit expansion is possible by utilizing multi-stage FRET. The slight increase in the output intensity in the cases of (0, 1, 1) and (1, 0, 1) may be due to the excitation of Alexa 546 and the FRET between FAM and Alexa 594, which can be improved by adjusting the fluorescent molecule composition and excitation wavelength.

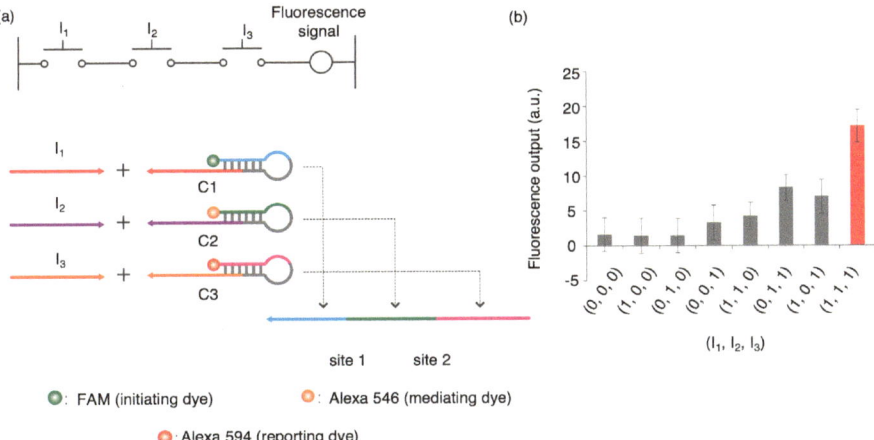

Fig. 12 **a** Design and **b** results for $I1 \wedge I2 \wedge I3$. Reproduced with permission from Appl. Phys. Lett. 101, 233703 (2012). Copyright 2012, AIP Publishing LLC

4 Optical Program of DNA Scaffold Logic

Optical input can be incorporated into the framework of the DNA scaffold logic. The use of fluorescent molecules that can reversibly change their fluorescence properties upon light irradiation is effective. In this subsection, we introduce light control using cyanine-based fluorescent molecules, as well as their applications in the optical control of the FRET pathway and optically programmable DNA scaffold logic.

4.1 Optical Control of FRET

According to Eq. (1), the energy transfer efficiency in FRET depends on the overlap between the fluorescence spectrum of the donor molecule and the absorption spectrum of the acceptor molecule. Photoresponsive proteins and fluorescent molecules are molecules whose absorption spectra change because of changes in their molecular structure induced by light irradiation or binding with functional groups [44]. Using these molecules, it is possible to switch the on/off state of signal transmission via FRET in response to light signals [45]. Here, we utilize optical control based on cyanine-based fluorescent molecules as an example of absorption spectrum modulation. Cyanine-based fluorescent molecules are light-controllable molecules that can control their fluorescent or bleached state through light irradiation [44]. This mechanism is based on a photochemical reaction that induces the binding and dissociation of the thiols of the cyanine dye [44]. Additionally, when other fluorescent molecules (activators) are present near the cyanine-based molecules, the dissociation of the cyanine-based molecules and thiols can be controlled by their excitation [46,

Fig. 13 Schematic illustration of optical control of FRET

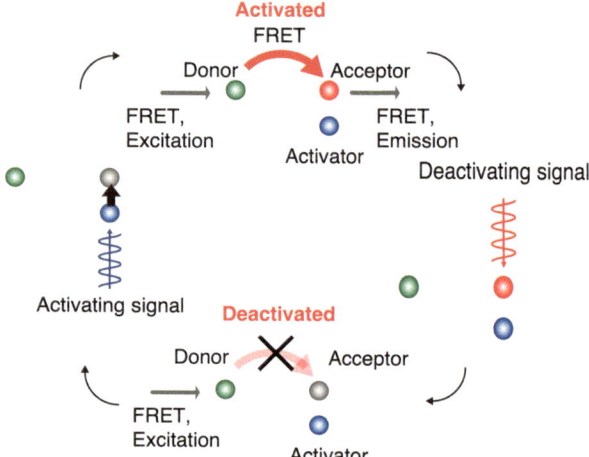

47]. This means that the wavelength of the activated light can be set by selecting the fluorescent molecules to be placed nearby. Without synthesizing special fluorescent molecules or fluorescent proteins, it is possible to control the fluorescence properties using light at various wavelengths by selecting fluorescent molecules to be placed nearby.

Figure 13 shows an overview of the optical control of FRET. Fluorescent molecules with three different functions (donors, acceptors, and activators) are prepared. Cyanine-based fluorescent molecules are used as acceptors, and activators are placed near the cyanine molecules. When the acceptor is strongly excited, it changes to a dark state. In this state, the excitation of the donor does not cause FRET to the acceptor. When the activator is excited, the acceptor recovers its fluorescent state, enabling FRET from the donor to the acceptor. Thus, FRET can be switched on and off using light irradiation.

4.2 Optical Control of FRET Pathway

FRET pathways are controlled by using cyanine-based fluorescent molecules, as shown in Fig. 14 [48]. Two sets of acceptor and activator molecules are arranged on the DNA scaffold. Because the absorption peak wavelengths of acceptors 1 and 2 are different, photoactivatable fluorescent molecules in these systems can be independently controlled using external light of different wavelengths. When only acceptor 1 is activated, FRET occurs only in pathway 1. Conversely, when only acceptor 2 is activated, FRET occurs only in pathway 2. Therefore, by independently switching each system, it is possible to change the energy transfer pathway. When the activation light for pathway 1 is irradiated, the fluorescence intensity of pathway 1 mainly increases, while that of pathway 2 decreases. This indicates that FRET occurs more

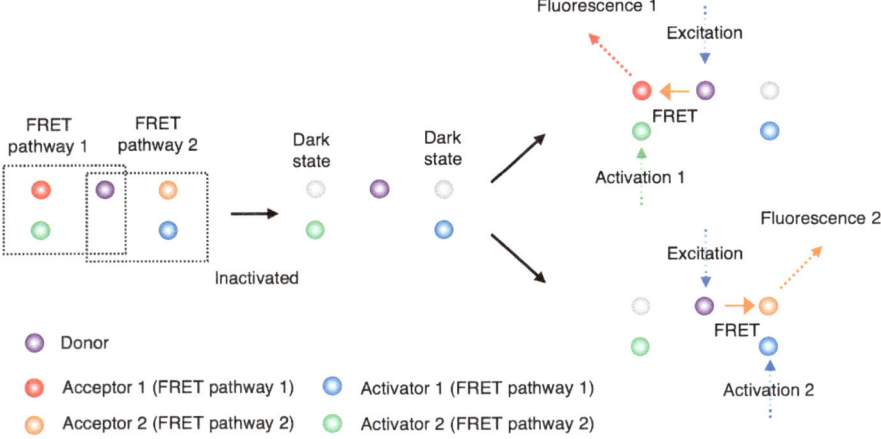

Fig. 14 Optical control of FRET pathway

frequently in pathway 1 than in pathway 2. In contrast, when applying the activation light for pathway 2, more FRET occurs in pathway 2 than in pathway 1. These results demonstrate that systems using different activators can be independently activated at the corresponding wavelengths and that FRET efficiency can be controlled.

4.3 Optically Programmable DNA Scaffold Logic

The optical control of a FRET molecular logic circuit has been demonstrated using a FRET pathway optical selection technique [49]. This enables the execution and timing of molecular logic operations to be controlled solely by light irradiation, without altering the solution environment. This approach is expected to enable the processing of spatiotemporal information from dynamically changing molecular information, which is difficult to implement in conventional DNA logic circuits that execute operations based solely on molecular reactions.

Molecular logic operations are executed by controlling the arrangement of fluorescent molecules using changes in the DNA structure in response to molecular inputs and utilizing the formation of the FRET pathway, as shown in Fig. 15. Fluorescent molecules are attached to the scaffold DNA based on self-organizing reactions using DNA complementarity. FRET pathways are designed to extend from the initiating dye to the reporting dye when the given logical conditions are satisfied. Multiple FRET pathways with different programming operations are prepared on the scaffold DNA. The selected pathway is then controlled by manipulating the activation or deactivation of the light-sensitive fluorescent molecules using light irradiation. Molecular logic operations are executed based on the selected pathway and its timing.

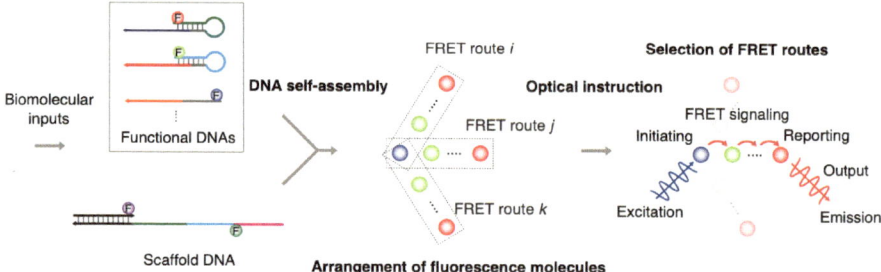

Fig. 15 Schematic illustration of optically programmable DNA scaffold logic. Reproduced with permission from Appl. Phys. Lett. 107, 013701 (2015). Copyright 2015, AIP Publishing LLC

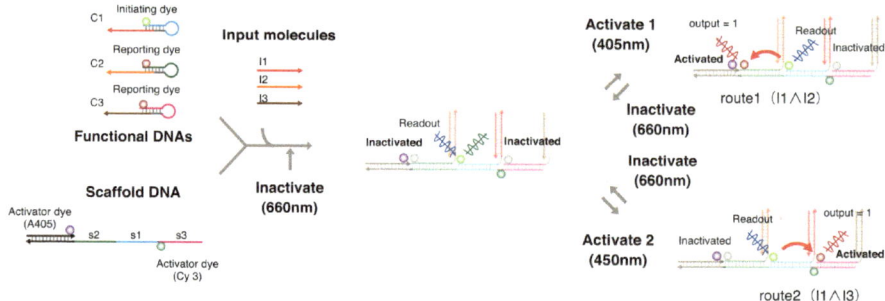

Fig. 16 Schematic illustration for optical switch I1∧I2 and I1∧I3 executions

To demonstrate time control of molecular information processing, we designed a FRET molecular logic circuit with light-controlled execution timing for I1∧I2 and I1∧I3 (Fig. 16). Functional DNA regulates the binding of fluorescent molecules to scaffold DNA according to the input molecular information. I1∧I2 and I1∧I3 are assigned to FRET pathways 1 and 2, respectively. To selectively induce FRET in each pathway, activator molecules (pathway 1: Alexa 405, pathway 2: Cy3) with different absorption spectra are pre-arranged on the scaffold DNA. By irradiating light at wavelengths corresponding to each activator molecule (pathway 1: 405 nm, pathway 2: 450 nm), the FRET pathway is selected, and the assigned operation is executed.

In the experiment, reporter molecules with different fluorescence spectra (pathway 1: Cy5, pathway 2: Cy5.5) were used to distinguish outputs. Input molecules I1, I2, and I3 were sequentially added to represent the time variation of the molecular information in the solution. The fluorescence intensity significantly increased only when the input molecular information satisfied the logical condition assigned to the light-selected pathway (Fig. 17). This result demonstrates that the timing of molecular logical operations can be controlled by light signals.

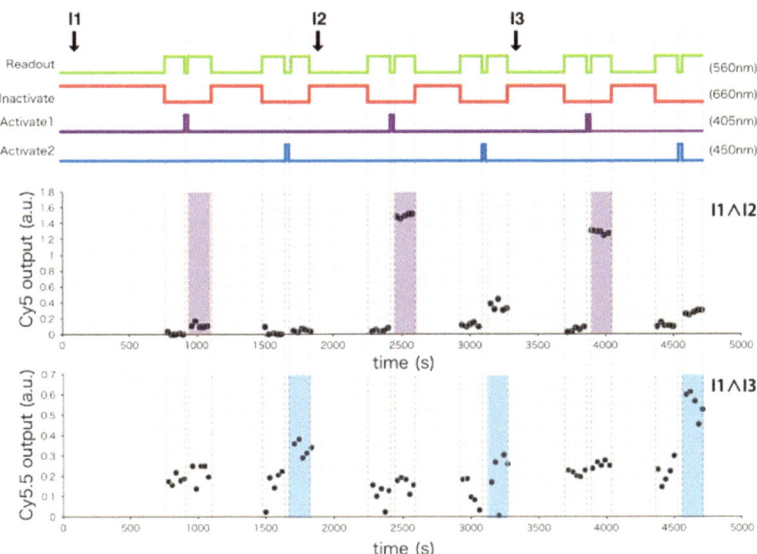

Fig. 17 Time courses of light irradiation signals and fluorescence output of I1∧I2 and I1∧I3, in response of inputs

5 FRET Network-Based Computation

Sections 3.3 and 3.4 described a FRET device capable of performing simple logic operations at the molecular level (below 10 nm). FRET pathways are controlled by the placement of particles and molecules, energy level saturation control, and molecular property changes, implemented by a pre-designed FRET computational device. Although scaling up FRET computing devices that apply microfabrication technologies, such as DNA nanotechnology, has been anticipated, the number of FRET cascades is limited owing to the red-shifted FRET. Large-scale computation using FRET devices has not yet been implemented.

As a potential solution to this issue, a signal processing device employing FRET networks has been proposed [30]. This device operates by learning the input/output characteristics of FRET network devices, which comprise a variety of quantum dots randomly arranged at a high density [28, 50]. The input signals are encoded as light signals using time and wavelength. Utilizing a high-density random FRET network device alone does not achieve efficient signal processing to provide the desired output. However, recent breakthroughs in machine learning have facilitated the estimation of object information and the attainment of desired optical responses from input/output data, even for cases in which optical modeling is challenging. By adopting this approach, the input–output characteristics of a densely aggregated FRET network device with the disorder can be learned, allowing for the implementation of the desired optical transformation, and thereby enabling micro-scale and large-scale optical signal processing, which was previously hindered by physical constraints in using the designed FRET pathways.

By using machine learning techniques to impart the desired function to the FRET network device, a wide range of signal processing techniques, including input light signal distribution, product operation, and amplification, can be implemented through the control of single-step FRET with light. Despite the random distribution of quantum dots within the FRET network device, they respond to light control and modulate the fluorescence signals [51]. Applying machine learning algorithms to the input/output response of the FRET network device allows programming of the FRET network device using a control light, enabling the desired signal modulation of input/output signals, such as processing for neural networks or reservoir computing.

The FRET network-based approach holds the potential to enable a fast and efficient optical implementation for predicting and recognizing spatiotemporal data using reservoir computing. This technology has wide-ranging applicability to various types of spatiotemporal data, including images, sounds, and videos, and is poised to contribute significantly to the development of new computing devices.

References

1. J.R. Lakowicz, *Principles of Fluorescence Spectroscopy* (Springer, Berlin, 2006)
2. B. Huang, M. Bates, X. Zhuang, Super-resolution fluorescence microscopy. Annu. Rev. Biochem. **78**, 993–1016 (2009)
3. S. Tyagi, F.R. Kramer, Molecular beacons: probes that fluoresce upon hybridization. Nat. Biotechnol. **14**(3), 303–308 (1996)
4. R.W. Wagner, J.S. Lindsey, A molecular photonic wire. J. Am. Chem. Soc. **116**(21), 9759–9760 (1994)
5. A.P. De Silva, S. Uchiyama, Molecular logic and computing. Nat. Nanotechnol. **2**(7), 399–410 (2007)
6. D.A. Parthenopoulos, P.M. Rentzepis, Three-dimensional optical storage memory. Science **245**(4920), 843–845 (1989)
7. M. Kuscu, O.B. Akan, The internet of molecular things based on fret. IEEE Internet Things J. **3**(1), 4–17 (2015)
8. J.W. Goodman, *Introduction to Fourier Optics* (Roberts and Company Publishers, 2005)
9. T. Nishimura, Y. Ogura, K. Yamada, H. Yamamoto, J. Tanida, A photonic dna processor: concept and implementation, in *Nanoengineering: Fabrication, Properties, Optics, and Devices VIII*, vol. 8102 (SPIE, 2011), pp. 33–40
10. R.M. Clegg, Fluorescence resonance energy transfer. Curr. Opin. Biotechnol. **6**(1), 103–110 (1995)
11. S. Weiss, Fluorescence spectroscopy of single biomolecules. Science **283**(5408), 1676–1683 (1999)
12. R. Roy, S. Hohng, T. Ha, A practical guide to single-molecule fret. Nat. Methods **5**(6), 507–516 (2008)
13. K.E. Sapsford, L. Berti, I.L. Medintz, Materials for fluorescence resonance energy transfer analysis: beyond traditional donor-acceptor combinations. Angewandte Chemie International Edition **45**(28), 4562–4589 (2006)
14. M.C. Dos Santos, W.R. Algar, I.L. Medintz, N. Hildebrandt, Quantum dots for förster resonance energy transfer (fret). TrAC Trends Anal. Chem. **125**, 115819 (2020)

15. K. Boeneman, D.E. Prasuhn, J.B. Blanco-Canosa, P.E. Dawson, J.S. Melinger, M. Ancona, M.H. Stewart, K. Susumu, A. Huston, I.L. Medintz, Self-assembled quantum dot-sensitized multivalent dna photonic wires. J. Am. Chem. Soc. **132**(51), 18177–18190 (2010)
16. T. Nishimura, Y. Ogura, J. Tanida, Multiplexed fluorescence readout using time responses of color coded signals for biomolecular detection. Biomed. Opt. Express **7**(12), 5284–5293 (2016)
17. T. Nishimura, Y. Ogura, K. Yamada, Y. Ohno, J. Tanida, Biomolecule-to-fluorescent-color encoder: modulation of fluorescence emission via dna structural changes. Biomed. Opt. Express **5**(7), 2082–2090 (2014)
18. T. Nishimura, Y. Ogura, J. Tanida, Reusable molecular sensor based on photonic activation control of dna probes. Biomed. Opt. Express **3**(5), 920–926 (2012)
19. J. Zheng, Fret and its biological application as a molecular ruler, in Biomedical Applications of Biophysics (Springer, Berlin, 2010), pp. 119–136
20. S. Preus, L.M. Wilhelmsson, Advances in quantitative fret-based methods for studying nucleic acids. ChemBioChem **13**(14), 1990–2001 (2012)
21. T. Nishimura, Y. Ogura, J. Tanida, A nanoscale set-reset flip-flop in fluorescence resonance energy transfer-based circuits. Appl. Phys. Express **6**(1), 015201 (2012)
22. C. LaBoda, C. Dwyer, A.R. Lebeck, Exploiting dark fluorophore states to implement resonance energy transfer pre-charge logic. IEEE Micro **37**(4), 52–62 (2017)
23. C.D. LaBoda, A.R. Lebeck, C.L. Dwyer, An optically modulated self-assembled resonance energy transfer pass gate. Nano Lett. **17**(6), 3775–3781 (2017)
24. H.M. Watrob, C.-P. Pan, M.D. Barkley, Two-step fret as a structural tool. J. Am. Chem. Soc. **125**(24), 7336–7343 (2003)
25. E. Petryayeva, W.R. Algar, I.L. Medintz, Quantum dots in bioanalysis: a review of applications across various platforms for fluorescence spectroscopy and imaging. Appl. Spectrosc. **67**(3), 215–252 (2013)
26. M. Heilemann, P. Tinnefeld, G. Sanchez Mosteiro, M. Garcia Parajo, N.F. Van Hulst, M. Sauer, Multistep energy transfer in single molecular photonic wires. J. Am. Chem. Soc. **126**(21), 6514–6515 (2004)
27. Q. Yang, L. Zhou, Y.-X. Wu, K. Zhang, Y. Cao, Y. Zhou, D. Wu, F. Hu, N. Gan, A two dimensional metal-organic framework nanosheets-based fluorescence resonance energy transfer aptasensor with circular strand-replacement dna polymerization target-triggered amplification strategy for homogenous detection of antibiotics. Anal. Chim. Acta **1020**, 1–8 (2018)
28. N. Tate, Y. Miyata, S. Sakai, A. Nakamura, S. Shimomura, T. Nishimura, J. Kozuka, Y. Ogura, J. Tanida, Quantitative analysis of nonlinear optical input/output of a quantum-dot network based on the echo state property. Opt. Express **30**(9), 14669–14676 (2022)
29. S. Sakai, A. Nakamura, S. Shimomura, T. Nishimura, J. Kozuka, N. Tate, Y. Ogura, J. Tanida, Verification of spatiotemporal optical properties of quantum dot network by regioselective photon counting, in *Photonics in Switching and Computing* (Optica Publishing Group, 2021), pp. Tu5A–2
30. M. Nakagawa, Y. Miyata, N. Tate, T. Nishimura, S. Shimomura, S. Shirasaka, J. Tanida, H. Suzuki, Spatiotemporal model for fret networks with multiple donors and acceptors: multicomponent exponential decay derived from the master equation. JOSA B **38**(2), 294–299 (2021)
31. S. Shimomura, T. Nishimura, J. Kozuka, N. Tate, S. Sakai, Y. Ogura, J. Tanida, Nonlinear-response neurons using a quantum-dot network for neuromorphic computing, in *Photonics in Switching and Computing* (Optica Publishing Group, 2021), pp. Tu5B–4
32. A. Kuzuya, M. Komiyama, Dna origami: fold, stick, and beyond. Nanoscale **2**(3), 309–321 (2010)
33. N.C. Seeman, H.F. Sleiman, Dna nanotechnology. Nat. Rev. Mater. **3**(1), 1–23 (2017)
34. C.M. Spillmann, S. Buckhout-White, E. Oh, E.R. Goldman, M.G. Ancona, I.L. Medintz, Extending fret cascades on linear dna photonic wires. Chem. Commun. **50**(55), 7246–7249 (2014)
35. S. Buckhout-White, C.M. Spillmann, W.R. Algar, A. Khachatrian, J.S. Melinger, E.R. Goldman, M.G. Ancona, I.L. Medintz, Assembling programmable fret-based photonic networks using designer dna scaffolds. Nat. Commun. **5**(1), 5615 (2014)

36. M. Ohtsu, T. Kawazoe, T. Yatsui, M. Naruse, Nanophotonics: application of dressed photons to novel photonic devices and systems. IEEE J. Sel. Top. Quantum Electron. **14**(6), 1404–1417 (2008)
37. E.R. Walter, Y. Ge, J.C. Mason, J.J. Boyle, N.J. Long, A coumarin-porphyrin fret break-apart probe for heme oxygenase-1. J. Am. Chem. Soc. **143**(17), 6460–6469 (2021)
38. M.D. Mottaghi, C. Dwyer, Thousand-fold increase in optical storage density by polychromatic address multiplexing on self-assembled dna nanostructures. Adv. Mater. (deerfield Beach, Fla.) **25**(26), 3593–3598 (2013)
39. C. Pistol, C. Dwyer, A.R. Lebeck, Nanoscale optical computing using resonance energy transfer logic. IEEE Micro **28**(6), 7–18 (2008)
40. A. Saghatelian, N.H. Völcker, K.M. Guckian, V.S.-Y. Lin, M.R. Ghadiri, Dna-based photonic logic gates: And, nand, and inhibit. J. Am. Chem. Soc. **125**(2), 346–347 (2003)
41. W. Yoshida, Y. Yokobayashi, Photonic boolean logic gates based on dna aptamers. Chem. Commun. (2), 195–197 (2007)
42. W.R. Algar, K.D. Krause, Developing fret networks for sensing. Annu. Rev. Anal. Chem. **15**, 17–36 (2022)
43. T. Nishimura, Y. Ogura, J. Tanida, Fluorescence resonance energy transfer-based molecular logic circuit using a dna scaffold. Appl. Phys. Lett. **101**(23), 233703 (2012)
44. G.T. Dempsey, M. Bates, W.E. Kowtoniuk, D.R. Liu, R.Y. Tsien, X. Zhuang, Photoswitching mechanism of cyanine dyes. J. Am. Chem. Soc. **131**(51), 18192–18193 (2009)
45. S. Uphoff, S.J. Holden, L. Le Reste, J. Periz, S. Van De Linde, M. Heilemann, A.N. Kapanidis, Monitoring multiple distances within a single molecule using switchable fret. Nat. Methods **7**(10), 831–836 (2010)
46. M. Bates, B. Huang, G.T. Dempsey, X. Zhuang, Multicolor super-resolution imaging with photo-switchable fluorescent probes. Science **317**(5845), 1749–1753 (2007)
47. M. Bates, B. Huang, X. Zhuang, Super-resolution microscopy by nanoscale localization of photo-switchable fluorescent probes. Curr. Opin. Chem. Biol. **12**(5), 505–514 (2008)
48. R. Fujii, T. Nishimura, Y. Ogura, J. Tanida, Nanoscale energy-route selector consisting of multiple photo-switchable fluorescence-resonance-energy-transfer structures on dna. Opt. Rev. **22**, 316–321 (2015)
49. T. Nishimura, R. Fujii, Y. Ogura, J. Tanida, Optically controllable molecular logic circuits. Appl. Phys. Lett. **107**(1), 013701 (2015)
50. S. Shimomura, T. Nishimura, Y. Miyata, N. Tate, Y. Ogura, J. Tanida, Spectral and temporal optical signal generation using randomly distributed quantum dots. Opt. Rev. **27**, 264–269 (2020)
51. M. Tanaka, J. Yu, M. Nakagawa, N. Tate, M. Hashimoto, Investigating small device implementation of fret-based optical reservoir computing, in *IEEE 65th International Midwest Symposium on Circuits and Systems (MWSCAS)* (IEEE, 2022), pp. 1–4

Quantum-Dot-Based Photonic Reservoir Computing

Naoya Tate

Abstract Reservoir computing is a novel computational framework based on the characteristic behavior of recurrent neural networks. In particular, a recurrent neural network for reservoir computing is defined as a reservoir, which is implemented as a fixed and nonlinear system. Recently, to overcome the limitation of data throughput between processors and storage devices in conventional computer systems during processing, known as the Von Neumann bottleneck, physical implementations of reservoirs have been actively investigated in various research fields. The author's group has been currently studying a *quantum dot reservoir*, which consists of coupled structures of randomly dispersed quantum dots, as a physical reservoir. The quantum dot reservoir is driven by sequential signal inputs using radiation with laser pulses, and the characteristic dynamics of the excited energy in the network are exhibited with the corresponding spatiotemporal fluorescence outputs. We have presented the fundamental physics of a quantum dot reservoir. Subsequently, experimental methods have been introduced to prepare a practical quantum dot reservoir. Next, we have presented the experimental input/output properties of our quantum dot reservoir. Here, we experimentally focused on the relaxation of fluorescence outputs, which indicates the characteristics of optical energy dynamics in the reservoir, and qualitatively discussed the usability of quantum dot reservoirs based on their properties. Finally, we have presented experimental reservoir computing based on spatiotemporal fluorescence outputs from a quantum dot reservoir. We consider that the achievements of quantum dot reservoirs can be effectively utilized for advanced reservoir computing.

1 Introduction

Reservoir computing [1, 2] is one of the most popular paradigms in recent machine learning and is especially well-suited for learning sequential dynamic systems. Even when systems display chaotic [3] or complex spatiotemporal phenomena [4],

N. Tate (✉)
Kyushu University, 744 Motooka, Nishi-Ku, Fukuoka 819-0395, Japan
e-mail: tate@ed.kyushu-u.ac.jp

© The Author(s) 2024
H. Suzuki et al. (eds.), *Photonic Neural Networks with Spatiotemporal Dynamics*,
https://doi.org/10.1007/978-981-99-5072-0_4

which are considered as exponentially complicated problems, an optimized reservoir computer can handle them efficiently.

On the other hand, von Neumann-type architecture is known as one of the most familiar computer architectures, which consists of a processing unit, a control unit, memory, external data storage, and input/output mechanisms. While such architecture is now widely utilized, in the conventional von Neumann-type architecture, bottlenecks between the processing unit and memory are inevitable when implementing the parallel operation of sequential processing [5]. Hence, the development of a physical reservoir using various physical phenomena that can act as a non-von Neumann-type architecture is required. Unlike other types of neural network models, reservoir models are expected to be suitable for physical implementation as their inner dynamics are fixed as a general definition and do not need to be modified during processing. Thus far, various methods for the physical implementation of reservoirs, such as electrochemical cells [5], analog VLSI [6], and memristive nanodevices [7], have been actively discussed.

We focused on the energy propagation between dispersed quantum dots (QDs) as a phenomenon for implementing a physical reservoir. QDs are nanometer-sized structures that confine the motion of charge carriers in all three spatial directions, leading to discrete energy levels based on quantum size effects. In general, the emission properties of QD can be adjusted by changing their sizes and structures. In addition, since QDs are typically fabricated using semiconductor materials, they exhibit good stability and durability. Recently, QDs have been incorporated into semiconductor devices, such as light-emitting diodes [8, 9], lasers [8, 10], and field-effect transistors [11]. Our QD reservoir (QDR) consists of randomly connected transfer paths of optical energy between the QDs and reveals the spatiotemporal variation in the fluorescence output. Recently, we experimentally demonstrated short-term memory capacity as a physical reservoir [12].

In this Chapter, we have discussed the fundamentals of QDR and the experimental protocol for preparing QDR samples. Moreover, the results of actual reservoir computing based on the spatiotemporal fluorescence outputs of QDR samples have been shown. Generally, a reservoir requires large amounts of computational resources during the learning process. Using the spatiotemporal data obtained from the nonlinear transformation from an optical input signal to the fluorescence output via QDR, learning without calculating the individual states of all the nodes in the reservoir layer is possible. As a result, by physically implementing the reservoir layer and learning based on spatiotemporal data, we expect to solve the target problem with a lower reservoir power consumption than other existing implementations.

2 Quantum-Dot-Based Physical Reservoir

2.1 Basics

As shown in Fig. 1, a QDR consists of randomly dispersed QDs. The optical input to the QDR was determined by the incidence of the laser pulse, and some QDs were excited by this incident light. While the excited electron energy behaves as a localized optical energy, it can be directly relaxed to the ground energy level with/without fluorescence radiation. Another typical phenomenon of such a localized optical energy transfer, based on the Förster resonance energy transfer (FRET) mechanism, is that a QD, initially in its electronic excited state, may transfer energy to a neighboring QD through nonradiative dipole–dipole coupling. Generally, the fluorescence and absorption spectra of QD partially overlap; thus, FRET is probabilistically allowed. Furthermore, if the FRET network consists of different types of QDs, each type of QDs is defined as a donor or acceptor. FRET from an acceptor-QD to a donor-QD is prohibited, and an acceptor-QD often acts as a destination in each network. During the energy transfer, optical energy percolates in the FRET network, which can be regarded as the optical input being partially memorized with subsequent forgetting.

In particular, in our scheme for realizing the optical input/output of QDR, the light pulses are spatially incident on the QD network in parallel and excite multiple QDs. For the excitation of the QDs, the optical energy of the input light must be greater than the bandgap energy of each QDs. The excited QDs probabilistically emit fluorescent photons with optical energy similar to the bandgap energy. In contrast, part of the optical energy is transferred from one excited QD to another QD based on the FRET

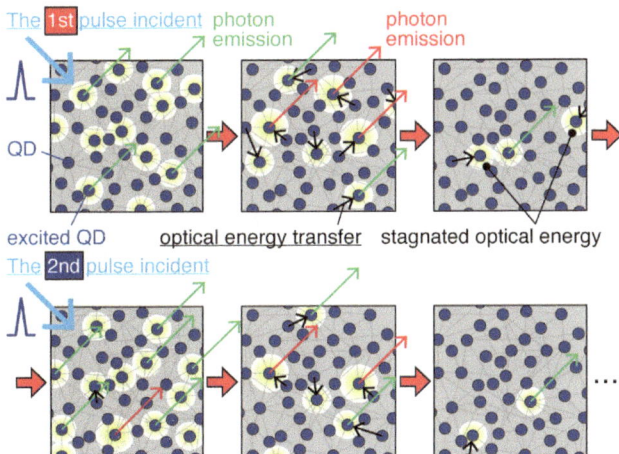

Fig. 1 Schematic of the inner dynamics of a QDR by sequential incidence of light pulses occurring due to localized transfer and temporal stagnation of optical energy during periodic irradiation of light pulses

mechanism. After single or multiple steps of energy transfer, some of the optical energy in the network is emitted from the destination QD. Generally, light emissions via multiple energy transfers necessarily occur later than that involving no energy transfer. Consequently, the fluorescence output from the QDR can be sequentially obtained by sparsely counting fluorescence photons.

Additionally, as shown in the lower half of Fig. 1, when the next light pulse is input to the QDR during the temporal stagnation of the inner optical energy, the transfer process and corresponding emission of fluorescence photons reveal a different tendency from the previous condition of the network because of the different internal state of the QDR from the previous condition; namely, unlike the upper half of Fig. 1, some QDs are already excited. Here, a saturated situation in which all the QDs are excited is not anticipated. Consequently, in such cases, the optical outputs in response to sequential optical inputs cannot be predicted using a simple linear sum of a single input/output. Therefore, the non-linearity of an input/output can be quantitatively evaluated by comparing the linear sum of a single input/output. In other words, such a setup works as a fusion-type setup of the processor and storage device, which maintains the inner states during multiple optical inputs and outputs and can directly read out its output as fluorescence via complicated signal processing in the network.

2.2 Experimental Demonstration: Randomly-Dispersed QDR

For the experimental preparation of QDR sample, we used two types of QDs as components of the QD network: CdS dots (NN-labs, CS460; peak wavelength of the emission: 465–490 nm, 3.0 nmol/mg, represented in catalog) and CdS/ZnS dots (NN-labs, CZ600; peak wavelength of the emission: 600–620 nm, 1.0 nmol/mg, represented in catalog) with toluene solutions. In this case, the CdS and CdS/ZnS dots acted as donors and acceptors, respectively. Additionally, polydimethylsiloxane resins (PDMS; Dow Corning, Sylgard184) were used as base materials to fix and protect the inner QDR. The basic procedure for preparing a QD sample using these materials is shown in Fig. 2a and is as follows.

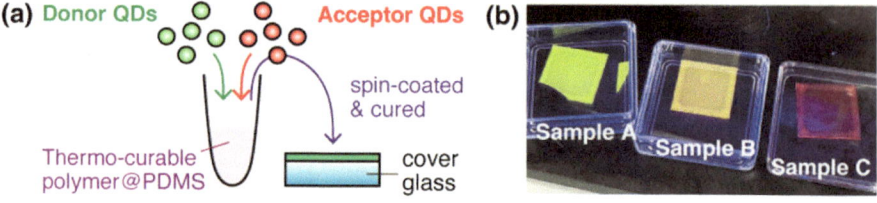

Fig. 2 a Schematic of experimental process. **b** Appearance of three QD samples under UV light illumination

First, the two QD solutions were mixed with 1,000 μL of PDMS base solution. To control the configuration of the QDR, the CdS and CdS/ZnS dots were mixed in ratios of 3:1, 1:1, and 0:1 for Samples A, B, and C, respectively. After mixing, each mixture was heated for evaporating toluene. Then, 100 μL of polymerization initiator PDMS was added to the mixture, and the resulting solution was dropped on the cover glass. The mixture was spread on a cover glass using a spin coater (MIKASA, MS-B100, rotational speed: 3,000 rpm) for 100 s to randomly disperse the QDs in the resin, and the respective QDR was expected to be formed. The assumed thickness of the samples was less than 1 μm. After the mixtures were thinned, to fix the alignments of the QDs in each mixture, the thinned samples were heated on a hot plate (EYELA, RCH-1000) at 150 °C for 600 s. The prepared samples appeared transparent under ambient light; however, they emitted fluorescence under UV illumination, as shown in Fig. 2b.

2.3 Experimental Demonstration: Electrophoresis

Electrophoretic deposition (EPD) was applied as another method to prepare QDR sample [13–16]. The EPD of the QD layer was accomplished by applying a voltage between two conductive electrodes suspended in a colloidal QD solution. The electric field established between the electrodes drives QD deposition onto the electrodes. EPD, as a manufacturing process, efficiently uses the starting colloidal QD solutions.

To start the EPD process, two GZO glasses coated with a Ga-doped ZnO layer were secured 1.0 cm apart with their conductive sides parallel and facing one another, as illustrated in Fig. 3a. An electric field of 15 V/cm was then applied, and the electrodes were placed in the QD solution. After a few minutes, the electrodes were QD-rich toluene droplets that left uneven QD deposits after being dried from the surface using compressed air. In this demonstration, we used CdSe/ZnS dots (Ocean Nanotech, QSR580; peak wavelength of emission: 580 nm, 10 mg/mL, represented in the catalog) with a toluene solution. Owing to their emission and absorption spectra, the CdSe/ZnS dots can function as both donors and acceptors. Figure 3b, c show the appearance of the negative and positive electrodes under UV light illumination, respectively. As the fluorescence of the QD layer can be observed by the eye only on the negative electrodes under UV light, the QD layer was confirmed to be deposited by the EPD process and not by any other phenomenon.

Here, we prepared three samples using the EPD method under 10, 20, and 60 min of deposition time, which we define as 10, 20, and 60 M samples, respectively. Figure 4a–c show fluorescence microscopic images under UV irradiation. As shown, the coverage rate of each sample by deposited-QDs was varied from 49.8 to 76.1% by increasing deposition time from 10 to 60 min. Atomic force microscopy (AFM) topography images of the QD layers on the GZO anodes are shown in Fig. 5a–c. The resulting QD layer apparently consisted of aggregated QD components, and the size of each unit component was 100–500 μm. The number of QD components increased with an increase in the deposition time.

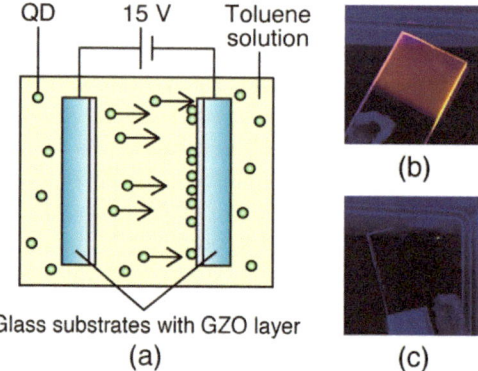

Fig. 3 a Schematic of the EPD process. A voltage applied between two parallel, conducting electrodes spaced 1.0 cm apart drives the deposition of the QDs. Appearance of **b** negative and **c** positive electrodes after 60 min of EPD process under UV light illumination

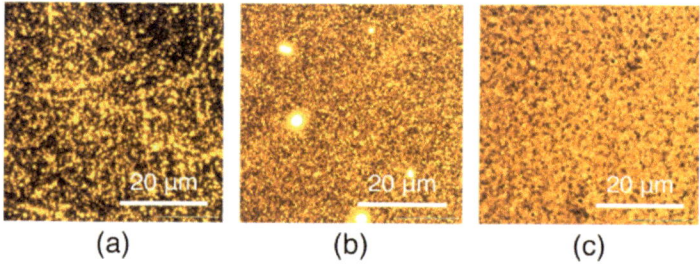

Fig. 4 Fluorescence microscopic images of the surface of an electrophoretically deposited QD layer on **a** 10 M sample, **b** 20 M sample, and **c** 60 M sample, respectively

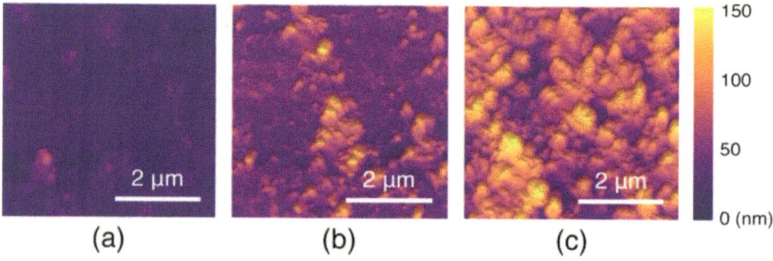

Fig. 5 AFM topography images of the surface of an electrophoretically deposited QD layer on **a** 10 M sample, **b** 20 M sample, and **c** 60 M sample, respectively

3 Non-linearity of the QD Reservoir

3.1 Experimental Setup

We focused on the fluorescence relaxation from acceptor QDs for verifying the non-linearity of our QDR. Here, photons emitted from each QD were necessarily output at various times, regardless of whether the photons are emitted via energy transfers. In other words, the results of photon counting, which are triggered by the timing of the optical input, indicate the characteristics of QDR.

For experimental verification, we used a Ti:Al$_2$O$_3$ laser (Spectra Physics, MaiTai), which emitted optical pulses with a pulse length of 100 fs, an optical parametric amplifier (Spectra Physics, TOPAS-prime), and a wavelength converter (Spectra Physics, NirUVis) as the light sources for irradiating the QD samples. The oscillation frequency and wavelength were set as 1 kHz and 457 nm, respectively. The laser power and polarization were appropriately controlled for exciting the QD samples and effectively counting the fluorescent photons, as shown in Fig. 6.

The delay line generated a time lag Δt between the first and second pulses incident on the QDR sample. The range of Δt was set at 0.64–7.4 ns, and the corresponding optical length was controlled by a stage controller driven by a stepping motor with a position resolution of 20 μm/step. Fluorescence photons induced by optical excitation of the QDR samples passed through the focusing lens again and were reflected in the detection setup using a Glan–Thompson polarizer. After passing through a bandpass filter (UQG Optics, G530, transmission wavelength: 620 ± 10 nm), the fluorescent photons were propagated to a photon detector (Nippon Roper, NR-K-SPD-050-CTE). Since the irradiated light passed through the polarizers, only the

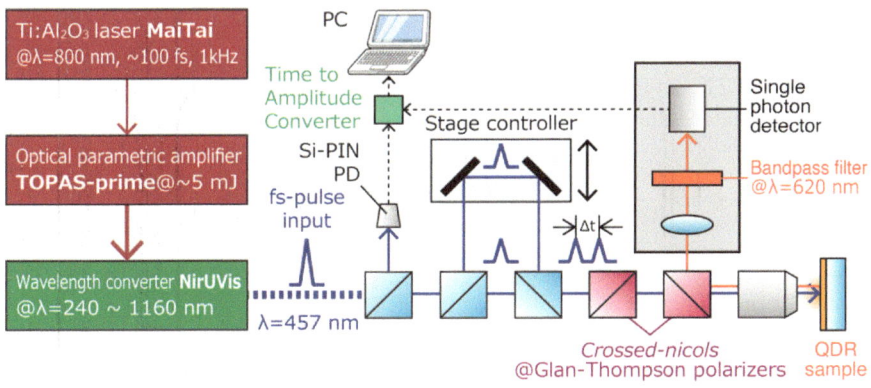

Fig. 6 Schematic of experimental photon-counting setup to identify characteristic of the QDR sample

fluorescence photons from the acceptor-QDs were selectively obtained by the single-photon detector. Then, by synchronizing with the trigger signal using a Si-PIN photodiode (ET-2030) on a time-to-amplitude converter (TAC; Becker and Hickl GMbH, SPC-130EMN), the time-resolved intensities were obtained, and the results were collated using PC software (Becker and Hickl GMbH, SPC-130 MN) as the lifetime of each fluorescence. Here, excitation power was set as 5.0 μW, which was enough to suppress and did not induce saturated situation of QD excitations.

Before verifying the non-linearity, the fluorescence relaxation due to multiple incident laser pulses was experimentally verified in the setup as the basic specification of our three samples: Samples A, B, and C. The left-hand side of Fig. 7 shows an example of the obtained photon-counting result from the acceptor QDs in response to double incident laser pulses, which were obtained at a certain area in Sample A. As shown, two rising phases were recognized, which were due to the first and second incident laser pulses. The time lag Δt between the two pulses was 7.4 ns. Under these conditions, the second pulse was irradiated before the induced optical energy was dissipated, and the corresponding photons were counted in response to the first incident pulse, which corresponded to the stagnation time of the QDR induced by the first incident pulse. Therefore, we focused on photon counting after the second incident pulse, which was extracted from the right side of Fig. 7.

To discuss the spatial variation of the QD networks in each sample, three individual areas were irradiated in the three samples, and the fluorescence photons were counted. To quantitatively compare the fluorescence relaxation of each sample, the results were fitted using an exponential equation:

$$C(t) = A + Be^{-t/\tau}, \tag{1}$$

where A and B are individual positive constants, and τ denotes the time constant of the fluorescence lifetime. The values of these parameters were selected to appropriately match the experimental results. As a result, τ was calculated to be 168–315, 129–257, and 99–212 ps for Samples A, B, and C, respectively. Clear differences were observed in the results for each sample. For Sample C, which contained no donor QDs

Fig. 7 Experimentally obtained photon counting result of a QDR sample under irradiation by double pulses with time difference Δt of 7.4 ns

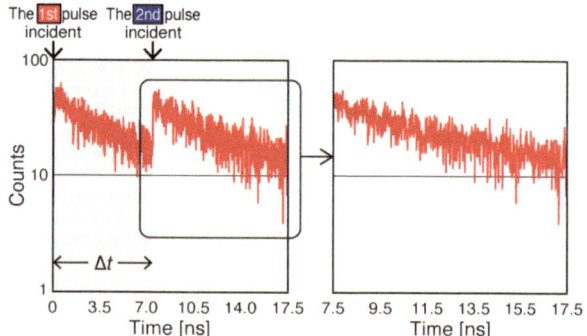

and a smaller total number of QDs, energy transfer between QDs rarely occurred, and the QDs excited by the laser pulse directly emitted photons without any energy transfer. Therefore, Sample C revealed the shortest τ. In contrast, since Sample A contained the largest number of donor QDs, frequent energy transfers were expected to occur. As a result, the excited optical energy was allowed to stagnate over a longer time in the QDR and obtained at various times owing to various energy transfers. Therefore, Sample A revealed the longest τ among the three. The results indicated a clear relationship between the configuration of the QD network and the extent of the echo state owing to FRET between the QDs.

3.2 Qualitative Non-linearity

To quantitatively evaluate the non-linearity of each photon-counting result, we employed their correlation analysis in response to double incident pulses with a linear sum of single inputs/outputs, which are the photon counts due to a single pulse incident on each sample. For correlation analysis, Pearson correlation coefficients, R, were calculated as follows:

$$R = \frac{\sum_{i=1}^{n}(x_i - \overline{x})(y_i - \overline{y})}{\sqrt{\sum_{i=1}^{n}(x_i - \overline{x})^2}\sqrt{\sum_{i=1}^{n}(y_i - \overline{y})^2}}, \tag{2}$$

where x and y represent the data obtained with the first and second incident pulses, respectively, n is the data size, x_i and y_i are individual data points indexed with i, and $\overline{x}$ and $\overline{y}$ represent the data means of x and y, respectively. While nonlinear inputs/outputs were expected to be difficult to approximate with a linear sum of the separately obtained single inputs/outputs, lower and higher correlation coefficients corresponded to the larger and smaller nonlinearities of each input/output, respectively. Furthermore, during irradiation with optical pulses, the length of the delay path in the optical setup, as shown in Fig. 6, was controlled to set the time difference Δt between the two pulses. Here, Δt was set to 0.64, 0.84, 1.6, and 7.4 ns, and the photon counts in response to the second incident pulse were determined for the three samples. Correlation coefficients were calculated from the photon counting results, which were obtained with several Δt values at three individual areas on the three samples.

Overall, as shown in Fig. 8, with increasing Δt, higher and more converged correlation coefficients were observed, implying a gradual dissipation of the echo states induced by the first incident optical pulse. Conversely, for Δt shorter than 1.0 ns, the second pulse was incident before dissipation of the echo state excited by the first incident pulse. Consequently, lower correlations and corresponding higher non-linearity were successfully revealed.

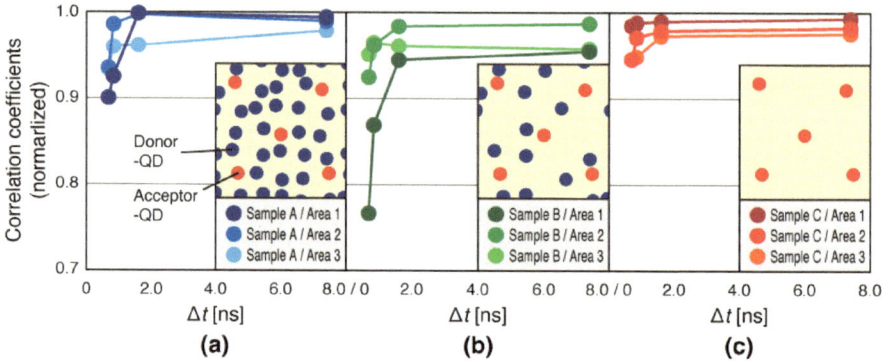

Fig. 8 Comparison of correlation coefficients corresponding to non-linearity of fluorescence of three samples, **a** Sample A, **b** Sample B, and **c** Sample C

Furthermore, the three lines in the results for Sample C revealed similar curves, and the correlation coefficients were greater than 0.95, which corresponded to a smaller non-linearity. These findings were attributed to the sparse alignment of the QDs, as schematically shown in the insets of Fig. 8. Specifically, the input/output varied from area to area, and FRET between the QDs was rarely allowed. As a result, the nonlinear input/output was not sufficiently revealed in Sample C. However, in the case of Sample A, since the number of QDs was sufficient in all areas, many paths for FRET were expected. Similar input/output tendencies were obtained in each area, and a higher non-linearity than that of Sample C was revealed. In the case of Sample B, tendencies showing the most variation in each area were observed, and higher non-linearity was often revealed in some areas. As a result, we verified the echo state properties of our QDR samples and quantitatively measured hold times of less than 1.0 ns with our experimental conditions. As shown in Fig. 8, the hold time and spatial variation of the echo state properties clearly depended on the composition of the QDR sample.

4 Spatio-Temporal Photonic Processing

4.1 Basics

Based on the FRET mechanism, as described in the previous section, after the laser irradiation of the QDR, the excited energy in some QDs was probabilistically transferred to the surroundings. After multistep transferences through several FRET paths, the energy was probabilistically irradiated as the fluorescence of the QDs at various times. Moreover, the fluorescence intensity varied spatially because of the random distribution of the QDs. Consequently, the fluorescence output of the QDR can be defined as two-dimensional spatiotemporal information, which is reflected as the

Fig. 9 Schematic of the spatiotemporal fluorescence output based on various FRETs in the QDR

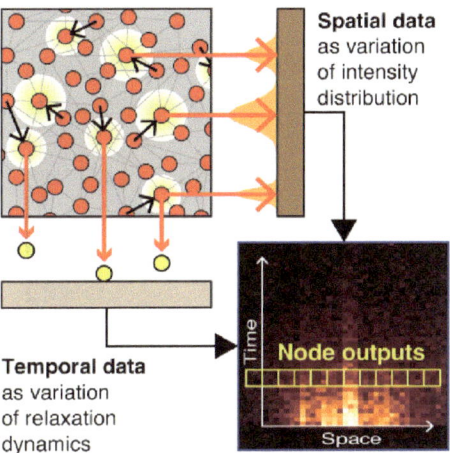

Spatial data as variation of intensity distribution

Temporal data as variation of relaxation dynamics

Node outputs

Time

Space

nonlinear input/output and short-term memory of the QDR, as conceptually shown in Fig. 9. Reservoir computing can be driven effectively using an appropriate readout of several calculation parameters for reservoir computing from fluorescence outputs.

4.2 Streak Measurement

As an experimental demonstration of QDR-based reservoir computing, we prepared QDR samples using the EPD method, which we defined as 10, 20, and 60 M samples in the previous section. The experimental setup for the fluorescence measurement is shown in Fig. 10. The fluorescence output of the QDR sample was detected using the time-correlated single-photon counting (TCSPC) method [17]. The setup and experimental conditions of the light source were the same as those for the previous setup shown in Fig. 6.

Several parameters of the reservoir model were identified. The QDR sample was irradiated with first, second, and double pulses. As shown in Fig. 10, the delay line generated 1 ns of the lag between the first and second incidents on the QDR sample. Corresponding streak images were obtained using a streak camera (Hamamatsu C10910) upon the insertion of an appropriate bandpass filter (Edmund Optics, #65–708, transmission wavelength: 600 ± 5 nm) to extract the fluorescent output. The streak camera mainly consisted of a slit, streak tube, and image sensor. The photons of fluorescence to be measured as outputs of the QDR sample were focused onto the photocathode of the streak tube through the slit, where the photons were converted into a number of electrons proportional to the intensity of the incident light. These electrons were accelerated and conducted toward the phosphor screen, and a high-speed voltage synchronized with the incident light was applied. The electrons were swept at a high speed from top to bottom, after which they collided against the

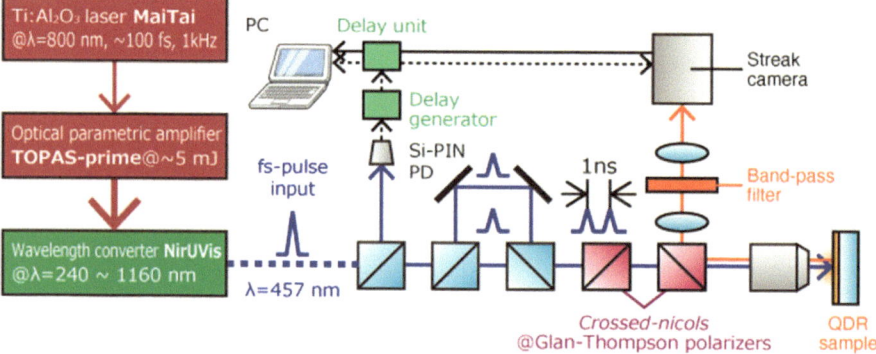

Fig. 10 Schematic of experimental setup for streak measurement based on TCSPC method

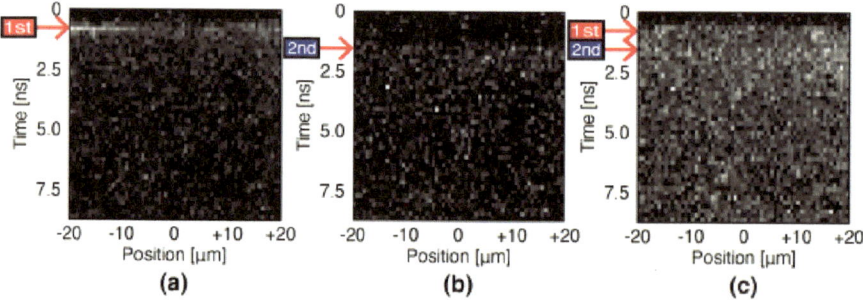

Fig. 11 Examples of obtained streak images as outputs of a QDR sample with **a** the first pulse, **b** the second pulse, and **c** double pulses

phosphor screen of the streak tube and were converted into a spatiotemporal image. Streak images of a QDR sample irradiated by the first, second, and double pulses are shown in Fig. 11a, b, c, respectively.

4.3 Spatiotemporal Reservoir Model

Referring to the echo state network [1], we define the updating formula for the state vector s of a reservoir as follows:

$$S_{t+1} = \tanh(\alpha(W_{res} S_t + W_{in} u_t)), \tag{3}$$

where s_t is the vector of the reservoir node at t, and W_{res} denotes the connection weight between nodes. The non-linearity of each node is revealed as a hyperbolic tangent function. Reservoir dynamics can be described using iterative applications

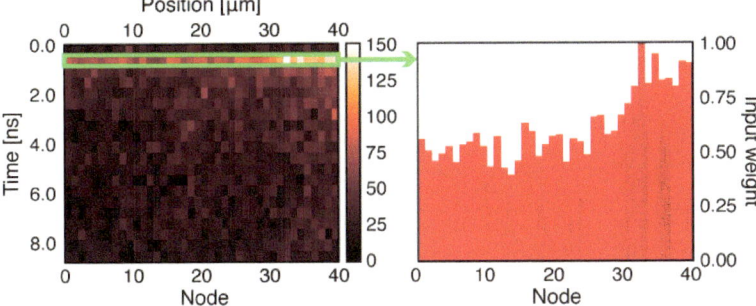

Fig. 12 One-dimensional fluorescence intensity for experimental determination of W_{in}

of the function. W_{in} is the vector of input to the reservoir, and u_t is the coefficient representing the sequential evolution of W_{in}.

Input weight matrix. W_{in} is defined as a linear mapping vector that projects the input vector u to the state vector space s, namely, reservoir nodes. In the QDR experiment, W_{in} was determined by the intensity distribution of the fluorescence, which corresponds to the distribution of dispersed QDs. Therefore, in streak images, the one-dimensional fluorescence intensity of all nodes immediately after irradiation must be read to determine the W_{in}, as shown in Fig. 12.

Leakage rate. *As the fundamentals of reservoir computing, the leakage rate of the respective nodes directly controls the retention of data from previous time steps. Therefore, the reservoir can act as a short-term memory. In the experiment of QDR, relaxation time τ of fluorescence output is directly related to the leakage rate of each node. In the streak images, we focused on the respective nodes and extracted the fluorescence relaxation, as shown in Fig. 13. The relaxations were fitted using* $N = N_0 exp(-t/\tau)$ *for determining the relaxation time* τ_i *for each node.*

Connection weight matrix. *The connection weights* between various nodes in the reservoir corresponded to the FRET efficiencies between the dispersed QDs in our

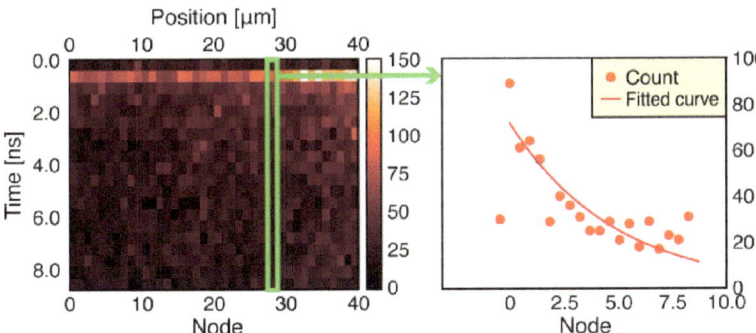

Fig. 13 Extraction of fluorescence relaxation for experimental determination of relaxation time τ

Fig. 14 Comparison of
fluorescence outputs
between all nodes for
determination of W_{res}

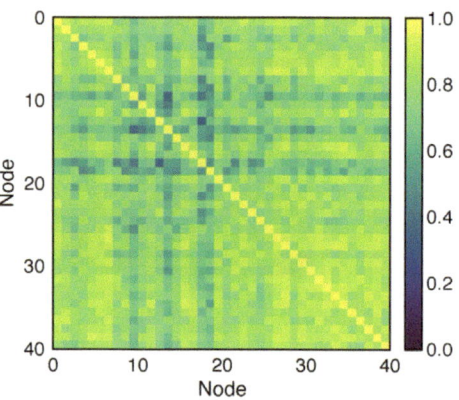

experiment. Based on these interactions, the inner state of a reservoir can sequentially
evolve to handle complex tasks. To search W_{res}, we assumed that multiple W_{res}s and
fluorescence relaxations of neighboring nodes were compared for each W_{res} as shown
in Fig. 14. For quantitative comparison, the respective correlation coefficients were
calculated. We then determined W_{res}, which revealed a similar variance σ_R^2 with the
experimental results.

Activate function. The change in the inner state s due to the variation in the input
vector u is defined by an active function. Here, we approximately set αtanh as the
active function of QDR and optimized its coefficient α. We varied α and respectively
compared with experimental results based on the calculation of Pearson's correlation
coefficients. Specifically, fluorescence relaxation by the first incident pulse and the
second pulse was added and compared with relaxation by double incident pulses to
identify non-linearity, as shown in Fig. 15. Then, a was determined, which revealed
similar non-linearity with the experimental results.

4.4 Experimental Demonstration

Based on the experimental conditions and obtained results, we demonstrated the
sequential prediction of XOR logic based on machine learning. As XOR is one of
the simplest tasks that is linearly nonseparable, it is often selected for experimentally
demonstrating machine learning using an artificial neural network. In our experiment,
the original streak image was arranged with 40 nodes in the spatial direction and 20
steps in the temporal direction. Each temporal step corresponded to 0.434 ns and each
size of node was 1 μm. Figure 16 shows some examples of data sequences predicted
by our reservoir models, which were constructed by utilizing streak images of 10,
20, and 60 M samples, respectively.

Furthermore, we successfully predicted sequential XOR logic using 1.0% of the
mean bit error rate (BER) with 100 trials, as shown in Fig. 17. The results clearly

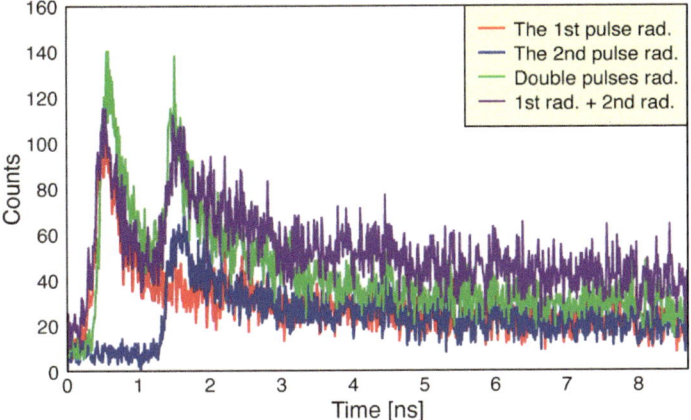

Fig. 15 Comparison of fluorescence relaxation by the first pulse incident, the second pulse incident, and the cumulative result of the two with relaxation by double pulses incident

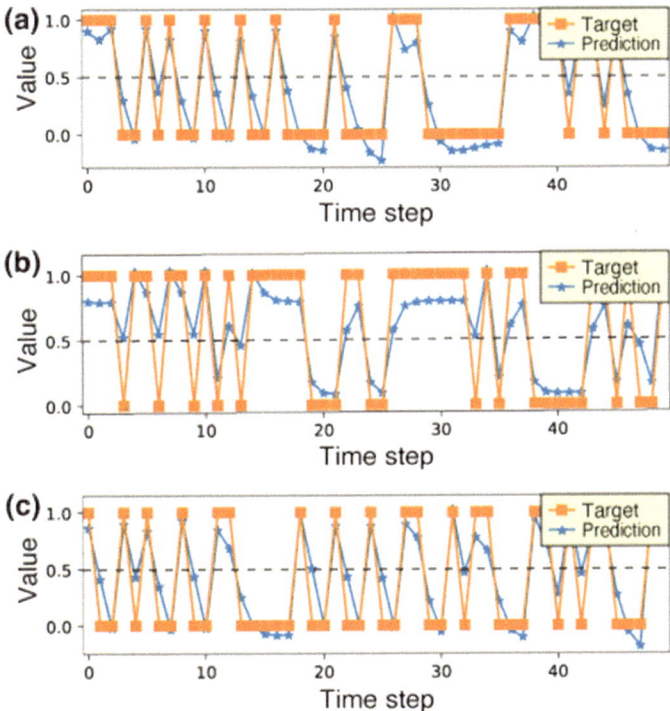

Fig. 16 Prediction results obtained by our reservoir models based on streak images of **a** 10 M, **b** 20 M, and **c** 60 M samples

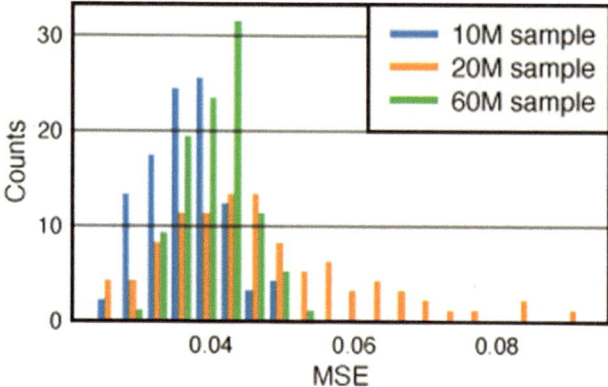

Fig. 17 Comparison of MSEs of reservoir models using 10, 20, and 60 M samples with 100 respective trials

show that each model exhibits its own performance owing to the variations in the QDR, as shown in Figs. 4 and 5. However, the theoretical relationship between each performance metric and the corresponding QDR is now under discussion.

5   Conclusion and Future Prospect

In this section, the results of an investigation into the applicability of QDR as a physical reservoir with an optical input/output is presented. Based on the idea that QDR is expected to reveal nonlinear input/output owing to the short-time memory of optical energy in the network, photon counting of the fluorescence outputs obtained in response to sequential short-light pulses was performed for verifying the optical input/output of our original QDR. Consequently, the non-linearity of the input/output, which is a fundamental requirement for the realization of effective machine learning, was qualitatively verified. Moreover, we demonstrated that reservoir computing based on the spatiotemporal fluorescence output of QDRs can learn the XOR problem and make correct predictions with a low BER. In future studies, we will extend the applicability of our idea to execute practical tasks for a larger amount of time-series data based on the outputs of our spatiotemporal fluorescence processing. In addition, larger spatial variation is also expected to be one of the fundamental requirements in QDR for physical implementations of reservoir computing because varied nonlinear input/outputs in a single reservoir are useful for effective machine learning based on QDR, that is, nanophotonic reservoir computing. The optimization of the composition required for targeted processes in nanophotonic reservoir computing remains an open topic for further investigation.

Acknowledgements The author would like to thank many colleagues and collaborators for illuminating the discussions over several years, particularly Y. Miyata, S. Sakai, A. Nakamura, S. Yamaguchi, S. Shimomura, T. Nishimura, J. Kozuka, Y. Ogura, and J. Tanida. This study was supported by JST CREST and JPMJCR18K2.

References

1. H. Jaeger, H. Haas, Harnessing nonlinearity: predicting chaotic systems and saving energy in wireless communication. Science **304**, 78–80 (2004)
2. W. Maass, T. Natschläger, H. Markram, Real-time computing without stable states: a new framework for neural computation based on perturbations. Neural Comput. **14**, 2531–2560 (2002)
3. J. Pathak, Z. Lu, B.R. Hunt, M. Girvan, E. Ott, Using machine learning to replicate chaotic attractors and calculate Lyapunov exponents from data. Chaos **27**, 121102 (2017)
4. J. Pathak, B. Hunt, M. Girvan, Z. Lu, E. Ott, Model-free prediction of large spatiotemporally chaotic systems from data: a reservoir computing approach. Phys. Rev. Lett. **120**, 24102 (2018)
5. B. Widrow, W.H. Pierce, J.B. Angell, Birth, life, and death in microelectronic systems. IRE Trans. Military Electron **5**, 191–201 (1961)
6. C. Mead, *Analog VLSI and Neural Systems* (Addison-Wesley, 1989)
7. G. Snider, Cortical computing with memristive nanodevices. Sci-DAC Review **10**, 58–65 (2008)
8. M. Sugawara (ed.), *Semiconductors and Semimetals*, vol. 60 (Academic, London, 1999)
9. J. Sabarinathan, P. Bhattacharya, P.-C. Yu, S. Krishna, J. Cheng, D.G. Steel, Appl. Phys. Lett. **81**, 3876 (2002)
10. D.L. Huffaker, G. Park, Z. Zou, O.B. Shchekin, D.G. Deppe, Appl. Phys. Lett. **73**, 2564 (1998)
11. H. Drexler, D. Leonard, W. Hansen, J.P. Kotthaus, P.M. Petroff, Phys. Rev. Lett. **73**, 2252 (1994)
12. Y. Miyata, et al., Basic experimental verification of spatially parallel pulsed-I/O optical reservoir computing. Inf. Photonics PA16 (2020)
13. O.O. van der Biest, L.J. Vandeperre, Annu. Rev. Mater. Sci. **29**, 327–352 (1999)
14. M.A. Islam, I.P. Herman, Appl. Phys. Lett. **80**, 3823–3825 (2002)
15. M.A. Islam, Y.Q. Xia, D.A. Telesca, M.L. Steigerwald, I.P. Herman, Chem. Mater. **16**, 49–54 (2004)
16. P. Brown, P.V. Kamat, J. Am. Chem. Soc. **130**, 8890–8891 (2008)
17. S. Sakai, et al., Verification of spatiotemporal optical property of quantum dot network by regioselective photon-counting experiment. Photonics Switch. Comput. Tu5A (2021)

Exploring Integrated Device Implementation for FRET-Based Optical Reservoir Computing

Masanori Hashimoto, Takuto Matsumoto, Masafumi Tanaka, Ryo Shirai, Naoya Tate, Masaki Nakagawa, Takashi Tokuda, Kiyotaka Sasagawa, Jun Ohta, and Jaehoon Yu

Abstract This chapter explores a reservoir computing (RC) device based on fluorescence resonance energy transfer (FRET) between quantum dots (QDs). We propose a compact structure in which optical input/output and quantum dots are adjacently placed without lenses or delay lines. The proposed structure exploits the QD-based optical reservoir as an intermediate layer and adopts memory to enable recurrent inputs. We evaluate the feasibility of the proposed structure by applying tasks that require nonlinearity. Simulation-based experimental results show that the proposed device can perform logistic map, time-series XOR, and NARMA10. A proof-of-concept implementation with a commercial image sensor demonstrates that the proposed structure can solve XOR and MNIST tasks. Also, we discuss the energy advantage over conventional digital circuit implementations.

1 Introduction

Reservoir computing simplifies model training by keeping the randomly generated middle layer unchanged and only training the output connection. Physical reservoirs, which use physical phenomena to replace the middle layer, are gaining popularity

M. Hashimoto (✉) · T. Matsumoto · R. Shirai
Department of Informatics, Kyoto University, Kyoto, Japan
e-mail: hashimoto@i.kyoto-u.ac.jp

M. Tanaka
Osaka University, Suita, Japan

N. Tate
Kyushu University, Fukuoka, Japan

M. Nakagawa
Fukuoka Institute of Technology, Fukuoka, Japan

T. Tokuda · J. Yu
Tokyo Institute of Technology, Yokohama, Japan

K. Sasagawa · J. Ohta
Nara Institute of Science and Technology, Ikoma, Japan

© The Author(s) 2024
H. Suzuki et al. (eds.), *Photonic Neural Networks with Spatiotemporal Dynamics*,
https://doi.org/10.1007/978-981-99-5072-0_5

as they may be more efficient than traditional digital circuits [1]. Physical reservoirs require nonlinearity, short-term memory, and high dimensionality to be effective. Various physical reservoirs have been studied, including light-based [2], oscillator-based [3], and mechanical reservoirs [4].

With the rise of IoT, edge computing is becoming more prevalent in applications where centralized data processing is not suitable. Deep learning requires significant computational power, making edge devices with limited hardware and energy resources challenging. Additionally, the demand for computing proximity to users necessitates the development of smaller devices.

FRET is a phenomenon where excited states are transferred between adjacent QDs based on the QD types and their distance as discussed in Chap. 4. The densely populated, randomly generated QD network inherently exposes several nonlinear relationships between the input excitation light and output fluorescence. The excitation state represents memory, making the FRET behavior promising for physical reservoirs. However, the tiny size of QDs makes it challenging to know which QD emits each photon by photodiode (PD) arrays or image sensors. Additionally, the short duration of state holding and fluorescence lifetime require repeated sensing. To overcome these issues, optical devices often use lenses and delay lines, but they increase the device size.

This study proposes a simple structure that utilizes FRET for computing without the use of lenses or delay lines and can be easily miniaturized to address the issues mentioned. The structure comprises tiny LEDs as excitation light sources, a sheet of QD network, a filter that can eliminate excitation light and transmit fluorescence, and a photodiode array. As a first step, this study mostly considers only a single type of QD network for FRET-based reservoir computing. Using a FRET simulator, we confirm the feasibility of the device by mapping several tasks requiring nonlinearity. A proof-of-concept (PoC) device is implemented using a commercial image sensor and a droplet sheet of QDs. Experimental validation shows that XOR and MNIST tasks can be performed using the PoC device. Finally, we discuss the advantage of computational energy.

2 Proposed Device Structure

2.1 Device Structure

We propose a device structure for FRET-based reservoir computing, which is shown in Fig. 1 [5]. The physical reservoir, which serves as the intermediate layer in reservoir computing, corresponds to the gray box in the figure. The device consists of a 2D-array light source (input), a sheet containing numerous randomly placed QDs, and a 2D PD array (output) arranged in a straight line. The light source provides the optical input that excites the QDs, and the PD measures the intensity of the fluorescence. To minimize the form factor of the device, lenses are not used in this structure. The light

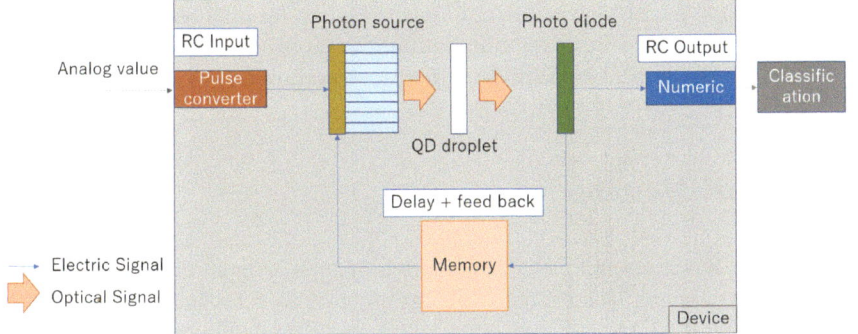

Fig. 1 Proposed device structure. ©[2022] IEEE. Reprinted, with permission, from [5]

source, QDs, and PDs are intended to be stacked and housed in a single package in the expected final implementation.

As mentioned earlier, the excited state memory of QDs is short, and detecting fluorescence on a per-QD basis is impractical due to the size mismatch between QDs and PDs. To address these, the proposed structure includes a digital memory to form a recurrent network. Additionally, single-photon detection is challenging due to the low sensitivity of typical PDs and isotropic photon emission. Therefore, stable reservoir output requires repeated input and accumulation. Taking into account this accumulation time, the reservoir operates discretely in time like a sequential circuit, and its recurrent behavior progresses via feedback through the memory. Finally, the digitized reservoir output is fed to a lightweight machine learning model, such as linear support vector machines or ridge regression, to obtain the final output.

A closer view of the light source, QDs, and PDs is shown in Fig. 2, where a filter is used to eliminate the excitation light. High rejection-ratio bandpass filters, such as those in [6], are recommended for this purpose. Since QDs are much smaller than

Fig. 2 Closer view near QDs. ©[2022] IEEE. Reprinted, with permission, from [5]

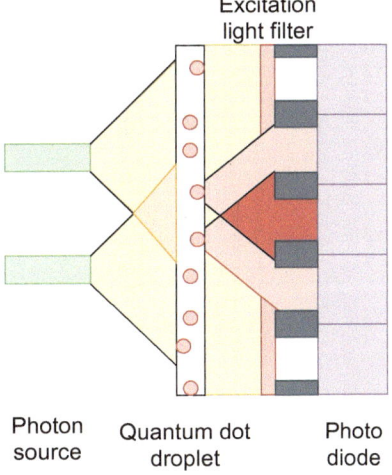

PDs and the photon emission is isotropic, each PD receives photons emitted by many QDs. Nonetheless, even in this configuration, the proposed structure can exploit a nonlinear input-output relationship suitable for reservoir computing.

2.2 Network Mapping

The proposed device structure enables a switching matrix function through the feedback memory, allowing for the selective mapping of an echo state network (ESN) onto the device. An ESN is a non-physical reservoir implementation consisting of artificial neurons whose recurrent connections and weights are randomly determined, with a nonlinear transformation performed using a nonlinear activation function [7].

This work proposes to adjust the memory switching matrix to selectively map a compact ESN achieving high performance onto the proposed device structure. Figure 3 illustrates the correspondence between the ESN and the network on the proposed device, where a set of 3×3 light source array and 3×3 PD array is considered equivalent to a node in an ESN. The spatial and temporal overlap of the light from the light sources can provide interaction between multiple light sources, and the arrows between the nodes from PD to the light sources represent delayed PD output and recurrent inputs from the light sources. The weights can be set by adjusting the relative positions of the light source and PD. In the device, the nodes are separated in time using external memory, while in space, they are sufficiently distant from each other to avoid unexpected FRETs between nodes. In addition to the conventional training of the output part, the feedback matrix also needs to be determined.

Fig. 3 Network construction. ©[2022] IEEE. Reprinted, with permission, from [5]

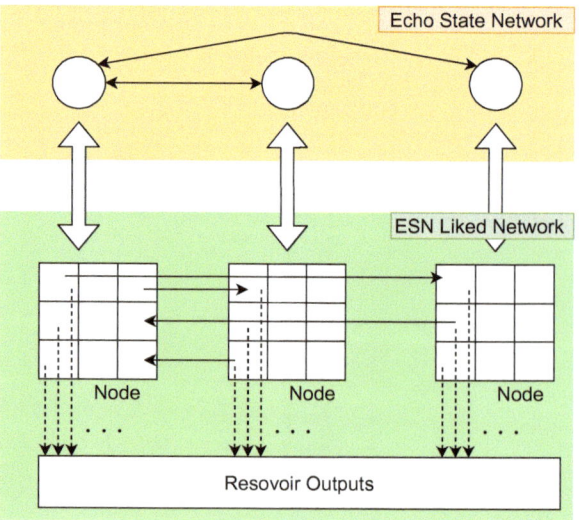

Table 1 Simulation setup

QD density	1000 dots/μm^2
# of MC trials	200
PD size	600 nm
PD pitch	1000 nm
QD-PD distance	1000 nm
LS-QD distance	1000 nm
Sim. step	100 ns

©[2022] IEEE. Reprinted, with permission, from [5]

2.3 Experiments

2.3.1 Setup

For the experiments, QD networks are generated randomly based on the conditions described in Table 1, assuming the use of QD585 [8]. Other device parameters are also listed in Table 1. Two types of light sources are used: DC and pulsed sources. The intensity of the DC source corresponds to the input value, while the input value is represented by the pulse count in unit time for the pulsed source. Specifically, the input light is pulsed with a period of 10/(input value) [ns], where the pulse width is constant at 1 ns.

In this study, we utilized a simulator [9] to simulate the behavior of QDs, which was introduced in Chap. 5. The simulator employs a Monte Carlo method to stochastically simulate the state transitions of QDs, including excitation, FRET, fluorescence, and inactivation, based on tRSSA [10]. Unlike the original FRET simulator, which only simulates the QD states, our simulation framework replicates the proposed device structure and simulates it as a complete device, with the FRET simulator as its core engine. The input light and fluorescence, for instance, decay based on the squared distance in our framework.

2.3.2 Memory-Unnecessary Tasks

We begin by evaluating two tasks that approximate nonlinear functions without the need for memory.

Logistic map

The logistic map is a chaotic system that is highly sensitive to small changes in initial conditions. It can be expressed by the equation $x_{t+1} = \alpha x_t(1 - x_t)$, where our experiment assumes a fixed value of $\alpha = 4.0$ and an initial value of $x_0 = 0.2$.

In this logistic mapping, the output value becomes the next input value, and the prediction process relies on the predictions made one after another. To learn the

Fig. 4 Prediction result of logistic map in time domain. ©[2022] IEEE. Reprinted, with permission, from [5]

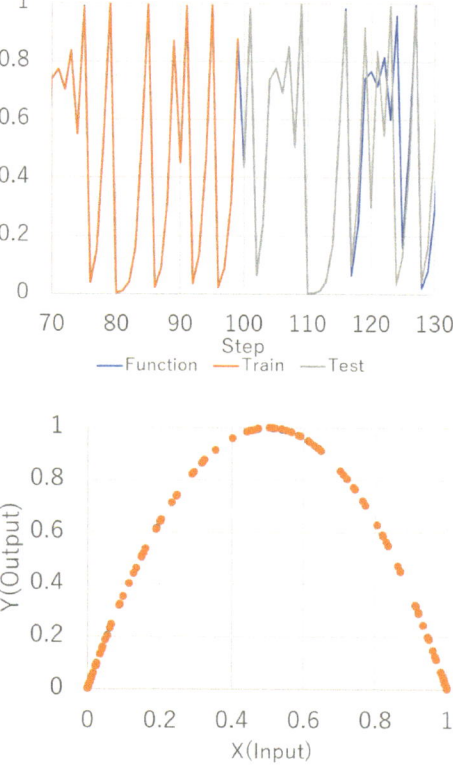

Fig. 5 Prediction result of logistic map in input-output domain. ©[2022] IEEE. Reprinted, with permission, from [5]

function, ridge regression is employed. The training process achieved a mean squared error (MSE) of 9.61×10^{-9}, using a 5×5 PD array and a DC light source.

The performance of the trained model is shown in Fig. 4, with the original function Y, the trained data *Train*, and the prediction *Test*. The training is conducted over the first 100 steps, and the prediction is performed after the 100-th step. Although the system is chaotic, the first 17 steps (i.e., 100–117) are well-approximated, indicating that the approximation is viable. In Fig. 5, we plot an X-Y diagram with the input on the horizontal axis and the output on the vertical axis. It is evident that the function and the prediction are nearly identical.

XOR

We conducted an experiment to test whether the proposed structure can derive non-linearity using a two-input XOR function, $y = \text{XOR}(x_1, x_2)$. To classify the output, we used a linear support vector machine (SVM), which cannot approximate XOR alone. We used 200 cases for training and an additional 50 cases for evaluation, with the inputs x_1 and x_2 given from two locations using pulsed light sources. For 0 and 1 inputs, the pulse frequencies were set to 50 and 100 MHz, respectively.

Fig. 6 Network structure for time-series XOR. ©[2022] IEEE. Reprinted, with permission, from [5]

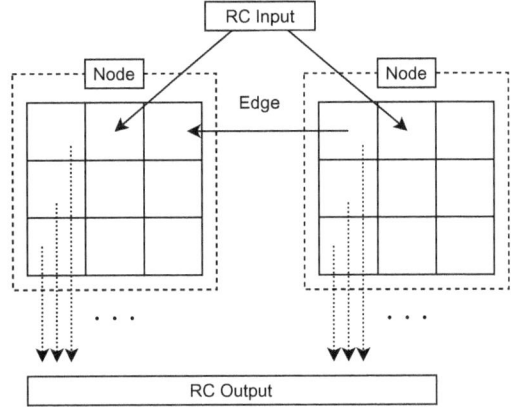

We tested two configurations: 2×2 and 3×3 PD arrays. In both cases, we gave the input lights to the most distant diagonal locations. The training and evaluation in the 3×3 PD array achieved 100% accuracy, whereas in the 2×2 case, both the training and evaluation accuracies were 75%, indicating poor approximation of the XOR function. We attribute this difference in accuracy to the variation in distance between the light source and the PD. In the 3×3 case, there were six variations in distance, while in the 2×2 case, there were only three.

2.3.3 Memory-Necessary Temporal Tasks

Next, we will evaluate tasks that require memory in the reservoir.

Time-series XOR

In the time-series XOR experiment, random 0 and 1 inputs are used to predict the XOR result of the current and previous inputs, i.e., $d(n) = \text{XOR}(u(n), u(n - 1))$. The input is generated by the same pulsed light source as in the previous XOR experiment. The feedback input is adjusted based on the amount of photons received by the associated PD, with the pulse period being inversely proportional to the amount of photons received.

Figure 6 shows one network configuration tested, where there is a one-step memory from the right node to the left node. The right node provides the previous input to the left node, which can then process both the current and previous inputs. Each node has 3×3 PDs, resulting in 18 outputs as a reservoir, with a linear support vector machine applied to the reservoir outputs. Both training and evaluation achieved 100% accuracy in the experiment.

NARMA10

Next, we evaluate the capability of reservoir computing using NARMA [11], which is a standard benchmark for testing reservoir performance. NARMA is expressed by the following equation:

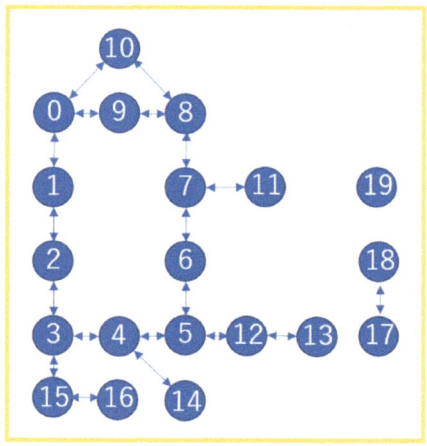

$$d(n+1) = a_1 d(n) + a_2 d(n) \sum_{i=0}^{m-1} d(n-i) + a_3 u(n-m+1)u(n) + a_4, \qquad (1)$$

where a_i are constants, and $u(n)$ is the input at time n. The output d depends on the inputs of the previous m steps, which means m-step memory is necessary. In this experiment, we evaluate the widely used NARMA10 with $m = 10$.

Due to the extensive number of steps in NARMA10, resulting in a long simulation time, only 100 FRET simulation trials were conducted. The experiment is designed for online training, where weights are sequentially updated in every step using the sequential least-squares method. The total number of training steps was 600 for three different sets of NARMA10 parameters, with 200 training steps for each set.

To find appropriate network structures, we generated multiple ESNs by varying the number of nodes and edges. The models were then trained and evaluated using the RMSE metric. Among them, we selected a 20-node network with cyclic structures depicted in Fig. 7 that provided accurate results. We then mapped this network onto the proposed device structure following the procedure explained in Sect. 2. Each node in the network was assumed to have a 3×3 PD array and a 3×3 light source array.

The results of the NARMA10 experiment are presented in Fig. 8, where the blue line corresponds to the target function $d(n)$ and the orange line represents the predicted values. At the beginning of the training, there is some distance between the blue and orange lines, but they gradually overlap as the training progresses. The root mean square error (RMSE) between the target and predicted values is 0.020.

Fig. 8 NARMA10 prediction result. ©[2022] IEEE. Reprinted, with permission, from [5]

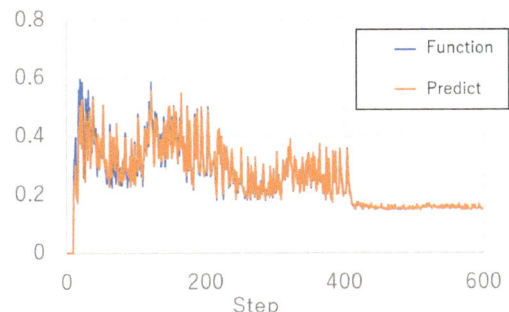

3 Proof-of-Concept Prototyping

To realize reservoir computing using QDs, it is necessary to confirm whether QD fluorescence can be observed in a small experimental system and whether its output can be used for learning. In this section, we construct a small experimental system using a commercial image sensor, observe the fluorescence of QDs, and evaluate the possibility of learning from the output.

3.1 Implementation

As part of the experimental setup, we installed a device incorporating an image sensor in a darkroom. Figures 9 and 10 depict the experimental systems used for the XOR and MNIST tasks, respectively, with the main difference being the used light source. Thinly coated QDs were applied to the cover glass, which was then placed closely on the image sensor. The image was obtained by illuminating the QDs with excitation light from above using a light source fixed to the XYZ stage substrate, and capturing the image with the image sensor. For the XOR task, two 430 nm LEDs were used as

Fig. 9 Experimental setup with LEDs for XOR task

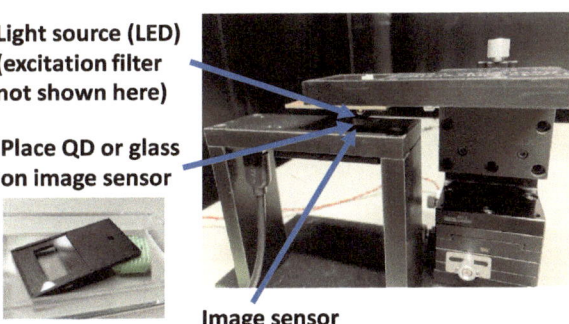

Light source (LED) (excitation filter not shown here)

Place QD or glass on image sensor

Image sensor

Fig. 10 Experimental setup with a projector for MNIST

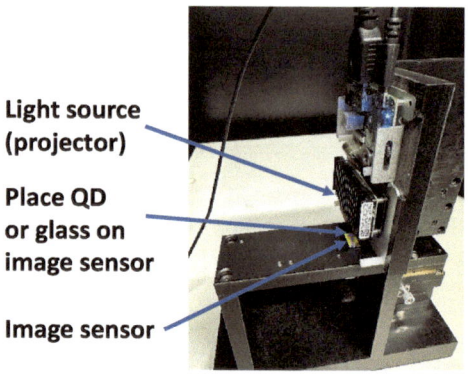

Light source (projector)

Place QD or glass on image sensor

Image sensor

the light sources, while for the MNIST task, a laser micro-projector with a resolution of 1280 × 720 (HD301D1, Ultimems, Inc.) was employed.

The QDs used in the XOR task are QDs (CdSe/ZnS, ALDRIICH) with a single center wavelength of 540 nm, and are fabricated in thin film form to realize a network structure. 100 μL of a solution of 30 μL of QDs and 270 μL of thermosetting resin (Sylgard 184) is poured into a cover glass. The film is deposited by spin coating. The resin is then heated to cure it. Regarding the QDs employed in the MNIST task, two types were used: CdSe/ZnS QDs with a 600 nm wavelength and CdSe/ZnS QDs with a 540 nm wavelength.

The performance of the filter used to separate the excitation light from the fluorescence is essential to observe fluorescence using a compact lensless image sensor. This image sensor is a commercial image sensor (CMV 20000, AMS) with a pixel count of 2048 × 1088 and a pixel size of 5.5 μm. We implemented a bandpass filter on the image sensor consisting of a long-pass interference filter, a fiber optic plate (FOP), a short-pass interference filter, and an absorption filter. Interference filters have a high rejection rate for perpendicular light but are angle-dependent, allowing scattered light to pass through. Therefore, by combining an absorption filter with angle-independent absorption with the interference filter, a high excitation light rejection is achieved, which is transmission of 10^{-8} at the assumed excitation light wavelength of 430–450 nm [6]. This allows the fluorescence of QDs whose wavelength is 540 nm and 600 nm to be transmitted through the filter, and only the excitation light is removed.

3.2 Evaluation

3.2.1 XOR

The nonlinearity of the reservoir layer is necessary to function as a reservoir computing device. To evaluate this nonlinearity, we conducted an experiment to check whether it is possible to solve the XOR problem using a linear learner when two

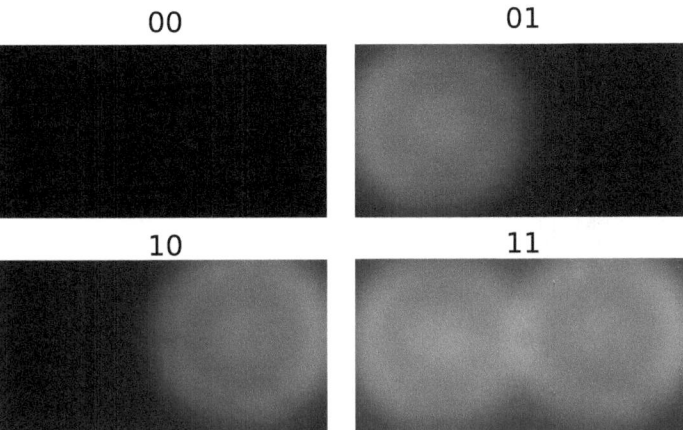

Fig. 11 Example of fluorescence pictures captured by the image sensor for XOR task

inputs are input from the light source. The captured images are used for training linear SVM. An example of the input image is shown in Fig. 11. Sixty images are taken for each of 00, 01, 10, and 11 (240 images in total), 40 for training and 20 for inference. Comparison is made between the case with QD and the case without QD (glass only, no excitation filter).

The experimental procedure involves determining the DC light intensity required to represent input values of 0 and 1, in order to achieve maximum task accuracy. Subsequently, the LEDs are turned off for input 0, while a constant current is applied for input 1. To ensure equal light intensity from both LEDs, the magnitude of the constant current for input 1 is adjusted to compensate for any spectral shift that may cause the two LEDs to exhibit different intensities at the same current.

The SVM utilized a fixed pixel range size of 32×32 pixels, and learning was performed at each location by shifting the location within the entire image (2048 $\times$ 1088 pixels). To address the issue of higher accuracy in areas with lower pixel values, images were captured with varying exposure times. This ensured that the pixel values in the bright areas of the image were approximately the same for both cases: with and without QDs. The accuracy for each location in the 32×32 pixel case is depicted in Fig. 12.

It was observed that using a shorter exposure time of 4 ms resulted in improved accuracy, both with and without QDs. Specifically, with a 4 ms exposure time, the QDs enhanced accuracy across a wider range of the image sensor, indicating their contribution to nonlinear computation. Additionally, the difference in accuracy between exposure times of 4 and 40 ms suggests that the image sensor itself may also exhibit nonlinearity. This finding underscores the necessity for image sensors dedicated to reservoir computing to possess nonlinear pixels, which may not be suitable for conventional image sensing applications.

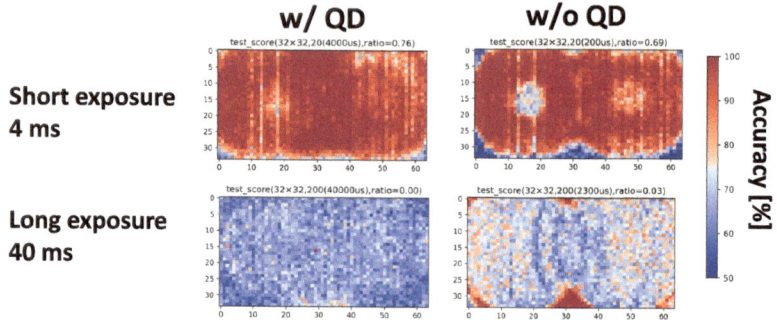

Fig. 12 Spatial distributions of XOR task accuracy across the image sensor. Each 32 × 32 pixels are given to SVM

3.2.2 MNIST

We employed the Newton conjugate gradient method to train the reservoir output for MNIST using logistic regression. Experiments were conducted on four different input image sizes: 28 × 28, 140 × 140, 280 × 280, and 600 × 600, as illustrated in Fig. 13. The fluorescence images shown in Fig. 13 serve as representative examples. For the 28 × 28 image size, we used 900 fluorescence pictures for training and 150 pictures for testing. For the other sizes, we used 1,500 pictures for training and 500 pictures for testing. Distant pixels that did not observe fluorescence were excluded from the training data. To prevent pixel value saturation in the image sensor, we modified the color of the projector light when QDs were not employed.

Table 2 presents the accuracy results, indicating that the accuracy decreases as the input image size decreases. For an input image size of 600 × 600, the accuracies with and without QDs were 88.8 and 87.6%, respectively. In comparison, the accuracy achieved using the original MNIST data was 87.0% under the same training conditions. Therefore, in both cases, the PoC implementation achieved higher accuracy than the linear regressor. Furthermore, the accuracy was higher when QDs

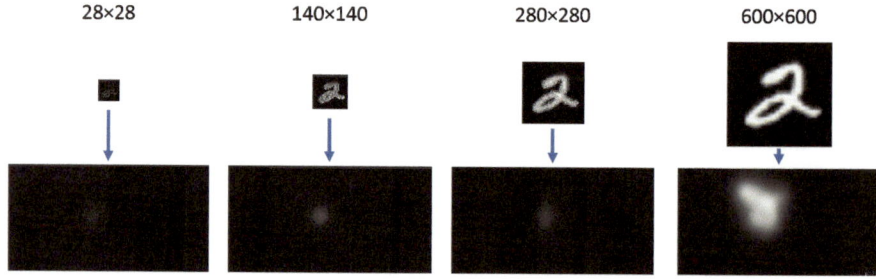

Fig. 13 Example of fluorescence pictures captured by the image sensor for MNIST task

Table 2 MNIST accuracy. The highest accuracy with linear regressor is 87.0%. The accuracies in bold are better than that of the linear regressor

Image size	w/QD		w/QD	
	Accuracy (%)	Exposure (ms)	Accuracy (%)	Exposure (ms)
28 × 28	23.3	500	38.0	500
140 × 140	40.6	100	62.0	50
280 × 280	71.0	16.7	81.0	50
600 × 600	88.8	16.7	87.6	50

were used compared to when they were not. This improvement in accuracy can be attributed to the nonlinearity of QDs and the image sensor. We will further investigate the conditions under which QDs can offer a more significant advantage.

4 Discussion on Energy Advantage

This section aims to investigate the potential power-saving benefits of the proposed reservoir computing device. To accomplish this, we conduct an analysis of the energy consumption of the physical reservoir computer and compare it with that of a digital circuit implementation.

4.1 Power Estimation Approach

Consider a structure consisting of a light source and a sensor directly below it, as shown in Fig. 14. Energy consumption in this structure is generated by the light source and sensor. For simplicity, we consider the energy consumed by the photodiode directly below the light source that receives the most light. Photons emitted from the light source excite the QD, from which photons stochastically enter the photodiode on the sensor at the origin directly below. The sensor part that receives the fluorescence and converts it into an output consists of a photodiode, a comparator, and an 8-bit counter. The comparator detects a certain voltage drop on the photodiode and converts it into a pulse, which is counted by the 8-bit counter. If the operating time is the time it takes for the sensor at the origin to count 256 pulses, this can be obtained from the number of photons incident on the photodiode per unit time, and the sensitivity and conversion efficiency of converting the photons to a voltage. The energy consumption is calculated by multiplying the power consumption of the sensor and the light source by the operation time.

Fig. 14 Assumed structure

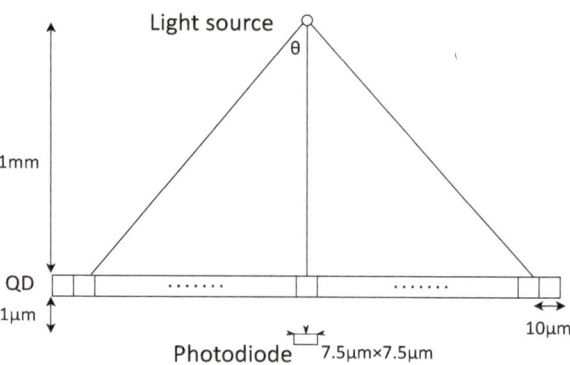

4.1.1 Probability of Photon Incidence

It is assumed that photons emitted from a light source follow a normal angular distribution with the divergence angle of the light source, which is the angle at which the intensity is halved, serving as the half-width. The photons are stochastically directed toward that angle. The probability of a photon entering a 10 μm × 10 μm section of the QD surface located 1mm away from the light source is illustrated in Fig. 15, where the probability density is integrated over each section of the QD surface to obtain the section probability. The photons emitted from the QD as fluorescence are presumed to travel in a random direction.

Figure 16 shows the probability of a photon incident on the photodiode at the origin, which is 1 μm away from the QD surface. For greater distances, photons enter the photodiode at the origin from a broad range of locations, but for 1 μm, photons primarily enter the photodiode from the QD directly above it.

Fig. 15 Incidence probability from the light source to the QD surface (Distance 1 mm)

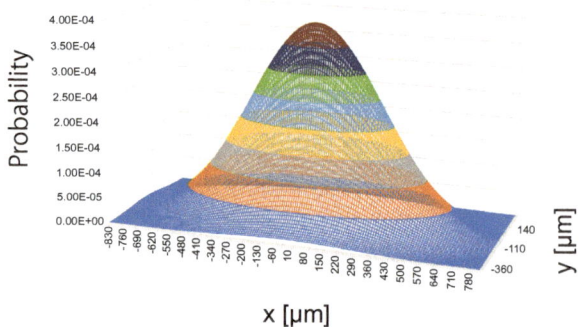

Fig. 16 Incidence probability from the QD surface to the origin PD (Distance 1 μm)

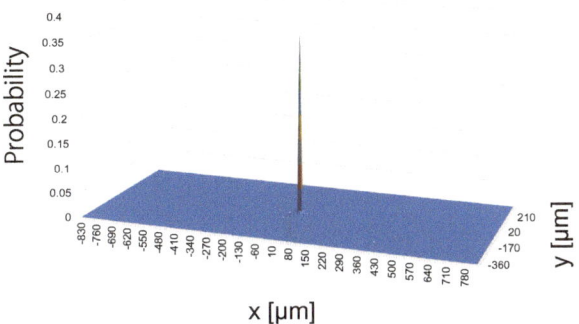

4.1.2   Photon Input/Output Ratio in QD network

Let us calculate the input-output ratio of photons in a QD. The decay of the fluorescence intensity $I(t)$ in a QD is expressed as an exponential decay as follows [12]:

$$\frac{dI(t)}{dt} = -\frac{1}{\tau_0} I(t), \tag{2}$$

$$I(t) = I_0 \exp(-\frac{t}{\tau_0}). \tag{3}$$

In this study, τ_0 is assumed to be constant, but it is known that the average fluorescence lifetime of the entire QD varies depending on the density and light intensity [13]. FRET phenomena are often observed between donors and acceptors, but they can also occur in a single QD. For simplicity, we consider fluorescence in a single type of QD.

The QDs that are newly excited by incoming photons are assumed to be QDs that are not currently excited. It is also assumed that the QDs are equally spaced lattice structures. Adding the excited term, we obtain the following equation:

$$\frac{dI(t)}{dt} = -\frac{1}{\tau_0} I(t) + (N_A - I(t)) \times \frac{\sigma_A}{S} \times N_{photon}, \tag{4}$$

where $I(t)$ is the number of excited QDs, N_A is the number of QDs in the region of interest, σ_A is absorption cross section of QD, S is region area, and N_{photon} is the number of photons injected into the region per unit time.

Since N_{photon} is a constant in the case of DC incidence, $\frac{dI(t)}{dt} = 0$ in equilibrium, we have

$$I(t) = \frac{N_A \times \frac{\sigma_A}{S} \times N_{photon}}{\frac{1}{\tau_0} + \frac{\sigma_A}{S} \times N_{photon}}. \tag{5}$$

In the case of pulse input, N_{photon} is changed to a square wave, and we use the transient response given by the following equation. For pulse input in this study, the period was set to 20 ns, with an on-time of 1 ns and an off-time of 19 ns:

$$\frac{dI(t)}{dt} = \begin{cases} -\frac{1}{\tau_0}I(t), & \text{(light source off)} \\ -\frac{1}{\tau_0}I(t) + (N_A - I(t)) \times \frac{\sigma_A}{S} \times N_{photon}. & \text{(light source on)} \end{cases} \quad (6)$$

At this time, the number of photons emitted from the QD surface as fluorescence per unit time is given by substituting $I(t)$ into Eq. (2) and multiplying it by the QD's emission efficiency.

4.1.3 Sensor Energy

The sensor part is supposed to consist of a photodiode, a comparator, and an 8-bit counter. The assumed structure is shown in Fig. 17. In the photodiode, the voltage across it under reverse bias gradually decreases depending on the number of incident photons. The overall operation of the sensor is as follows.

1. Photodiode: A photodiode converts an incident photon into a voltage by storing it for a certain period of time, and the voltage is reduced from the supply voltage.
2. Source follower: The voltage drop of the photodiode is reduced to a voltage in the range appropriate for the operation of the next-stage amplifier.

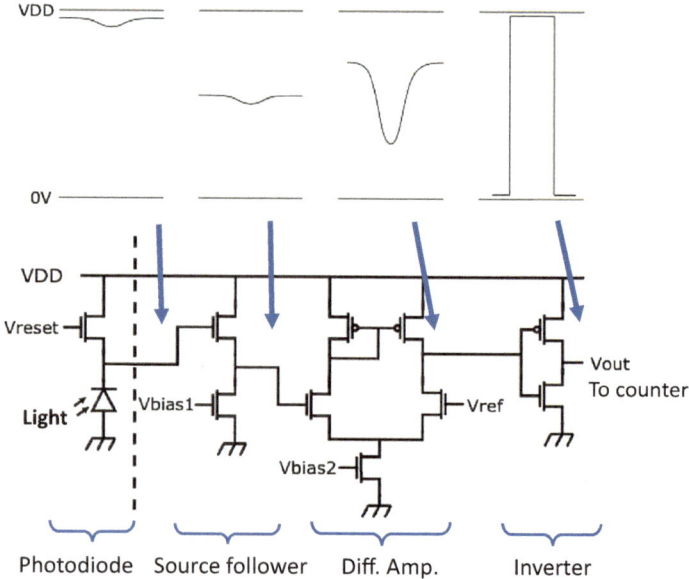

Fig. 17 Sensor structure and waveforms in it

3. Differential amplifier: The voltage change at the photodiode is amplified.
4. Inverter: The output voltage is converted into pulses.
5. 8-bit counter: Up to 256 8-bit pulses are counted.

The input voltage to the differential amplifier is determined by the photons incident on the photodiode multiplied by the voltage conversion efficiency. In this evaluation, the voltage drop required to output one pulse is 100 mV. Therefore, the operating time is the time required for this 100 mV voltage drop with the highest photon intensity multiplied by 256. In addition, the power consumption of the sensor unit is calculated by simulating the consumption energy when a triangular wave is assumed to be applied to the sensor unit as a voltage drop.

4.2 Result

4.2.1 Energy Dissipation

Using the probability of incidence on the photodiode at the origin and the assumed conversion efficiency, we calculate the time required for the comparator to output one pulse, and multiply it by the power consumption of the sensor and the light source to obtain the energy consumption, as shown in Table 3. The parameters used in the calculation are shown in Table 4. In this study, a laser diode (NDV4316, NICHIA) is assumed as the light source, and the sensitivity and conversion efficiency are based on the values of existing image sensors (S11639, Hamamatsu Photonics).

As shown in Table 3, using pulsed input light results in longer operation time. The light source does not consume energy when off, but the sensor continues to operate. Therefore, the energy consumption of the sensor increases roughly proportional to the operation time even when the light source is off. From an energy standpoint, DC input is better than pulsed input for the light source, but the pulsed input might be better in terms of utilizing the short-term memory of the QD.

Table 3 Operating time and energy consumption of DC light and pulsed light

	DC light	Pulsed light
Necessary time for pulse (ns)	56.9	1250
Operating duration (μs)	14.5	320
Energy dissipated by light source (μJ)	8.4	9.14
Energy dissipated by sensor (nJ)	11.4	237.6

Table 4 Parameters used for energy calculation

Parameter	Value
PD conversion efficiency (μV/e)	25
Sensitivity (A/W)	0.3
Section width (μm)	10
Input wavelength (nm)	400
Output wavelength (nm)	600
QD absorption cross section (cm^2)	1.51×10^{-17}
QD fluorescence lifetime (ns)	20
Rated power of light source (mW)	576

4.2.2 Energy Comparison with Digital Circuit Implementation

To assess the energy efficiency of the computed energy discussed in the previous section, we compare it with the energy consumption in a digital circuit implementation. In neural networks, the energy consumption in multiplication and accumulation (MAC) operations and memory access becomes a concern as the complexity grows. The energy consumption for 32-bit floating-point operations and memory access is presented in Table 5 [14]. We will utilize these values for the comparative evaluation in this section.

Let m be the number of light sources and n be the number of PDs. Consider adding up weighted inputs in a fully connected layer. We assume that the input transformation in the QD network is equivalent to the calculation in a fully connected layer. If we read weights from RAM (DRAM, SRAM), we need to perform addition, multiplication, and weight reading $m \times n$ times each. On the other hand, the energy consumption of the light sources and sensors is proportional to their respective numbers. Therefore, the energy consumption for digital circuit implementation and physical reservoir computing implementation can be calculated as follows:

- SRAM: $(0.9 + 3.7 + 5) \times mn = 9.6$ pJ $\times mn$
- DRAM: $(0.9 + 3.7 + 640) \times mn = 644.6$ pJ $\times mn$
- DC light source + sensor: 8.4 μJ $\times m + 11.4$ nJ $\times n$
- pulse light source + sensor: 9.14 μJ $\times m + 238$ nJ $\times n$.

Table 5 Energy consumption of 32-bit floating-point arithmetic and memory access

Operation	Energy (pJ)
Addition	0.9
Multiplication	3.7
SRAM Read (8 KB)	5
DRAM Read (8 KB)	640

Table 6 m, n values achieving lower power dissipation than digital implementation (DC light)

	SRAM	DRAM
100 mV	$m:10^4, n:10^6$	$m:10^2, n:10^5$
10 mV	$m:10^3, n:10^5$	$m:10, n:10^4$

Table 7 m, n values achieving lower power dissipation than digital implementation (pulsed light)

	SRAM	DRAM
100 mV	$m:10^6, n:10^6$	$m:10^3, n:10^5$
10 mV	$m:10^5, n:10^5$	$m:10^2, n:10^4$

Tables 6 and 7 show the values of m and n where the energy consumption of the light sources and the sensors is smaller than that of digital circuit implementation. We compared the energy consumption of SRAM read and DRAM read in digital circuit implementation. In this simulation, we assume that the comparator outputs 1 pulse at 100 mV, and when 10 mV corresponds to 1 pulse, the time becomes 1/10. From Tables 6 and 7, we can see that increasing the number of light sources m and PDs n results in lower energy consumption than digital circuit implementation. This is because in the proposed device, the energy consumption only increases by the sum of the energy consumption of the light sources and sensors when m and n are increased. Comparing DC and pulse light sources, the DC light source gives smaller m value, meaning that the DC light source is more energy efficient.

In this evaluation, we assumed that the conversion of the input in the QD network corresponds to the computation in the fully connected layer, but more energy-saving operation can be expected when more complex conversions are performed.

5 Summary

In this chapter, we explored the viability of a compact implementation for FRET-based optical reservoir computing. The proposed device can be integrated into a single package containing the light source, QDs, filters, photodetectors, and digital signal processing capabilities. Our evaluations, which included simulation-based and proof-of-concept-based assessments, demonstrated that the proposed device is capable of performing tasks and is energy efficient for large-input large-output computation. Moving forward, we plan to develop a dedicated chip for FRET-based reservoir computing, which will be integrated into a single package.

References

1. G. Tanaka, T. Yamane, J.B. Héroux, R. Nakane, N. Kanazawa, S. Takeda, H. Numata, D. Nakano, A. Hirose, Recent advances in physical reservoir computing: a review. Neural Netw. **115**, 100–123 (2019)
2. S.S. Vatsavai, I. Thakkar, Silicon photonic microring based Chip-Scale accelerator for delayed feedback reservoir computing (2021). arXiv:2101.00557
3. J. Torrejon, M. Riou, F.A. Araujo, S. Tsunegi, G. Khalsa, D. Querlioz, P. Bortolotti, V. Cros, K. Yakushiji, A. Fukushima, H. Kubota, S. Yuasa, M.D. Stiles, J. Grollier, Neuromorphic computing with nanoscale spintronic oscillators. Nature **547**(7664), 428–431 (2017)
4. K. Nakajima, H. Hauser, T. Li, R. Pfeifer, Information processing via physical soft body. Sci. Rep. **5**, 10487 (2015)
5. M. Tanakal, J. Yu, M. Nakagawa, N. Tate, M. Hashimoto, Investigating small device implementation of FRET-based optical reservoir computing, in *Proceedings International Midwest Syposium on Circuits and Sytems (MWSCAS)* (2022), pp. 1–4
6. K. Sasagawa, Y. Ohta, M. Kawahara, M. Haruta, T. Tokuda, J. Ohta, Wide field-of-view lensless fluorescence imaging device with hybrid bandpass emission filter. AIP Adv. **9**(3), 035108 (2019)
7. H. Jaeger, *The "echo state" approach to analysing and training recurrent neural networks* (Tech. rep, GMD - German National Research Institute for Computer Science, 2001)
8. U. Utzinger, *Spectra Database Hosted at the University of Arizona* (2011)
9. M. Nakagawa, Y. Miyata, N. Tate, T. Nishimura, S. Shimomura, S. Shirasaka, J. Tanida, H. Suzuki, Spatiotemporal model for FRET networks with multiple donors and acceptors: multicomponent exponential decay derived from the master equation. J. Opt. Soc. Am. B **38**(2), 294 (2021)
10. V.H. Thanh, C. Priami, Simulation of biochemical reactions with time-dependent rates by the rejection-based algorithm. J. Chem. Phys. **143**(5), 054104 (2015)
11. H. Jaeger, Adaptive nonlinear system identification with echo state networks, in *Proceedings of the 15th International Conference on Neural Information Processing Systems*, NIPS'02 (MIT Press, Cambridge, MA, USA, 2002), pp. 609–616
12. J.R. Lakowicz, *Principles of Fluorescence Spectroscopy* (Springer Science & Business Media, 2013)
13. S. Shimomura, T. Nishimura, Y. Miyata, N. Tate, Y. Ogura, J. Tanida, Spectral and temporal optical signal generation using randomly distributed quantum dots. Opt. Rev. **27**(2), 264–269 (2020)
14. M. Horowitz, 1.1 computing's energy problem (and what we can do about it), in *IEEE International Solid-State Circuits Conference Digest of Technical Papers (ISSCC)* (IEEE 2014), pp. 10–14

FRET Networks: Modeling and Analysis for Computing

Masaki Nakagawa

Abstract FRET networks, which refer to energy transfer networks between nanopar-
ticles due to Förster resonance energy transfer (FRET), are promising physical phenomena for realizing high-speed, efficient, and compact information processing.
These networks can generate rich spatiotemporal signals that help in information
processing and are capable of function approximation, time-series prediction, and
pattern recognition. This chapter presents a mathematical model and analysis for
FRET networks, including some simulation methods for the model, and demonstrates the power of FRET networks for information processing.

1 Introduction

The energy transfer caused by dipole–dipole interactions between fluorescent
molecules is known as Förster resonance energy transfer (FRET). Förster theory
[1] states the energy transfer rate (the expected number of energy transferred per unit
time) to be such that

$$k^{\text{FRET}} = \frac{3}{2}\frac{\kappa^2}{\tau}\left(\frac{R_0}{r}\right)^6,$$

(1)

where κ is an orientation factor, τ is a natural excited-state lifetime, R_0 is the Förster
distance, and r is the distance between fluorescent molecules. This relation shows
that the energy transfer rate sensitively depends on the distance r between fluorescent
molecules as it is proportional to r to the power of -6.

Consider randomly distributing a large number of fluorescent molecules. The
fluorescent molecules are separated by various distances. Therefore, the energies
on the network are transferred through diverse pathways. If time-series signals are
input as excitation light in the network, we can expect high-dimensionalized and
nonlinearized time-series signals to be produced as fluorescence. Furthermore, we

M. Nakagawa (✉)
Fukuoka Institute of Technology, 3-30-1 Wajiro-higashi, Higashi-ku, Fukuoka 811-0295, Japan
e-mail: m-nakagawa@fit.ac.jp

© The Author(s) 2024
H. Suzuki et al. (eds.), *Photonic Neural Networks with Spatiotemporal Dynamics*,
https://doi.org/10.1007/978-981-99-5072-0_6

can expect some memory of previous input to be left in the network because input energy cycles to some extent in the network. Therefore, this energy transfer network due to FRET can be a prominent phenomenon toward a novel information processing device.

FRET has been used as local nanoscale signaling without molecular diffusion processes. For example, local signaling by FRET between fluorescent molecules on DNA substrates is used as photonic logic gates [2–5]. As previously stated, the kinetics of single-step FRET is well explained by Förster theory, even in special environments such as membranes and solutions (see Chaps. 13–15 in [1]). Furthermore, the kinetics of multistep FRET (a cascade of FRET) can also be understood in principle using Förster theory. Multistep FRET has been demonstrated experimentally on linear DNA scaffolds using heterogeneous fluorescent dyes [6] and even homogeneous fluorescent dyes [7]. Furthermore, multistep FRET occurs also in hetero- and homogeneous quantum dots (QDs) [8, 9]. If spatially distributed fluorescent molecules are excited simultaneously, multistep FRET can occur over multiple locations and times, where fluorescent molecules act as both donors and acceptors depending on temporally changing situations. This multistep FRET network is the one we will consider in the study.

The spatiotemporal dynamics of FRET networks are very important from an information processing perspective. For example, the spatiotemporal dynamics of multistep FRET have been used to design intelligent system components, such as unclonable physical keys [10] and photonic logic gates [2–5]. Some FRET networks on spatially distributed QDs are shown to generate diverse spatiotemporal signals that can be used for information processing [11]. The key to designing information-processing applications for FRET networks is to understand the spatiotemporal behavior of FRET networks. In our previous paper [12], we developed a spatiotemporal mathematical model for FRET networks and revealed its temporal characteristic behavior. We emphasize that our model applies to any fluorescent molecule. However, we concentrate on QD-based FRET networks because QDs are expected to be important fundamental elements in realizing compact and energy-efficient information-processing systems [13–16].

The rest of this chapter is organized as follows: In Sect. 2, we introduce a spatiotemporal model for FRET networks, called the multiple-donor model, and show various analytical (theoretical) results. Section 3 presents some FRET network simulation methods, from deterministic to stochastic ones, and compares deterministic and stochastic methods to reveal the pros and cons of both methods. In Sect. 4, we show the power of FRET networks for information processing by simulations, particularly nonlinear function approximation, chaotic time-series prediction, and handwritten digit recognition. Finally, we summarize this chapter and mention some future works in Sect. 5.

2 Spatiotemporal Model for FRET Networks

In this section, we first introduce a spatiotemporal mathematical model for FRET networks. Then, we show some analytical results for the model. The first part of this section almost follows our previous paper [12].

2.1 Multiple-Donor Model

Wang et al. [17, 18] developed a mathematical model to describe the dynamics of multistep FRET that assumes networks with *no more than one* excited molecule, which we refer to as the "single-donor model." The single-donor model is a continuous-time Markov chain (CTMC) with a finite or countable state space where the time spent in each state is exponentially distributed. The single-donor model assumes that the system has only one excited molecule and hence, cannot consider the "level occupancy effect," which means that already excited molecules are effectively forbidden from energy absorption. However, the level occupancy effect is essential for FRET networks because they involve multiple excited molecules (donors) and non-excited molecules (acceptors). The "multiple-donor model" [12] is an extended version of the single-donor model that assumes networks with multiple excited molecules and non-excited molecules. Although similar models that consider the level occupancy effect already exist [19, 20], their approaches differ from ours in the following points: (i) Their main aim is to present a Monte Carlo simulation algorithm using their models. On the contrary, our main aim is to produce the theoretical results using our model and thus understand the spatiotemporal behavior fundamentally. (ii) They introduce the level occupancy effect as the complete exclusion of already excited molecules from their roles as acceptors, whereas our model incorporates such roles by considering the Auger recombination. (iii) Their models mainly handle the decay processes, whereas our model additionally covers the light-induced excitation process.

In the multiple-donor model for the system consisting of N QDs, we first represent the system state by an element in $\{0, 1\}^N$, where each QD is assigned either a ground state "0" or an excited state "1." Then, we consider the system state probability $P_{i_1 \ldots i_N}(t)$ such that the system is in $(i_1, i_2, \ldots, i_N) \in \{0, 1\}^N$ at time t, where i_n represents the nth QD's state, 0 or 1. Evidently, $0 \le i_1 + i_2 + \cdots + i_N \le N$. It should hold that $\sum_{(i_1, \ldots, i_N) \in \{0,1\}^N} P_{i_1 \ldots i_N}(t) = 1$ for all $t \in \mathbb{R}$. As we consider the whole system state at each time, we can consider the level occupancy effect in FRET between QDs, as shown later.

Next, we define the state transition rules between system states $(i_1, \ldots, i_N)$. The following symbols are used:

$$S_n = (i_1, \ldots, i_{n-1}, 0, i_{n+1}, \ldots, i_N),$$
$$S_n^* = (i_1, \ldots, i_{n-1}, 1, i_{n+1}, \ldots, i_N).$$

The state transition rules consist of the following five rules in the multiple-donor model:

$$S_n^* \xrightarrow{k_n^F} S_n, \tag{2a}$$

$$S_n^* \xrightarrow{k_n^N} S_n, \tag{2b}$$

$$S_n^* + S_m \xrightarrow{k_{nm}^{FRET}} S_n + S_m^*, \tag{2c}$$

$$S_n^* + S_m^* \xrightarrow{k_{nm}^{FRET}} S_n + S_m^*, \tag{2d}$$

$$S_n \xrightarrow{k_n^E(t)} S_n^*, \tag{2e}$$

where k_n^F and k_n^N denote the rate constants of radiative (fluorescence) decay and nonradiative decay for the nth QD, respectively, and k_{nm}^{FRET} denotes the rate constant of FRET from the nth QD to the mth QD. $k_n^E(t)$ denotes the rate constant of the excitation process with irradiation of time-dependent excitation light for the nth QD. The time spent in each state is exponentially distributed with each rate constant, k_n^F, k_n^N, k_{nm}^{FRET}, or $k_n^E(t)$. We also illustrate the state transition rules (2a–2e) in Fig. 1.

The rate constants in the state transition rules (2a–2e) are given using the fundamental physical constants as follows:

$$k_n^F = Q_n/\tau_n, \tag{3a}$$

$$k_n^N = (1 - Q_n)/\tau_n, \tag{3b}$$

$$k_{nm}^{FRET} = (3/2)\left(\kappa_{nm}^2/\tau_n\right)\left(R_{nm}/r_{nm}\right)^6, \tag{3c}$$

$$k_n^E(t) = \sigma_n I_{ex,n}(t), \tag{3d}$$

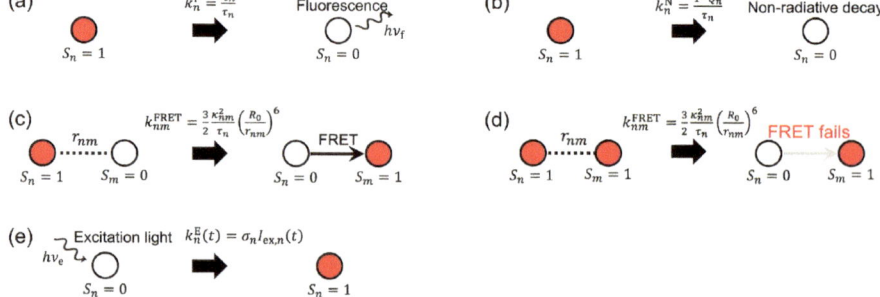

Fig. 1 State transitions in the multiple-donor model: (**a**) Deactivation due to fluorescence, (**b**) Nonradiative deactivation, (**c**) Excitation–deactivation due to FRET, (**d**) Deactivation due to the level occupancy effect, and (**e**) Excitation due to light-induced excitation. Note that $h\nu_{(f)}$, $h\nu_{(e)}$ denote a fluorescence photon and an excitation photon, respectively

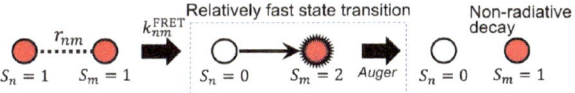

Fig. 2 Auger recombination in the process of two excited-QDs interactions. Since Auger recombination is a fast process, the resulting state transition follows Rule (2d)

where Q_n, τ_n denote the quantum yield and the natural excited-state lifetime for the nth QD, respectively. κ_{nm}^2 is the orientation factor between the nth and mth QDs. R_{nm} and r_{nm} denote the Förster and physical distances from the nth QD to the mth QD, respectively. σ_n denotes the collision cross section for the nth QD, and $I_{\text{ex},n}(t)$ denotes the irradiation photon density for the nth QD.

We note that the transition rules (2a–2c) are essentially equivalent to the ones of the single-donor model [17, 18]. On the other hand, the transition rules (2d–2e) are originally introduced in our study [12]. Rule (2d) describes an energy transfer by FRET and the subsequent Auger recombination. The Auger recombination is a nonradiative decay process from a higher energy excitation state $S_m^{**} = (i_1, \ldots, i_{m-1}, 2, i_{m+1}, \ldots, i_N)$ to the first level excitation state S_m^*. Although the Auger recombination can be modeled as several interactions [21], we will assume here for simplicity that this decay process is relatively rapid, i.e., $S_m^{**} \xrightarrow{\infty} S_m^*$ (see Fig. 2). Therefore, the resulting state transition follows Rule (2d). Finally, Rule (2e) is the state transition due to the light-induced excitation process.

For simplicity, we assume that the orientation factors κ_{nm}^2 and physical distances r_{nm} are constant in time, i.e., QDs have low anisotropies and minimal lateral motions during their excited-state lifetimes. Our model can also be applied to such situations for (i) orientation factor values other than the commonly assumed $2/3$ or even dynamic ones and (ii) diffusion of QDs during their excited states. However, when a considerably faster rotation or diffusion compared to their excited-state lifetimes is considered, one may need to use simpler models (see Chap. 4 in [22] or [23] for the dynamic averaging regime and [24] for the rapid-diffusion limit).

Considering the inflow and outflow of probability, the master equation of the multiple-donor model defined by the state transition rules (2a–2d) is given as follows:

$$\frac{d}{dt} P_{i_1 \cdots i_N}(t) = -\sum_{n=1}^{N} i_n (k_n^{\text{F}} + k_n^{\text{N}}) P_{i_1 \cdots i_N}(t)$$

$$+ \sum_{n=1}^{N} \bar{i}_n (k_n^{\text{F}} + k_n^{\text{N}}) S_n^+ P_{i_1 \cdots i_N}(t) - \sum_{n,m=1}^{N} i_n k_{nm}^{\text{FRET}} P_{i_1 \cdots i_N}(t)$$

$$+ \sum_{n,m=1}^{N} \bar{i}_n i_m k_{nm}^{\text{FRET}} S_n^+ P_{i_1 \cdots i_N}(t) + \sum_{n,m=1}^{N} \bar{i}_n i_m k_{nm}^{\text{FRET}} S_n^+ S_m^- P_{i_1 \cdots i_N}(t)$$

$$-\sum_{n=1}^{N} \bar{i}_n k_n^{\mathrm{E}}(t) P_{i_1 \cdots i_N}(t) + \sum_{n=1}^{N} i_n k_n^{\mathrm{E}}(t) S_n^{-} P_{i_1 \cdots i_N}(t), \tag{4}$$

where $\bar{i}_n$ denotes the inverted binary for the nth QD's state i_n, i.e., $\bar{i}_n = 1 - i_n$, and $S_n^{\pm}$ denotes the shift operator for states $(i_1, \ldots, i_N)$, i.e., $S_n^{\pm} P_{i_1 \cdots i_N}(t) = P_{i_1 \cdots i_n \pm 1 \cdots i_N}(t)$. Notably, if the state $(i_1, \ldots, i_n \pm 1, \ldots, i_N)$ is improper, i.e., not in $\{0, 1\}^N$, $P_{i_1 \cdots i_n \pm 1 \cdots i_N}(t) = 0$. In addition, we set $k_{nn}^{\mathrm{FRET}} = 0$. The time-dependent fluorescence intensity $I(t)$ is expressed as

$$I(t) = \sum_{(i_1, \ldots, i_N) \in \{0,1\}^N} \left[\sum_{n=1}^{N} i_n k_n^{\mathrm{F}} P_{i_1 \cdots i_N}(t) \right]. \tag{5}$$

Note that the master equation (4) includes spatial information through the rate constants k_{nm}^{FRET} (i.e., the network structure of QDs). In the following sections, we focus on the temporal behavior of the FRET network and analyze our model and also the single-donor model.

As mentioned earlier, the single-donor model assumes networks with *at most one* excited molecule [17, 18]. Namely, only the state transition rules described in (2a–2c) are considered. Therefore, the master equation of the single-donor model is as follows:

$$\frac{\mathrm{d}}{\mathrm{d}t} P_n(t) = -(k_n^{\mathrm{F}} + k_n^{\mathrm{N}}) P_n(t) - \sum_{m=1}^{N} k_{nm}^{\mathrm{FRET}} P_n(t) + \sum_{m=1}^{N} k_{mn}^{\mathrm{FRET}} P_m(t), \tag{6}$$

where $P_n(t)$ denotes the probability that the nth QD is in an excited state.

Now, we will show that the master equation (4) of the multiple-donor model coincides with the master equation (6) of the single-donor model if we assume networks with no excitation light and only one excited molecule, i.e., $k_n^{\mathrm{E}} = 0$ and $i_1 + \cdots + i_N \le 1$. To show this, let us assume $i_1 + \cdots + i_N \le 1$ and put $P_{0 \cdots 1 \cdots 0}^{\;n}(t) = P_n(t)$ in (4). Then, one can easily transform each term in the right-hand side of (4) as follows: (the first term) $= -(k_n^{\mathrm{F}} + k_n^{\mathrm{N}}) P_n(t)$, (the second term) $= 0$, (the third term) $= -\sum_{m=1}^{N} k_{nm}^{\mathrm{FRET}} P_n(t)$, (the fourth term) $= 0$, (the fifth term) $= \sum_{m=1}^{N} k_{mn}^{\mathrm{FRET}} S_m^{+} S_n^{-} P_n(t) = \sum_{m=1}^{N} k_{mn}^{\mathrm{FRET}} P_m(t)$,   (the sixth term) $= 0$, (the seventh term) $= 0$.

2.2 Analytical Results

2.2.1 Multicomponent Exponential Decay

Here, we present the fundamental temporal property of the decay process of the fluorescence intensity derived from the multiple-donor model with $k^{\mathrm{E}}(t) = 0$. We show that nontrivial network-induced properties of fluorescence intensity decay occur when multiple donors are considered. The derivation will almost follow our previous paper [12], except for the expression of some formulae.

Now consider the simplest situation where the network consists of only one type of QDs, i.e., $k_n^F = k^F$, $k_n^N = k^N$, $k^F + k^N = 1/\tau$ for all n, and $k_{nm}^{FRET} = k_{mn}^{FRET}$ for all n, m.

First, we will show that the single-donor model (6) implies the single-exponential decay. Because the fluorescence intensity in the single-donor model (6) becomes $I(t) = k^F \sum_{n=1}^{N} P_n(t)$, the derivative of $I(t)$ becomes

$$\frac{d}{dt} I(t) = -\frac{1}{\tau} I(t) - \sum_{n,m=1}^{N} k_{nm}^{FRET} k^F P_n(t) + \sum_{n,m=1}^{N} k_{mn}^{FRET} k^F P_m(t). \qquad (7)$$

The sum of the second and third terms on the right-hand side of (7) is zero because of the symmetricity of k_{nm}^{FRET}. Therefore, the single-donor model (6) implies the single-exponential decay, i.e., $I(t) = I(0) \exp(-t/\tau)$.

Next, we will show that the multiple-donor model (4) implies the multicomponent exponential decay, i.e., $I(t) = \sum_j \alpha_j \exp(-t/\tau_j)$. To show this, we define the l-excited states as

$$\Lambda_l = \{(i_1, \ldots, i_N) \in \{0, 1\}^N : i_1 + \cdots + i_N = l\}$$

and the time-dependent fluorescence intensity from the l-excited states as follows:

$$I_l(t) = \sum_{(i_1,\ldots,i_N)\in\Lambda_l} \left[\sum_{n=1}^{N} i_n k_n^F P_{i_1\ldots i_N}(t) \right]. \qquad (8)$$

Obviously, $I(t) = \sum_{l=1}^{N} I_l(t)$ holds from (5) and (8). The following expression for the all-excited-state probability $P_{1\ldots1}(t)$ can be easily derived from (4) in the case where the network consists of only one type of QDs:

$$\frac{d}{dt} P_{1\ldots1}(t) = -\frac{N}{\tau} P_{1\ldots1}(t) - \left(\sum_{n,m=1}^{N} k_{nm}^{FRET} \right) P_{1\ldots1}(t). \qquad (9)$$

Therefore, the all-excited-state probability $P_{1\ldots1}(t)$ shows the single-exponential decay as follows:

$$P_{1\ldots1}(t) = P_{1\ldots1}(0) \exp\left(-\frac{t}{\tau_*^{(N)}}\right), \quad \tau_*^{(N)} = \left(\frac{N}{\tau} + \sum_{n,m=1}^{N} k_{nm}^{FRET}\right)^{-1}. \qquad (10)$$

Because $I_N(t) = Nk^F P_{1...1}(t)$, the resulting fluorescence intensity $I_N(t)$ also shows the single-exponential decay $I_N(t) = I_N(0) \exp\left(-t/\tau_*^{(N)}\right)$. Similarly, the following expression for the $(N-1)$-excited-state probability $P_{1...\overset{n}{0}...1}(t)$ can be derived from (4) by a straightforward calculation:

$$\frac{d}{dt} P_{1...\overset{n}{0}...1}(t) = \sum_{m=1}^{N} A_{nm} P_{1...\overset{m}{0}...1}(t) + \left(\frac{N}{\tau} + \sum_{m=1}^{N} k_{nm}^{FRET}\right) P_{1...1}(t), \qquad (11)$$

where the elements of the matrix A are

$$A_{nm} = -\left(\frac{N-1}{\tau} + \sum_{\substack{n',m'=1 \\ (n' \neq n)}}^{N} k_{n'm'}^{FRET}\right)\delta_{nm} + k_{nm}^{FRET}(1 - \delta_{nm}). \qquad (12)$$

δ_{nm} denotes the Kronecker delta. Since matrix A is real symmetric, it has real eigenvalues and can be diagonalized. Let $\lambda_1, \ldots, \lambda_N$ be the real eigenvalues (with multiplicity) of the matrix A. One can show that the matrix A is negative definite, i.e., $\mathbf{x}^T A \mathbf{x} < 0$ for all nonzero $\mathbf{x} \in \mathbb{R}^N$, by a straightforward calculation and rearrangement of the terms:

$$\sum_{n,m=1}^{N} x_n A_{nm} x_m$$

$$= -\frac{N-1}{\tau}\sum_{n=1}^{N} x_n^2 - \sum_{n=1}^{N}\left(\sum_{\substack{n',m'=1 \\ (n' \neq n)}}^{N} k_{n'm'}^{FRET}\right) x_n^2 + \sum_{n,m=1}^{N} k_{nm}^{FRET} x_n x_m$$

$$= -\frac{N-1}{\tau}\sum_{n=1}^{N} x_n^2 - \sum_{n=1}^{N}\left(\sum_{\substack{n'<m' \\ (n',m' \neq n)}}^{N} 2k_{n'm'}^{FRET}\right) x_n^2 - \sum_{n<m} k_{nm}^{FRET}(x_n - x_m)^2 \quad < 0,$$

$$\tag{13}$$

where we frequently used the symmetricity of k_{nm}^{FRET}. Therefore, all of the eigenvalues λ_j are strictly negative. Thus, one can see that the solution of (11), $P_{1...\overset{n}{0}...1}(t)$, is a linear sum of $\exp\left(-t/\tau_*^{(N-1,1)}\right), \ldots, \exp\left(-t/\tau_*^{(N-1,N)}\right)$ and $\exp\left(-t/\tau_*^{(N)}\right)$, where the decay times are $\tau_*^{(N-1,j)} = |\lambda_j|^{-1}$ labeled in ascending order. Because $I_{N-1}(t) = (N-1)k^F \sum_{n=1}^{N} P_{1...\overset{n}{0}...1}(t)$, the resulting fluorescence intensity $I_{N-1}(t)$ shows the multicomponent exponential decay, including these exponential decay components. Finally, we will show that the fluorescence intensity $I_1(t)$ includes the exponential decay component $\exp(-t/\tau)$. The following expression for the 1-excited-state probability $P_{0...\overset{n}{1}...0}(t)$ can be derived from (4) by a straightforward calculation:

$$\frac{d}{dt} P_{0\ldots\underset{n}{1}\ldots0}(t) = -\frac{1}{\tau} P_{0\ldots\underset{n}{1}\ldots0}(t)$$

$$+ \sum_{\substack{m=1 \\ (m\neq n)}}^{N} \left(\frac{1}{\tau} + k_{mn}^{\mathrm{FRET}}\right) P_{0\ldots\underset{n}{1}\ldots\underset{m}{1}\ldots0}(t). \tag{14}$$

Therefore, the 1-excited-state probability $P_{0\ldots\underset{n}{1}\ldots0}(t)$ includes the exponential decay $\exp(-t/\tau)$ as $P_{0\ldots\underset{n}{1}\ldots\underset{m}{1}\ldots0}(t)$ goes asymptotically to zero faster than $P_{0\ldots\underset{n}{1}\ldots0}(t)$. Because $I_1(t) = k^{\mathrm{F}} \sum_{n=1}^{N} P_{0\ldots\underset{n}{1}\ldots0}(t)$, the resulting fluorescence intensity $I_1(t)$ also includes the exponential decay component $\exp(-t/\tau)$. Note that we used a physical insight in the argument for (14); hence, it is not rigorous proof. In summary, the multiple-donor model (4) implies that the fluorescence intensity $I(t)$ shows the multicomponent exponential decay, including at least $N + 2$ exponential decay components, i.e., $\exp\left(-t/\tau_*^{(N)}\right)$, $\exp\left(-t/\tau_*^{(N-1,1)}\right)$, ..., $\exp\left(-t/\tau_*^{(N-1,N)}\right)$, and $\exp\left(-t/\tau\right)$ if $P_{1\ldots1}(0) \neq 0$ and $P_{1\ldots\underset{n}{0}\ldots1}(0) \neq 0$ for some n. We expect that each fluorescence intensity $I_l(t)$ from the l-excited states has potentially up to $\binom{N}{l}$ exponential decay components. Therefore, the resulting fluorescence intensity $I(t)$ has potentially up to $\sum_{l=1}^{N} \binom{N}{l} = 2^N - 1$ exponential decay components. Note that the observable number of exponential decay components can be smaller if fewer QDs are initially excited.

The above theoretical result, i.e., the appearance of multicomponent exponential decay even in single-type QDs, is qualitatively supported by experimental results obtained from spatially distributed single-type CdSe/ZnS QDs. See Sect. 3B in our previous study [12] for further details.

2.2.2 Some Analytical Results for Small QD Systems

We will show some analytical formulae for the fluorescence intensity in specific cases that assume equidistant QD systems, as shown in insets of Fig. 3. For other points, we continue to treat networks consisting of only one type of QDs and no excitation light. Let us introduce

$$\mathbf{P}(t) = \begin{bmatrix} \mathbf{P}_{\Lambda_N}^{\mathsf{T}} & \mathbf{P}_{\Lambda_{N-1}}^{\mathsf{T}} & \cdots & \mathbf{P}_{\Lambda_0}^{\mathsf{T}} \end{bmatrix}^{\mathsf{T}},$$

where $\mathbf{P}_{\Lambda_l}$ denotes a column vector consisting of $P_{i_1\ldots i_N}(t)$ for $(i_1, \ldots, i_N) \in \Lambda_l$, in which subscripts are in descending order as binary numbers. For example, when $N = 2$, $\mathbf{P}(t) = [P_{11}, P_{10}, P_{01}, P_{00}]^{\mathsf{T}}$, and when $N = 3$, $\mathbf{P}(t) = [P_{111}, P_{110}, P_{101}, P_{011}, P_{100}, P_{010}, P_{001}, P_{000}]^{\mathsf{T}}$. The master equation (4) in the multiple-donor model can also be expressed as

$$\frac{d}{dt} \mathbf{P}(t) = M(t)\mathbf{P}(t), \tag{15}$$

where $M(t)$ denotes the transition matrix with 2^N rows and 2^N columns for a N QD system.

First, consider the simplest situation where the network consists of only one type of QDs and no excitation light. Then, the matrix M in the case of $N = 2$ (see inset in Fig. 3(a)) becomes

$$M = \frac{1}{\tau} \begin{bmatrix} -2(\rho_r + 1) & 0 & 0 & 0 \\ \rho_r + 1 & -(\rho_r + 1) & \rho_r & 0 \\ \rho_r + 1 & \rho_r & -(\rho_r + 1) & 0 \\ 0 & 1 & 1 & 0 \end{bmatrix} \quad \text{for } N = 2, \qquad (16)$$

where $\rho_r = (R_0/r)^6$, and r is the distance between two QDs.

Eigenvalue analysis using Maxima for the master equation (15) with the matrix (16) derives the fluorescence intensity decay of case $N = 2$, which has two exponential components:

$$I(t) = 2k^F \frac{A_r}{2\rho_r + 1} \exp\left(-\frac{2\rho_r + 2}{\tau}t\right) + k^F \frac{B_r}{2\rho_r + 1} \exp\left(-\frac{1}{\tau}t\right) \quad \text{for } N = 2, \qquad (17)$$

where $A_r = \rho_r P_{\Lambda_2}(0)$ and $B_r = 2(\rho_r + 1)P_{\Lambda_2}(0) + (2\rho_r + 1)P_{\Lambda_1}(0)$. We introduce initial state probabilities for l-excited states Λ_l:

$$P_{\Lambda_l}(0) = \sum_{(i_1,\dots,i_N)\in\Lambda_l} P_{i_1\dots i_N}(0).$$

We show the shape of (17) in Fig. 3(a).

In the same way, the matrix M in the case of $N = 3$ (see inset in Fig. 3(b)) becomes

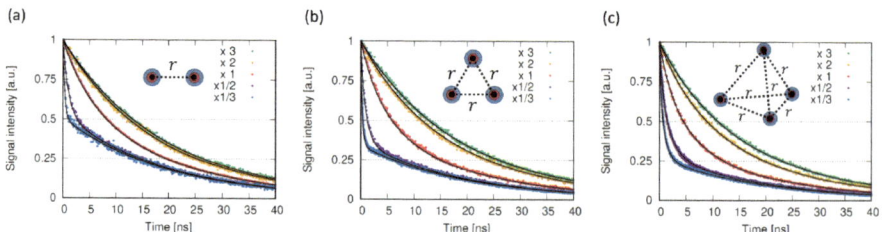

Fig. 3 Analytical and numerical fluorescence intensity decay for equidistant QD systems: (**a**) 2QD, (**b**) 3QD, and (**c**) 4QD systems (see insets). Each color dot corresponds to each distance $r = c^{1/n} \times R_0$, $c = 3, 2, 1, 1/2$, and $1/3$, respectively, where Förster distance $R_0 = 6.18$ nm, and Dimension $n = 2$ for (**a**) and (**b**), $n = 3$ for (**c**). The solid black lines are analytical results obtained using (17), (19), and (20) for (**a**), (**b**), and (**c**), respectively. The dots are numerical results obtained from stochastic simulation, tRSSA, described in the next section

$$M = \frac{1}{\tau} \begin{bmatrix} A_{1\times1} & \cdots & & \cdots & O \\ B_{3\times1} & C_{3\times3} & \ddots & & \vdots \\ \vdots & & D_{3\times3} & E_{3\times3} & \vdots \\ O & & \cdots & F_{1\times3} & \vdots \end{bmatrix} \quad \text{for } N = 3, \tag{18}$$

where

$$A = -(6\rho_r + 3), \qquad\qquad B = \begin{bmatrix} 2\rho_r + 1 & 2\rho_r + 1 & 2\rho_r + 1 \end{bmatrix}^{\mathsf{T}},$$

$$C = \begin{bmatrix} -(4\rho_r + 2) & \rho_r & \rho_r \\ \rho_r & -(4\rho_r + 2) & \rho_r \\ \rho_r & \rho_r & -(4\rho_r + 2) \end{bmatrix}, \quad D = \begin{bmatrix} \rho_r + 1 & \rho_r + 1 & 0 \\ \rho_r + 1 & 0 & \rho_r + 1 \\ 0 & \rho_r + 1 & \rho_r + 1 \end{bmatrix},$$

$$E = \begin{bmatrix} -(2\rho_r + 1) & \rho_r & \rho_r \\ \rho_r & -(2\rho_r + 1) & \rho_r \\ \rho_r & \rho_r & -(2\rho_r + 1) \end{bmatrix}, \quad F = \begin{bmatrix} 1 & 1 & 1 \end{bmatrix}.$$

Eigenvalue analysis using Maxima for the master equation (15) with the matrix (18) derives the fluorescence intensity decay of case $N = 3$, which has three exponential components:

$$
\begin{aligned}
I(t) = {} & 6k^{\mathrm{F}} \frac{A_r}{(3\rho_r + 1)(4\rho_r + 1)} \exp\left(-\frac{6\rho_r + 3}{\tau} t\right) \\
& + 2k^{\mathrm{F}} \frac{B_r}{(2\rho_r + 1)(4\rho_r + 1)} \exp\left(-\frac{2\rho_r + 2}{\tau} t\right) \\
& + k^{\mathrm{F}} \frac{C_r}{(2\rho_r + 1)(3\rho_r + 1)} \exp\left(-\frac{1}{\tau} t\right) \quad \text{for } N = 3,
\end{aligned}
\tag{19}
$$

where $A_r = \rho_r^2 P_{\Lambda_3}(0)$, $B_r = 3\rho_r(2\rho_r + 1)P_{\Lambda_3}(0) + \rho_r(4\rho_r + 1)P_{\Lambda_2}(0)$, and $C_r = 3(\rho_r + 1)(2\rho_r + 1)P_{\Lambda_3}(0) + 2(\rho_r + 1)(3\rho_r + 1)P_{\Lambda_2}(0) + (2\rho_r + 1)(3\rho_r + 1)P_{\Lambda_1}(0)$. We show the shape of (19) in Fig. 3(b).

Furthermore, eigenvalue analysis using Maxima for the master equation (15) with the matrix M for $N = 4$ (see inset in Fig. 3(c)) derives the fluorescence intensity decay of case $N = 4$ (where we avoid the long explicit formula for M), which has four exponential components:

$$
\begin{aligned}
I(t) = {} & 24k^{\mathrm{F}} \frac{A_r}{(4\rho_r + 1)(5\rho_r + 1)(6\rho_r + 1)} \exp\left(-\frac{12\rho_r + 4}{\tau} t\right) \\
& + 6k^{\mathrm{F}} \frac{B_r}{(3\rho_r + 1)(4\rho_r + 1)(6\rho_r + 1)} \exp\left(-\frac{6\rho_r + 3}{\tau} t\right) \\
& + 2k^{\mathrm{F}} \frac{C_r}{(2\rho_r + 1)(4\rho_r + 1)(5\rho_r + 1)} \exp\left(-\frac{2\rho_r + 2}{\tau} t\right) \\
& + k^{\mathrm{F}} \frac{D_r}{(2\rho_r + 1)(3\rho_r + 1)(4\rho_r + 1)} \exp\left(-\frac{1}{\tau} t\right) \quad \text{for } N = 4,
\end{aligned}
\tag{20}
$$

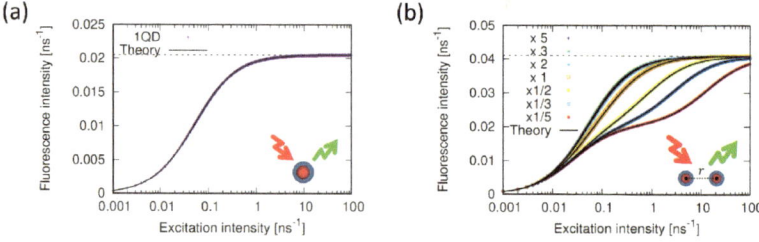

Fig. 4 Nonlinearity in the multiple-donor model: (**a**) single-QD, and (**b**) two-QD situation. (**b**) Each color dot corresponding to each distance $r = \sqrt{c} \times R_0$, $c = 5, 3, 2, 1, 1/2, 1/3, 1/5$, respectively, where Förster distance $R_0 = 6.18$ nm

where $A_r = \rho_r^3 P_{\Lambda_4}(0)$, $B_r = 4\rho_r^2(3\rho_r + 1)P_{\Lambda_4}(0) + \rho_r^2(6\rho_r + 1)P_{\Lambda_3}(0)$, $C_r = 6\rho_r$ $(2\rho_r + 1)(3\rho_r + 1)P_{\Lambda_4}(0) + 3\rho_r(2\rho_r + 1)(5\rho_r + 1)P_{\Lambda_3}(0) + \rho_r(4\rho_r + 1)$ $(5\rho_r + 1)P_{\Lambda_2}(0)$, and $D_r = 4(\rho_r + 1)(2\rho_r + 1)(3\rho_r + 1)P_{\Lambda_4}(0) + 3(\rho_r + 1)(2\rho_r + 1)(4\rho_r + 1)P_{\Lambda_3}(0) + 2(\rho_r + 1)(3\rho_r + 1)(4\rho_r + 1)P_{\Lambda_2}(0) + (2\rho_r + 1)(3\rho_r + 1)$ $(4\rho_r + 1)P_{\Lambda_1}(0)$. We show the shape of (20) in Fig. 3(c).

From the above specific results (17, 19, 20), we can infer the general case for $N \in \mathbb{N}$ that $\tau_*^{(N)} < \tau_*^{(N-1,1)} \leq \cdots \leq \tau_*^{(N-1,N)} < \tau$, that is, the network-induced decay times are shorter than the natural excited-state decay time τ.

We show the nonlinearities of the multiple-donor model in the stationary excitation situation at the end of this subsection. First, considering the single QD case (see inset in Fig. 4(a)) with stationary excitation (the rate constant k^E), the matrix M in the case of $N = 1$ becomes

$$M = \frac{1}{\tau} \begin{bmatrix} -\tau k^E & 1 \\ \tau k^E & -1 \end{bmatrix} \quad \text{for } N = 1. \tag{21}$$

Eigenvalue analysis using Maxima for the master equation (15) with the matrix (21) derives the stationary fluorescence intensity of case $N = 1$:

$$I_f = k^F \frac{I_e}{I_e + (\sigma\tau)^{-1}} \quad \text{for } N = 1, \tag{22}$$

where $I_e = \sigma^{-1}k^E$ is the excitation intensity for the collision cross section σ. We show this nonlinearity between the fluorescence intensity I_f and the excitation intensity I_e in Fig. 4(a).

Next, considering the two-QD case (see inset in Fig. 4(b)) with stationary uniform excitation (the rate constant k^E for each QD), the matrix M in the case of $N = 2$ becomes

$$M = \frac{1}{\tau} \begin{bmatrix} -2(\rho_r + 1) & \tau k^E & \tau k^E & 0 \\ \rho_r + 1 & -(\rho_r + 1 + \tau k^E) & \rho_r & \tau k^E \\ \rho_r + 1 & \rho_r & -(\rho_r + 1 + \tau k^E) & \tau k^E \\ 0 & 1 & 1 & -2\tau k^E \end{bmatrix} \quad \text{for } N = 2,$$

(23)

where $\rho_r = (R_0/r)^6$, and r is the distance between two QDs. Eigenvalue analysis using Maxima for the master equation (15) with the matrix (23) derives the stationary fluorescence intensity of case $N = 2$:

$$I_f = 2k^F \frac{I_e^2 + \sigma^{-1}(\tau^{-1}\rho_r + \tau^{-1})I_e}{I_e^2 + 2\sigma^{-1}(\tau^{-1}\rho_r + \tau^{-1})I_e + (\sigma\tau)^{-2}\rho_r + (\sigma\tau)^{-2}},$$

(24)

where $I_e = \sigma^{-1}k^E$ is the excitation intensity for the collision cross section σ. We show this nonlinearity between the fluorescence intensity I_f and the excitation intensity I_e in Fig. 4(b). Note that the nonlinearity in Fig. 4(b) depends on the distance r between two QDs and reveals an intermediate step as the distance r decreases. The two-QD case result of (24) suggests that in a general QD network, the nonlinearity between fluorescence intensity (output) and excitation intensity (input) depends on the network structure complexity and has multiple distinct intermediate steps.

3 Simulation Methods

We can use either deterministic or stochastic approaches to simulate the multiple-donor model. Each method has its advantages and disadvantages.

3.1 Deterministic Simulation

The deterministic approach numerically solves the master equation (4) as an initial value problem for a 2^N-dimensional system of ordinary differential equations, for example, the Euler method and the Runge–Kutta method:

$$\frac{d}{dt}\mathbf{P}(t) = \mathbf{F}(\mathbf{P}(t), t)$$

$$\text{or} \quad \frac{d}{dt}P_{i_1...i_N}(t) = F_{i_1...i_N}(\{P_{i_1...i_N}(t)\}, t),$$

(25)

where $\mathbf{P}(t) = (P_{1...1}(t), P_{1...0}(t), \ldots, P_{0...0}(t))^T$. The calculation time in the deterministic simulation increases exponentially with 2^N for the number of QDs, N. In the simulation with large N, the calculation of $F_{i_1...i_N}$ dominates the whole calculation

Algorithm 1 Calculate the value of $F_{i_1...i_N}$ in the right-hand side of (4)

Input: t, $(i_1, \ldots, i_N)$, and $\{P_{1...1}(t), \ldots, P_{0...0}(t)\}$
Output: $F_{i_1...i_N}(\{P_{i_1,...,i_N}(t)\}, t)$
1: set $sum = 0$
2: **for** $n = 1, \ldots, N$ **do**
3:   **if** $i_n = 1$ **then**
4:     $sum = sum - (k_n^{\mathrm{F}} + k_n^{\mathrm{N}}) P_{i_1...i_N}(t) + k_n^{\mathrm{E}}(t) S_n^{-} P_{i_1...i_N}(t)$
5:     **for** $m = 1, \ldots, N$ **do**
6:       $sum = sum - k_{nm}^{\mathrm{FRET}} P_{i_1...i_N}(t)$
7:     **end for**
8:   **else**
9:     $sum = sum + (k_n^{\mathrm{F}} + k_n^{\mathrm{N}}) S_n^{+} P_{i_1...i_N}(t) - k_n^{\mathrm{E}}(t) P_{i_1...i_N}(t)$
10:     **for** $m = 1, \ldots, N$ **do**
11:       **if** $i_m = 1$ **then**
12:         $sum = sum + k_{nm}^{\mathrm{FRET}} \{ S_n^{+} P_{i_1...i_N}(t) + S_n^{+} S_m^{-} P_{i_1...i_N}(t) \}$
13:       **end if**
14:     **end for**
15:   **end if**
16: **end for**

Note: This optimized algorithm was created by Dr. Jaehoon Yu (former associate professor at the Tokyo Institute of Technology).

---

time. Therefore, optimizing the calculation of $F_{i_1...i_N}$ is important to accelerate the deterministic simulation. We show an optimized algorithm for calculating $F_{i_1...i_N}$ in Algorithm 1, which reduces "if" conditional branches, "for" loops, and zero multiplications.

It is best to avoid using "if" conditional branches for GPU parallel computing. We transform the formula $F_{i_1...i_N}$ as follows for GPU parallelization:[1]

$$
F_{i_1...i_N}(\{P_{i_1,...,i_N}(t)\}, t) =
$$

$$
2 \sum_{n=1}^{N} \left(\frac{1}{2} - i_n \right) \left[\left(k_n^{\mathrm{F}} + k_n^{\mathrm{N}} + \sum_{m=1}^{N} (i_n \oplus i_m) k_{nm}^{\mathrm{FRET}} \right) P_{i_1...1...i_N}(t) - k_n^{\mathrm{E}}(t) P_{i_1...0...i_N}(t) \right]
$$

$$
+ \sum_{n,m=1}^{N} (1 - i_n) i_m S_n^{+} S_m^{-} P_{i_1...i_N}(t). \tag{26}
$$

This formula enables us to create an optimized algorithm that calculates $F_{i_1...i_N}$ without "if" conditional branches, which is suitable for GPU parallel computing. An algorithm optimized for GPU parallel computing will be created in future work.

---

[1] This transformed formula (26) was also created by Dr. Jaehoon Yu (former associate professor at the Tokyo Institute of Technology).

3.2 Stochastic Simulation

The stochastic approach generates a sample path of a stochastic process that obeys the master equation (4) by Gillespie's direct method (DM) [25], the first reaction method (FRM) [26], and the next reaction method (NRM) [27]. Consider one of the state transitions (2) in the multiple-donor model, say event i, and assume its rate constant is k_i. Then, the occurrence frequency per unit time, a_i, called "propensity," for the event i is

$$a_1^{(n)} = i_n k_n^{\mathrm{F}} \quad (n = 1, \ldots, N), \tag{27a}$$

$$a_2^{(n)} = i_n k_n^{\mathrm{N}} \quad (n = 1, \ldots, N), \tag{27b}$$

$$a_3^{(n,m)} = i_n (1 - i_m) k_{nm}^{\mathrm{FRET}} \quad (n, m = 1, \ldots, N; n \neq m), \tag{27c}$$

$$a_4^{(n,m)} = i_n i_m k_{nm}^{\mathrm{FRET}} \quad (n, m = 1, \ldots, N; n \neq m), \tag{27d}$$

$$a_5^{(n)} = i_n k_n^{\mathrm{E}}(t) \quad (n = 1, \ldots, N). \tag{27e}$$

The total number of the above propensities a_i is $3N + 2(N^2 - N) = 2N^2 + N$. However, we can reduce the total net number of propensities by integrating the third and fourth events:

$$a_1^{(n)} = i_n k_n^{\mathrm{F}} \quad (n = 1, \ldots, N), \tag{28a}$$

$$a_2^{(n)} = i_n k_n^{\mathrm{N}} \quad (n = 1, \ldots, N), \tag{28b}$$

$$a_{3 \wedge 4}^{(n,m)} = i_n k_{nm}^{\mathrm{FRET}} \quad (n, m = 1, \ldots, N; n \neq m), \tag{28c}$$

$$a_5^{(n)} = i_n k_n^{\mathrm{E}}(t) \quad (n = 1, \ldots, N), \tag{28d}$$

where we can determine which third or fourth event occurs according to i_m being 0 or 1. Therefore, the total number of propensities is reduced to $3N + (N^2 - N) = N^2 + 2N$. In the following, we write the propensities as a_i ($i = 1, \ldots, N^2 + 2N$) by flattening the above propensities $a_i^{(*)}$ of (28).

Gillespie-type algorithms are based on the fact that the waiting time τ_i until a subsequent event i occurs follows an exponential distribution with a propensity, $P(\tau_i) = a_i e^{-a_i \tau_i}$ ($\tau_i > 0$). The DM first generates a waiting time τ until *some* event occurs by an exponential distribution $P(\tau) = a e^{-a\tau}$ ($\tau > 0$), where a is the sum of propensities such that $a = \sum_i a_i$. After generating a waiting time τ, which event occurred is determined according to the ratio of propensities a_i (event i is selected with probability a_i/a). Depending on the event that has occurred, the state $(i_1, \ldots, i_N)$ is changed, and the propensities a_i are updated. We show a DM algorithm in Algorithm 2. We note that the DM does not assume time-dependent rate constants. Therefore, the DM cannot be adopted for the case of modulated excitation light, which includes time-dependent rate constants $k_n^{\mathrm{E}}(t)$.

The time-dependent rejection-based stochastic simulation algorithm (tRSSA) [28], one of the Gillespie-type algorithms, can handle time-dependent rate constants.

Algorithm 2 Direct method (DM) [25] for the multiple-donor model (2)

Input: an initial state $\mathbf{x} = (i_1, \ldots, i_N)$
Output: a sample path $\{(t, \mathbf{x})\}$
1: initialize $t = 0$ with initial state $\mathbf{x}$
2: compute propensities $\{a_i\}$ according to the state $\mathbf{x}$
3: **while** $t < t_{\max}$ **do**
4:　　compute $a = \sum_i a_i$
5:　　generate a random number $r_1 \sim U(0, 1)$
6:　　compute waiting time $\tau = (-1/a) \ln r_1$
7:　　update time $t = t + \tau$
8:　　generate a random number $r_2 \sim U(0, 1)$
9:　　select minimum index j s.t. $\sum_{i=1}^{j} a_i > r_2 a$
10:　　update state $\mathbf{x}$ depending on the selected event j
11:　　compute propensities $\{a_i\}$ according to the state $\mathbf{x}$
12: **end while**

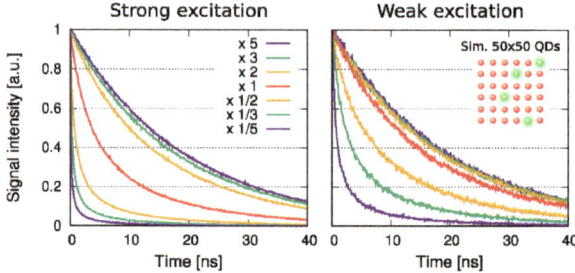

Fig. 5 Simulation results from tRSSA [28] for slightly large FRET networks: 50×50 lattice arrangements. The solid lines represent the intensity data accumulated every 0.1 ns. The theoretically predicted features are found as follows: (i) the multicomponent exponential decay, (ii) fast decay in earlier times and slow decay in later times, and (iii) slower decay during higher dilution or lower excitation

The modified NRM (MNRM) [29], another Gillespie-type algorithm, can also handle time-dependent rate constants. However, the generation of the waiting time in the MNRM relies on the tractable calculation of the integration of the time-dependent rate constants and the solution of the inverse problem (see [28] for details). Furthermore, the tRSSA we adopted here is a computationally efficient and versatile Gillespie-type algorithm that does not rely on such tractable calculations and inverse problem solutions. In the following, we adopted the tRSSA for the case of modulated excitation light and also the case of constant excitation light. We show a tRSSA algorithm for the multiple-donor model in Algorithm 3.

Figure 5 shows the simulation results obtained from tRSSA described above for slightly large FRET networks. The simulation was conducted in the following settings: QDs were located on a 50×50 lattice, and the lattice spacing was $\sqrt{c} \times R_0$, where $c = 5$ to $1/5$, as shown in the legend. The parameters of QDs were set to $Q = 0.40$ (quantum yield), $\tau = 19.5$ ns (natural excited-state lifetime), and $R_0 = 6.18$ nm (Förster distance), assuming QD585. These QD parameters are used repeatedly

Algorithm 3 The time-dependent RSSA (tRSSA) [28] for the multiple-donor model (2)

Input: an initial state $\mathbf{x} = (i_1, \ldots, i_N)$
Output: a sample path $\{(t, \mathbf{x})\}$
1: initialize $t = 0$ with initial state $\mathbf{x}$
2: define the bound $[\underline{\mathbf{x}}, \overline{\mathbf{x}}]$ as $\underline{x}_i = 0$ and $\overline{x}_i = x_i$ for $i = 1, \ldots, N$
3: discretize $[0, t_{\max}]$ to k intervals $0 < t_1 < \cdots < t_k = t_{\max}$
4: set $i = 1$
5: compute propensity bounds $\{\underline{a}_j\}$ and $\{\overline{a}_j\}$ according to $\underline{\mathbf{x}}$ and $\overline{\mathbf{x}}$, respectively
6: compute $\overline{a} = \sum_j \overline{a}_j$
7: **while** $t < t_{\max}$ **do**
8:     generate a random number $r_1 \sim U(0, 1)$
9:     compute waiting time $\tau = (-1/a) \ln r_1$
10:    update time $t = t + \tau$
11:    **if** $t > t_i$ **then**
12:        set $t = t_i$
13:        update $i = i + 1$
14:        compute propensity bounds $\{\underline{a}_j\}$ and $\{\overline{a}_j\}$ according to $\underline{\mathbf{x}}$ and $\overline{\mathbf{x}}$, respectively
15:        go to 7
16:    **end if**
17:    generate two random numbers $r_2, r_3 \sim U(0, 1)$
18:    select minimum index j s.t. $\sum_{k=1}^{j} \overline{a}_k > r_2 \overline{a}$
19:    set $accept = \mathbf{false}$
20:    **if** $r_3 \le \underline{a}_j / \overline{a}_j$ **then**
21:        set $accepted = \mathbf{true}$
22:    **else**
23:        compute propensities $\{a_j\}$ according to current state $\mathbf{x}$
24:        **if** $r_3 \le a_j / \overline{a}_j$ **then**
25:            set $accepted = \mathbf{true}$
26:        **end if**
27:    **end if**
28:    **if** $accepted = \mathbf{true}$ **then**
29:        update state $\mathbf{x}$ depending on the selected event j
30:        **if** $\mathbf{x} \in [\underline{\mathbf{x}}, \overline{\mathbf{x}}]$ **then**
31:            define a new bound $[\underline{\mathbf{x}}, \overline{\mathbf{x}}]$ around current state $\mathbf{x}$
32:            compute propensity bounds $\{\underline{a}_j\}$ and $\{\overline{a}_j\}$ according to $\underline{\mathbf{x}}$ and $\overline{\mathbf{x}}$, respectively
33:        **end if**
34:    **end if**
35: **end while**

in the following and are listed in Table 1. Furthermore, κ^2 (orientation factors) were set to $2/3$, assuming that our QD-experimental system is in the dynamic averaging regime for the three-dimensional spatial and orientational case (see Chap. 4 in [22] or [23] for effective kappa-squared values). In Fig. 5, "Strong excitation" and "Weak excitation" denote the initially excited QDs that account for 90% and 10% of the total amount, respectively. For simplicity, we assumed that QDs are points without volume in the simulation. We performed 10^4 independent simulation trials and averaged the results. We accumulated photons in a time interval of 0.1 ns at each time point to evaluate the fluorescence intensity. We confirmed that the simulation results show

Table 1 Simulation parameters assumed for QD585

Q Quantum yield	τ Natural excited-state lifetime	R_0 Förster distance	κ^2 Orientation factor
0.40	19.5 ns	6.18 nm	2/3

multicomponent exponential decay. Specifically, the decays were fast in earlier times, slow in later times, and finally, with the natural decay time τ, as stated above as the inference from the case $N = 2, 3, 4$. Moreover, the result shows faster decays as the density of excited QDs increases or the excitation becomes stronger, as expected due to the level occupancy effect. The effect promotes the emission of the transferred and saturated energy between excited QDs through a nonradiative process such as heat dissipation. As a result, the radiative energy dissipation becomes faster as the density of excited QDs increases or the excitation becomes strong (see also Sect. 4.1 in [11] for a more intuitive explanation).

3.3 Comparison Between Deterministic and Stochastic Simulation

We compare deterministic and stochastic simulations to understand the difference in characteristics. We assume common QD parameters, as shown in Table 1. We further assume a 4QD 2×2 lattice arrangement system, where the nearest neighbor distance is R_0, and the 4QD system consists of single-type QDs. In the following, the time step of the Runge–Kutta method in the deterministic simulation is set to 0.01 ns. Meanwhile, in the stochastic simulation, the fluorescence photon accumulation time is set to 0.1 ns, and the sampling number repeated for averaging is set to 10^6.

Figure 6 compares deterministic and stochastic simulations for the 4QD system with no excitation light situation. The obtained normalized fluorescence intensity decays are nearly identical, but the stochastic one includes small noises due to the intrinsic probabilistic nature of FRET. Figure 7 compares deterministic and stochastic simulations for the 4QD system with sinusoidal excitation light: $k^E(t) = A(1 + \varepsilon \sin(2\pi t/T))$, $A = 1$ nm^{-1}, $\varepsilon = 0.8$, and $T = 5$ ns. The obtained modulated fluorescence intensities are nearly identical even under excitation light, and the stochastic one includes small noises due to the FRET intrinsic probabilistic nature. Figure 8 compares deterministic and stochastic simulations for the 4QD system with rectangular excitation light: $k^E(t) = A\theta(t_w - (t \mod T))$, $A = 1$ nm^{-1}, $t_w = 1$ ns, and $T = 5$ ns, where $\theta(t)$ is the Heaviside function. The obtained modulated fluorescence intensities are also almost identical to the above cases, and the stochastic one again includes small noises due to the FRET intrinsic probabilistic nature.

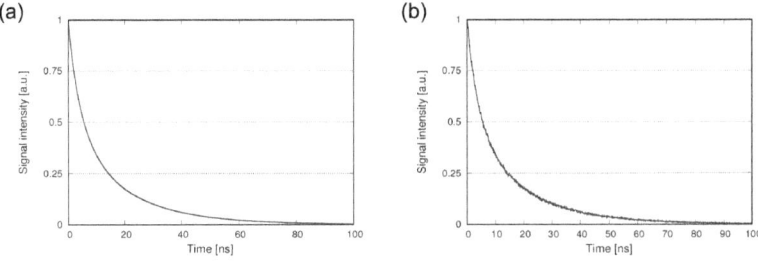

Fig. 6 Comparison of (**a**) deterministic and (**b**) stochastic simulations in the no excitation light situation for the 4QD 2 × 2 lattice arrangement system

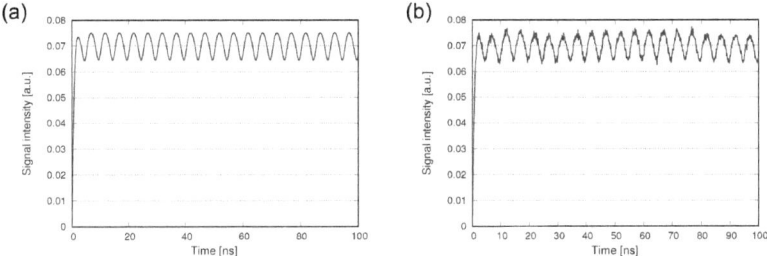

Fig. 7 Comparison of (**a**) deterministic and (**b**) stochastic simulations in the sinusoidal excitation light situation for the 4QD 2 × 2 lattice arrangement system

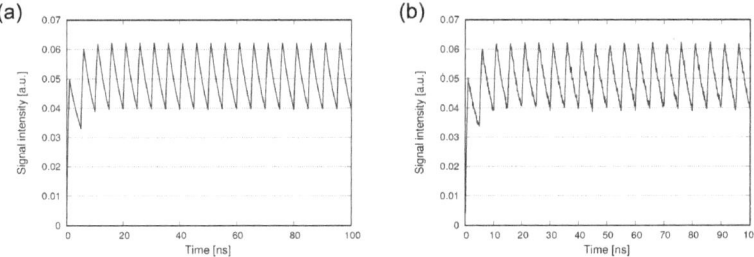

Fig. 8 Comparison of (**a**) deterministic and (**b**) stochastic simulations in the rectangular excitation light situation for the 4QD 2 × 2 lattice arrangement system

The above three cases suggest that the stochastic simulation results converge to the deterministic simulation results in the limit of large sampling numbers. A large sampling number is required for a clear result in stochastic simulations. Nonetheless, noises in stochastic simulations are faithful to actual observations. Therefore, the sampling number in the stochastic simulation is determined by the considered situations.

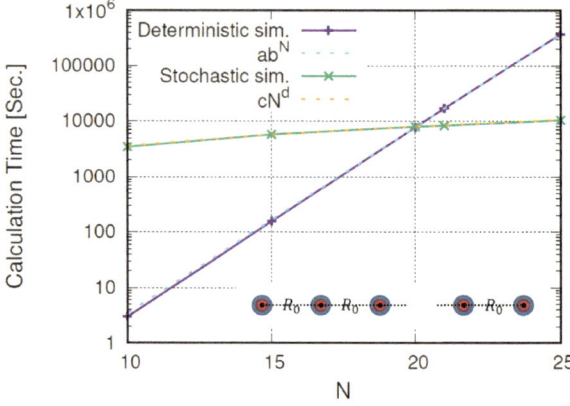

Fig. 9 Comparison of calculation times between deterministic and stochastic simulations. The network structure is chain-like, as shown in the bottom inset

Table 2 Pros and cons of deterministic and stochastic simulations

Deterministic simulation	Stochastic simulation
Merits	Merits
• Fewer QDs means faster calculations • Clear results are obtained	• Calculation time does not increase significantly even if the number of QDs increases • Noises are faithful to actual observations
Demerits	Demerits
• Calculation time increases exponentially if the number of QDs increases • It is necessary to investigate the influence of intrinsic noise separately	• Fewer QDs take longer calculation time than deterministic simulation • Evaluation of the results must be careful since inherent noise is always included

Finally, we show the trade-off nature between deterministic and stochastic simulations. Figure 9 represents the calculation times of deterministic and stochastic simulations in the case of chain-like networks. As shown in Fig. 9, deterministic simulations are appropriate for small-number situations. However, the calculation times of deterministic simulations increase exponentially as the number of QDs increases. On the other hand, the calculation times of stochastic simulations increase polynomially as the number of QDs increases. Therefore, the calculation times between stochastic and deterministic simulations reverse at some number of QDs. Table 2 summarizes the pros and cons of deterministic and stochastic simulations.

4   Information Processing Using FRET Networks

In this section, we show the power of FRET networks for information processing. This section follows our previous studies [30, 31]. As shown in Fig. 10(a), our information processing scheme has a standard structure with an input layer, a single hidden layer (consisting of FRET networks), and an output layer. The parameters to be trained are only output weights (the part of readout) connecting the hidden and output layers, such as reservoir computing. We sometimes call this hidden layer the "FRET-network reservoir." Feedback in Fig. 10(a) is optional for an autonomous signal generation or increased memory. The learning method used for the output weights is linear regression, particularly ridge regression, which is similar to reservoir computing.

Here, we assume a working hypothesis for the simulator's limited ability:

1. Each node consists of infinitely many two-QD pairs.
2. There are no interactions between such pairs.

The working hypothesis is not essential for information processing. It is only due to the simulator's limited memory, processing speed, etc. The working hypothesis implies that fluorescence from each node is free from inherent noise, which enables us to perform the deterministic simulation using Algorithm 1 with $N = 2$ for each node. Therefore, the simulation becomes low-cost and suitable for CPU-thread or GPU parallelization. We assume that FRET networks consist of QD585 and the simulation parameters of FRET networks are listed in Table 1.

We further assume a standard input scheme called the time-piecewise constant excitation, as shown in Fig. 10(b). The time-piecewise constant excitation has stationary excitation intensity within an excitation duration $[t_p, t_p + \Delta t)$ and switches to another excitation intensity just before the next excitation duration $[t_p + \Delta t, t_p + 2\Delta t)$. The fluorescence of each node is briefly observed just before the excitation switches. Therefore, the excitation switching time Δt equals the fluorescence sampling time. As shown in Fig. 10(b), the smaller the excitation switching time Δt, the more past inputs reflect the output. Thus, the excitation switching time and the memory in the reservoir are negatively correlated. We finally assume that

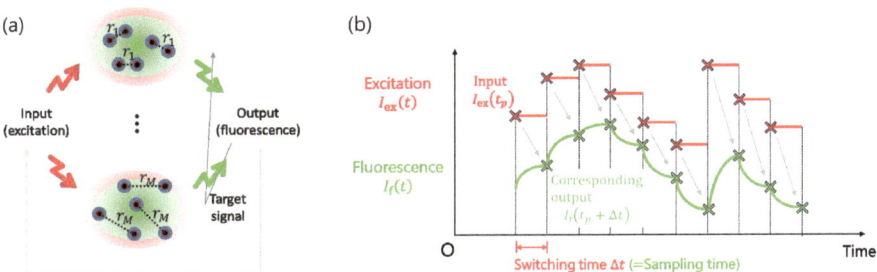

Fig. 10 FRET-network reservoir: (**a**) Fundamental network structure with optional feedback, (**b**) Input scheme by time-piecewise constant excitation

the excitation intensity is created from an exponentially enhanced input signal as follows:

$$I_{\text{ex}} = 10^{su_{\text{input}}+b}, \tag{29}$$

where s and b are a scale factor and a bias constant, respectively. The scale factor s and bias constant b are set such that the dynamic range of the excitation intensity I_{ex} is within a significant change region in the nonlinearity, approximately $[0.01, 100]$, as seen in Fig. 4(b).

4.1 Nonlinear Function Approximation

We show the ability of FRET networks for nonlinear function approximation [30]. The goal of the task is to learn the nonlinear transformation $u(t) \mapsto y(t) = \frac{1}{2}\{u(t)\}^7$ through input–output relation data $\{(\sin(t/5), \frac{1}{2}\sin^7(t/5))\}_{t=0,\Delta t,2\Delta t,\cdots,n\Delta t}$. The training and prediction phases are performed with no feedback. Since this task needs no memory, the excitation switching time Δt should be set large. Here we set the excitation switching time $\Delta t = 100$ ns.

Figure 11 represents the result of this nonlinear function approximation. Training is done by the first half of 100 input–output data (until $t = 10000$ ns), and prediction is performed by the following 100 input data (from $t = 10000$ ns until $t = 20000$ ns). Other simulation settings are as follows: the number of nodes is set to $N_{\text{net}} = 1000$, the distance between each two-QD pair is chosen between minimum $r_{\text{min}} = 0.3R_0$ nm and maximum $r_{\text{max}} = 2.0R_0$ nm, the scale factor and bias constant is set to $s = 1.0$ and $b = 0.0$, respectively, and the regularization factor for the ridge regression is set to $\lambda = 10^{-10}$. We set the transient duration to 80 ns to ignore the transient behaviors of the reservoir.

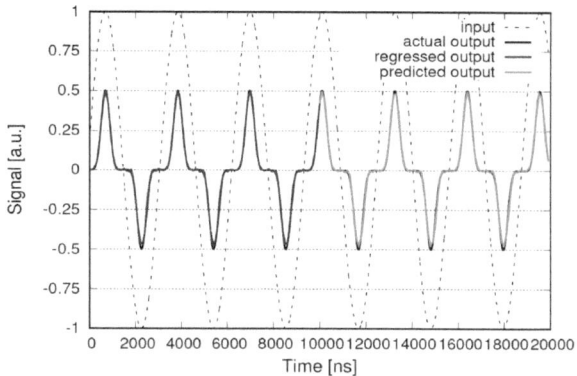

Fig. 11 Nonlinear function approximation: input $u(t) = \sin(t/5) \mapsto$ output $y(t) = \frac{1}{2}\sin^7(t/5)$

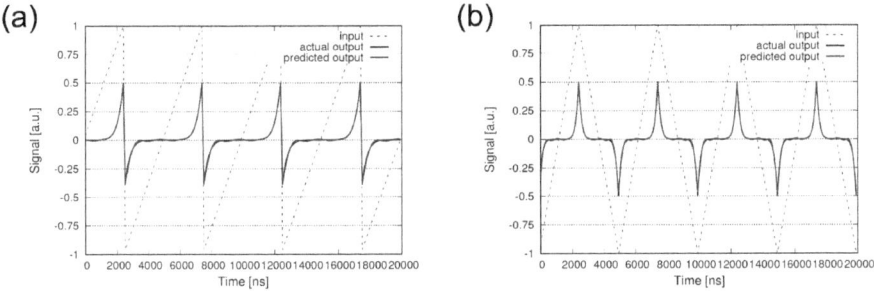

Fig. 12 Generalization of nonlinear function approximation $\tilde{u}(t) \mapsto \tilde{y}(t) = \frac{1}{2}\tilde{u}(t)^7$ using different input data from the original one $\sin(t/5)$, (**a**) a sawtooth wave and (**b**) a triangle wave. The prediction is performed with the pre-trained weight by $u(t) = \sin(t/5) \mapsto y(t) = \frac{1}{2}\sin^7(t/5)$

Furthermore, we check the generalization of this nonlinear function approxima-tion using different input data from the original one. Figure 12 shows the results. The prediction is performed with the pre-trained weight by the original input–output data. The predicted outputs become $\tilde{y}(t) = \frac{1}{2}\tilde{u}(t)^7$ in both cases, (a) sawtooth input wave and (b) triangle input wave. These results mean that the nonlinear function approximation is certainly generalized.

4.2 Chaotic Time-Series Prediction

We demonstrate the capability of FRET networks for chaotic time-series prediction with minimal memory requirements [30]. The goal of the task is to predict the next step of the Hénon map, $x_{n+1} = 1 - 1.4x_n^2 + 0.3x_{n-1}$, from the present step x_n with a memory of the past x_{n-1} left in the reservoir. This task imposes the use of memory in the reservoir. Therefore, the prediction phase is performed with feedback, as shown in Fig. 10(a). On the other hand, the training phase is performed with no feedback. Since this task needs some memory, the excitation switching time Δt should be moderately small. Here, we set the excitation switching time $\Delta t = 0.5$ ns.

Figure 13 represents the result of this chaotic time-series prediction. Training is done by the first half of 100 input–output data (until $t = 50$ ns), and prediction is performed by the following 100 input data (from $t = 50$ ns to $t = 100$ ns). Other simulation settings are as follows: the number of nodes is set to $N_{net} = 1000$, the distance between each two-QD pair is chosen between minimum $r_{min} = 0.3R_0$ nm and maximum $r_{max} = 2.0R_0$ nm, the scale factor and bias constant is set to $s = 1/3 + 0.1\xi$ and $b = 0.5 + 0.1\eta$ (where ξ, η are uniform random numbers in $[-1, 1]$), respectively, and the regularization factor for the ridge regression is set to $\lambda = 10^{-10}$. We set the transient duration to 40 ns to ignore the transient behaviors of the reservoir.

Furthermore, we check the attractor reconstruction made from the above chaotic time-series prediction. Figure 14 shows the result. Each cross point (green) in the

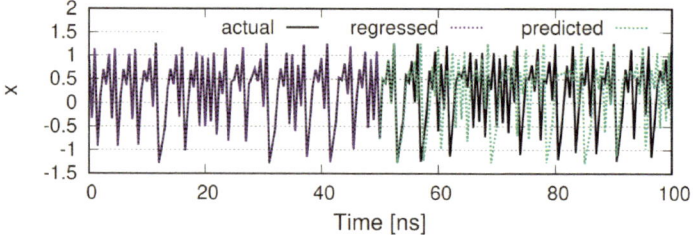

Fig. 13 Chaotic time-series prediction for the Hénon map

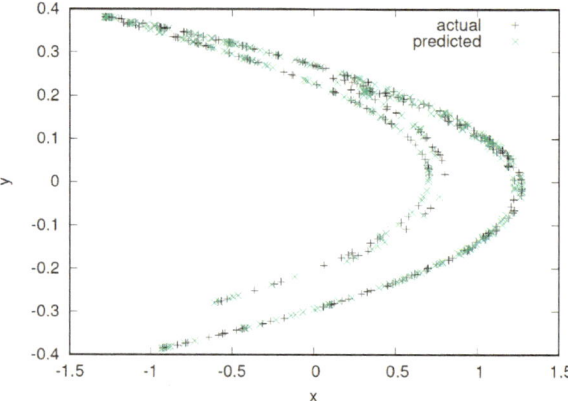

Fig. 14 Reconstructed strange attractor of the Hénon map from the predicted time series

figure denotes (x_n, x_{n+1}) made from the predicted time series in the above task. This result means that the Hénon map's strange attractor is certainly reconstructed.

Finally, we evaluate the performance of the chaotic time-series prediction by the root mean square error (RMSE), which is defined as $\text{RMSE} = \sqrt{\frac{1}{N} \sum_{n=1}^{N} \langle (x_n - \tilde{x}_n)^2 \rangle}$, where $x_n, \tilde{x}_n$ are actual and predicted time series, respectively, and $\langle \cdot \rangle$ is the average with respect to different initial conditions (note: initial conditions for x_n and $\tilde{x}_n$ are set to equal). Figure 15 shows RMSE versus (a) step n from the start of prediction and (b) the excitation switching time Δt. The excitation switching time $\Delta t = 0.5$ ns provides the best performance in this task as it has the smallest RMSE. This result means that this task needs appropriate memory. Thus, a large excitation switching time Δt reduces available memory, whereas a small excitation switching time Δt disrupts prediction by introducing unnecessary memory.

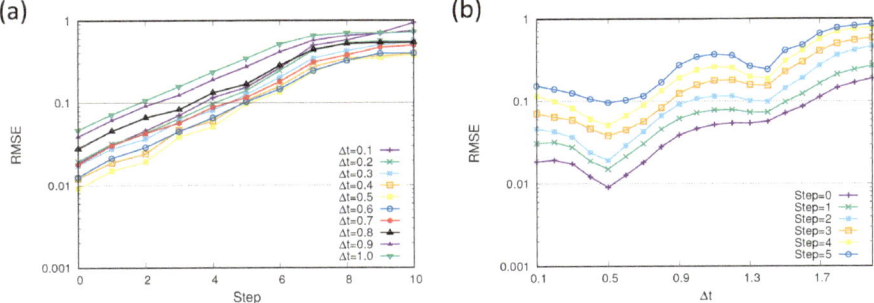

Fig. 15 Root mean square error (RMSE) versus (**a**) step n from the start of prediction and (**b**) excitation switching time Δt

4.3 Handwritten Digit Recognition

We show the ability of FRET networks for pattern recognition, particularly handwritten digit recognition [31]. The goal of the task is to classify handwritten digit images to correct digits using the MNIST handwritten digit dataset. Figure 16 depicts our pattern recognition scheme. Since this task needs no memory, the training and prediction phases are performed with no feedback, and the excitation switching time Δt should be set large. Here, we set the excitation switching time $\Delta t = 100$ ns as in the nonlinear function approximation.

We note that an input weight matrix is needed to transform an image vector $\mathbf{x}$ to an excitation intensity vector $\mathbf{I}_{ex}$. The image vector $\mathbf{x}$ and excitation intensity vector $\mathbf{I}_{ex}$ are N_i ($= 784$) and N_{net} dimensional, respectively. Therefore, the input weight matrix V has N_{net} rows and N_i columns such that $\mathbf{I}_{ex} = 10^{sV\mathbf{x}+b}$. The input

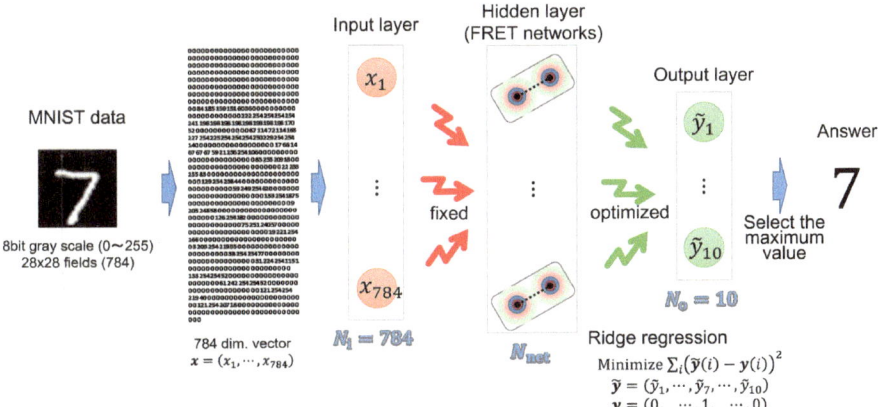

Fig. 16 Pattern recognition scheme using FRET networks

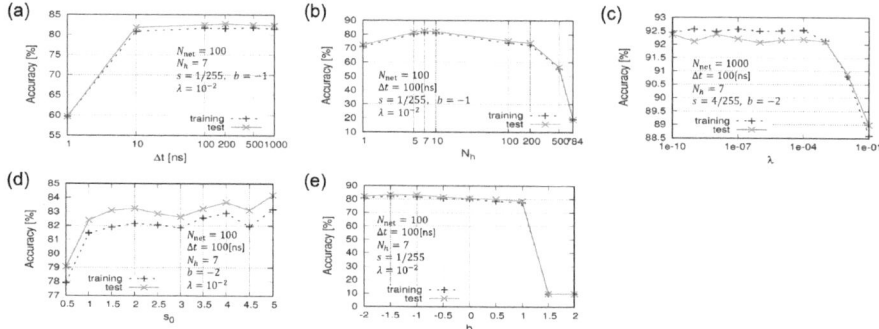

Fig. 17 Investigation of appropriate hyperparameters for MNIST handwritten digit recognition. Investigated hyperparameters are (**a**) excitation switching times Δt, (**b**) the number of nonzero elements N_h per row in the input weight matrix, (**c**) regularization parameters for ridge regression λ, (**d**) scale factors for excitation intensity s_0, and (**e**) bias constants for excitation intensity b

weight matrix V is better sparse. Therefore, the number of nonzero elements N_h per row is also better small. We randomly select the nonzero elements and set the sum of elements per row to one in the input weight matrix.

We first investigate appropriate hyperparameters, as shown in Fig. 17. The evaluation uses accuracy for the MNIST handwritten digit dataset with training data of 60000 and test data of 10000. Investigated hyperparameters are (a) excitation switching times Δt, (b) the number of nonzero elements N_h per row in the input weight matrix, (c) regularization parameters for ridge regression λ, (d) scale factors for excitation intensity s_0, and (e) bias constants for excitation intensity b. This investigation shows that (a) Δt should be large to some extent, (b) N_h should be approximately $1/100$ of the total, (c) λ should be small to some extent, and (d, e) s and b should be set such that the dynamic range of the excitation intensity is within the nonlinear region. The adopted hyperparameters are as follows: the excitation switching time is set to $\Delta t = 100$ ns, the number of nonzero elements per row in the input weight matrix is set to $N_h = 7$, the regularization factor for the ridge regression is set to $\lambda = 10^{-4}$, and the scale factor and bias constant is set to $s = 4/255$ and $b = -2$, respectively. The distance between each two-QD pair is chosen between minimum $r_{\min} = 0.3R_0$ nm and maximum $r_{\max} = 1.0R_0$ nm. In the task, we set no transient duration.

Figure 18 shows the main result of MNIST handwritten digit recognition: accuracy versus the number of FRET-network nodes. The accuracy is 92% in the case of $N_{\text{net}} = 1000$, 94% in the case of $N_{\text{net}} = 2000$, and finally reaches approximately 95% in the case of $N_{\text{net}} = 3000$. This accuracy is almost as good as the ELM-based MNIST handwritten digit recognition accuracy of 94% in the case of 1000 nodes (see Table 1 in [32]).

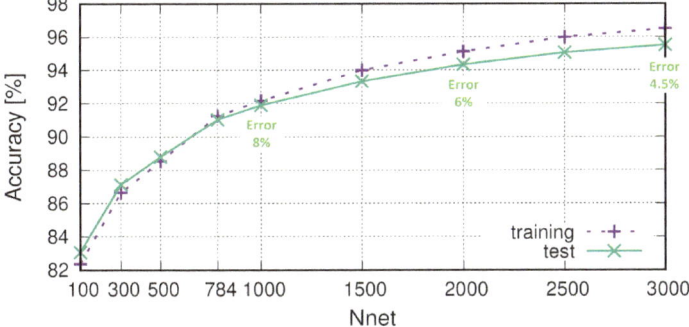

Fig. 18 Dependence of accuracy on the number of FRET-network nodes in MNIST handwritten digit recognition

5 Conclusions and Future Works

In this chapter, we first introduced a spatiotemporal model for FRET networks, called the multiple-donor model, and showed various analytical (theoretical) results. The derivation of network-induced multicomponent exponential decay and the analytical results for small QD systems, including the nonlinear relation between input excitation intensity and output fluorescence intensity, are demonstrated. We then presented the deterministic and stochastic simulation methods for FRET networks and compared their advantages and disadvantages. In general, deterministic simulations are appropriate for only a few QDs situations, whereas stochastic simulations are appropriate for many QDs situations in terms of computational costs. We finally showed the power of FRET networks for information processing by simulations, particularly nonlinear function approximation, chaotic time-series prediction, and MNIST handwritten digit recognition.

We are considering future work on reinforcement learning using FRET networks and spatial network design theory. Reinforcement learning is one of the important applications for FRET networks. We believe that the ability of FRET networks to recognize patterns and predict chaotic time series (with some memory) provides the power for reinforcement learning. On the other hand, the power of FRET networks would be maximized if the spatial network (spatial arrangement of QDs) were appropriately optimized. We currently lack a design theory for the spatial network of QDs. Therefore, developing spatial network design theory is an urgent issue for realizing novel information processing devices.

Acknowledgements I would like to thank my colleagues, especially Dr. Naoya Tate (Kyushu University) and Dr. Jaehoon Yu (former Tokyo Institute of Technology), for their valuable comments. This study is supported by JST CREST Grant Number JPMJCR18K2, Japan. Finally, I would like to express my deepest gratitude and love to my late mother, Junko Nakagawa (March 1953–January 2023), who gave me birth, raised me, and watched over my life.

References

1. J.R. Lakowicz, *Principles of Fluorescence Spectroscopy*, 3rd edn. (Springer, Boston, MA, 2006). https://doi.org/10.1007/978-0-387-46312-4
2. T. Nishimura, Y. Ogura, J. Tanida, Fluorescence resonance energy transfer-based molecular logic circuit using a DNA scaffold. Appl. Phys. Lett. **101**, 233703 (2012). https://doi.org/10.1063/1.4769812
3. C.D. LaBoda, A.R. Lebeck, C.L. Dwyer, An optically modulated self-assembled resonance energy transfer pass gate. Nano Lett. **17**, 3775–3781 (2017). https://doi.org/10.1021/acs.nanolett.7b01112
4. M. Massey, I.L. Medintz, M.G. Ancona, W.R. Algar, Time-gated FRET and DNA-based photonic molecular logic gates: AND, OR, NAND, and NOR. ACS Sens. **2**, 1205–1214 (2017). https://doi.org/10.1021/acssensors.7b00355
5. J. Inoue, T. Nishimura, Y. Ogura, J. Tanida, Nanoscale optical logic circuits by single-step FRET. IEEE Photonics J. **12**, 6500112 (2020). https://doi.org/10.1109/JPHOT.2020.2976489
6. C.M. Spillmann, S. Buckhout-White, E. Oh, E.R. Goldman, M.G. Anconac, I.L. Medintz, Extending FRET cascades on linear DNA photonic wires. Chem. Commun. **50**(55), 7246–7249 (2014). https://doi.org/10.1039/c4cc01072h
7. S. Vyawahare, S. Eyal, K.D. Mathews, S.R. Quake, Nanometer-scale fluorescence resonance optical waveguides. Nano Lett. **4**(6), 1035–1039 (2004). https://doi.org/10.1021/nl049660i
8. K.F. Chou, A.M. Dennis, Förster resonance energy transfer between quantum dot donors and quantum dot acceptors. Sensors **15**, 13288–13325 (2015). https://doi.org/10.3390/s150613288
9. N. Kholmicheva, P. Moroz, H. Eckard, G. Jensen, M. Zamkov, Energy transfer in quantum dot solids. ACS Energy Lett. **2**, 154–160 (2017). https://doi.org/10.1021/acsenergylett.6b00569
10. V. Nellore, S. Xi, C. Dwyer, Self-assembled resonance energy transfer keys for secure communication over classical channels. ACS Nano **9**, 11840–11848 (2015). https://doi.org/10.1021/acsnano.5b04066
11. S. Shimomura, T. Nishimura, Y. Miyata, N. Tate, Y. Ogura, J. Tanida, Spectral and temporal optical signal generation using randomly distributed quantum dots. Opt. Rev. (2020). https://doi.org/10.1007/s10043-020-00588-7. (March)
12. M. Nakagawa, Y. Miyata, N. Tate, T. Nishimura, S. Shimomura, S. Shirasaka, J. Tanida, H. Suzuki, Spatiotemporal model for fret networks with multiple donors and acceptors: multicomponent exponential decay derived from the master equation. J. Opt. Soc. Am. B **38**(2), 294–299 (2021). https://doi.org/10.1364/josab.410658
13. M. Naruse, N. Tate, M. Aono, M. Ohtsu, *Nanointelligence: Information Physics Fundamentals for Nanophotonics*, Chapter 1 (Springer, Berlin, Heidelberg, 2014), pp. 1–39 . https://doi.org/10.1007/978-3-642-40224-1_1
14. N. Tate, M. Naruse, W. Nomura, T. Kawazoe, T. Yatsui, M. Hoga, Y. Ohyagi, Y. Sekine, H. Fujita, M. Ohtsu, Demonstrate of modulatable optical near-field interactions between dispersed resonant quantum dots. Opt. Express **19**, 18260–18271 (2011). https://doi.org/10.1364/OE.19.018260
15. M. Naruse, P. Holmström, T. Kawazoe, K. Akahane, N. Yamamoto, L. Thylén, M. Ohtsu, Energy dissipation in energy transfer mediated by optical near-field interactions and their interfaces with optical far-fields. Appl. Phys. Lett. **100**, 241102 (2012). https://doi.org/10.1063/1.4729003
16. M. Naruse, M. Aono, S.-J. Kim, T. Kawazoe, W. Nomura, H. Hori, M. Hara, M. Ohtsu, Spatiotemporal dynamics in optical energy transfer on the nanoscale and its application to constraint satisfaction problems. Phys. Rev. B **86**, 125407 (2012). https://doi.org/10.1103/PhysRevB.86.125407
17. S. Wang, A.R. Lebeck, C. Dwyer, Nanoscale resonance energy transfer-based devices for probabilistic computing. IEEE Micro **35**, 72–84 (2015). https://doi.org/10.1109/MM.2015.124

18. S. Wang, R. Vyas, C. Dwyer, Fluorescent taggants with temporally coded signatures. Opt. Express **24**(14), 15528–15545 (2016). https://doi.org/10.1364/OE.24.015528
19. C. Berney, G. Danuser, FRET or No FRET: a quantitative comparison. Biophys. J. **84**, 3992–4010 (2003). https://doi.org/10.1016/S0006-3495(03)75126-1
20. C. Dwyer, A. Rallapalli, M. Mottaghi, S. Wang, *DNA Self-Assembled Nanostructures for Resonance Energy Transfer Circuits*, Chapter 2 (Springer, Berlin, Heidelberg, 2014), pp. 41–65. https://doi.org/10.1007/978-3-642-40224-1_2
21. M.T. Trinh, R. Limpens, T. Gregorkiewicz, Experimental investigations and modeling of auger recombination in silicon nanocrystals. J. Phys. Chem. C **117**, 5963–5968 (2013). https://doi.org/10.1021/jp311124c
22. I.L. Medintz, N. Hildebrandt, *FRET-Förster Resonance Energy Transfer: From Theory to Applications*, 1st edn. (Wiley, New York, 2013). https://doi.org/10.1002/9783527656028
23. R.E. Dale, J. Eisinger, W.E. Blumberg, The orientational freedom of molecular probes. the orientation factor in intramolecular energy transfer. Biophys. J. **26**(2), 161–193 (1979). https://doi.org/10.1016/S0006-3495(79)85243-1
24. D.D. Thomas, W.F. Carlsen, L. Stryer, Fluorescence energy transfer in the rapid-diffusion limit. Proc. Natl. Acad. Sci. **75**(12), 5746–5750 (1978). https://doi.org/10.1073/pnas.75.12.5746
25. D.T. Gillespie, Exact stochastic simulation of coupled chemical reactions. J. Phys. Chem. **81**(25), 2340–2361 (1977). https://doi.org/10.1021/j100540a008
26. D.T. Gillespie, General method for numerically simulating stochastic time evolution of coupled chemical-reactions. J. Comput. Phys. **22**(4), 403–434 (1976). https://doi.org/10.1016/0021-9991(76)90041-3
27. M.A. Gibson, J. Bruck, Efficient exact stochastic simulation of chemical systems with many species and many channels. J. Phys. Chem. A **104**(9), 1876–1889 (2000). https://doi.org/10.1021/jp993732q
28. V.H. Thanh, C. Priami, Simulation of biochemical reactions with time-dependent rates by the rejection-based algorithm. J. Chem. Phys. **143**, 054104 (2015). https://doi.org/10.1063/1.4927916
29. F. Anderson, A modified next reaction method for simulating chemical systems with time dependent propensities and delays. J. Chem. Phys. **127**(21), 214107 (2007). https://doi.org/10.1063/1.2799998
30. M. Nakagawa, Y. Miyata, N. Tate, T. Nishimura, S. Shimomura, S. Shirasaka, J. Tanida, H. Suzuki, Modeling and analysis of fret networks: toward development of novel information processing devices, in *Proceedings of the 2020 International Symposium on Nonlinear Theory and its Applications (NOLTA2020), The Institute of Electronics, Information and Communication Engineers (IEICE)* (Tokyo, Japan, 2020), pp. 217–220
31. M. Nakagawa, Pattern recognition using fret networks: a preliminary study, in *Proceedings of the 2022 International Symposium on Nonlinear Theory and its Applications (NOLTA2022), The Institute of Electronics, Information and Communication Engineers (IEICE)* (Tokyo, Japan, 2022), p. 131
32. M.D. McDonnell, M.D. Tissera, T. Vladusich, A. van Schaik, J. Tapson, Fast, simple and accurate handwritten digit classification by training shallow neural network classifiers with the "extreme learning machine algorithm. PLoS ONE **10**(8), e0134254 (2015). https://doi.org/10.1371/journal.pone.0134254

Quantum Walk on FRET Networks

Michihisa Takeuchi

Abstract In this section, we introduce the basics of quantum walk algorithm and its applications. Quantum walk is a natural extension of the concept of random walk in quantum way; therefore, the results obtained from the discussion are considered as the results by Quantum Computation. In principle, we can expect a certain type of the computation would be boosted. There are a variety of phenomena in Quantum walks, and much broader outcomes are often obtained than those from classical random walk. Such famous examples include quantum search algorithms and quantum simulations. In this article, we introduce a quantum simulation of QCD parton shower algorithm appearing in particle physics.

1 Introduction of Quantum Walk

In this section, we introduce the Quantum Walk algorithm as an extended version of Classical Random Walk.

1.1 Classical Random Walk

Let us begin with a traditional classical random walk on the integer points of a 1-dimensional line. We call the object which will move around on the line as "walker" and the movement of the "walker" is determined step-by-step randomly according to the given probability p. At each step, the "walker" can move to the left or the right integer points next to the current point with the probability of p and $1 - p$, respectively. We consider the same procedure t times repeated, then although we cannot predict the location of the "walker" n at the time t we can compute the probability of it.

M. Takeuchi (✉)
School of Physics and Astronomy, Sun Yat-sen University, Zhuhai, China
e-mail: takeuchi@mail.sysu.edu.cn

Graduate School of Information Science and Technology, Osaka University, Osaka, Japan

© The Author(s) 2024
H. Suzuki et al. (eds.), *Photonic Neural Networks with Spatiotemporal Dynamics*,
https://doi.org/10.1007/978-981-99-5072-0_7

Suppose at $t = 0$ the "walker" starts at $n = 0$, then the probability that he is at n at the time t is given by

$$P(t, n) = {}_t C_{\frac{t+n}{2}} 2^{-t}, \tag{1}$$

where ${}_t C_n = t!/(t - n)!n!$.

An important consideration here is that we can consider the "path", in other words, "history", the set of the positions of the "walker" at all the time steps before the ending time t, $\{n_{t'} | 0 \leq t' \leq t\}$, and that we can compute the probability of the appearance of a certain "path". For this example, each "path" appears at the probability of 2^{-t}. The above probability is given by counting the number of possible "paths" ending at position n. For the generic p, the appearance probability of each "path" becomes $p^{n_+}(1 - p)^{n_-}$, where $n = n_+ - n_-$ and $t = n_+ + n_-$.

For the random walk, we can compute the average $\mu = \langle n \rangle$ and the variance $\sigma^2 = \langle n^2 \rangle - \mu^2$ as $\mu = (2p - 1)t$ and $\sigma^2 = p(1 - p)t$. Especially, the standard deviation σ scales as $O(\sqrt{t})$. The asymptotic probability distribution becomes

$$\lim_{t \to \infty} P(X_t/\sqrt{t} \leq x) = \int_{-\infty}^{x} f(y) \mathrm{d}y, \quad \text{where,} \quad f(x) = \frac{1}{\sqrt{2\pi\sigma^2}} \exp\left[-\frac{(x - \mu)^2}{2\sigma^2} \right].$$

1.2 Quantum Walk

In classical random walk, we can predict the probability of the position of the walker at time t by considering all the possible "paths" of the walker and computing the probability of the "path". We want to consider the quantum version of the corresponding system. The most important property of the quantum system is that we can consider the superposition of the states. Thus, we in the end want to consider the superposition of the "paths".

The dynamics of a quantum walk can be described using a quantum mechanical formalism [1, 2]. The state of the particle on several nodes can be represented by a quantum state vector, which evolves according to a unitary operator. One can imagine that each node is lined on a 1-dimensional line, labeled with an integer n. Note that the following discussion is not restricted to the nodes lined in a line, but are valid as long as we can label the node with n, for example, in the case that the nodes are vertices on a graph. The position of the "walker" is described by the quantum state $|n\rangle$, which spans the position Hilbert space $\mathcal{H}_P = \{|n\rangle | n \in \mathbb{Z}\}$. Furthermore, for each node, we assume there are two discrete states, like spin up and down. This Hilbert space is denoted as $\mathcal{H}_C$, and we can label the two states with $\{| \uparrow \rangle, | \downarrow \rangle\}$, $\{|0\rangle, |1\rangle\}$, $\{|L\rangle, |R\rangle\}$, $\{|+\rangle, |-\rangle\}$, or often $\{|H\rangle, |T\rangle\}$, which means "head" and "tail". The coined operator is acting on this Hilbert space. The whole Hilbert space considered is the product of the two Hilbert spaces $\mathcal{H} = \mathcal{H}_P \otimes \mathcal{H}_C$, where the dimension of the Hilbert space is the product of the dimensions of $\mathcal{H}_P$ and $\mathcal{H}_C$. The quantum state $|\psi(t)\rangle$ at each time step t is given by

$$|\psi(t)\rangle = \sum_n \left[\psi_{n,+}(t)|n, +\rangle + \psi_{n,-}(t)|n, -\rangle\right],$$

where

$$\sum_n \left[|\psi_{n,+}(t)|^2 + |\psi_{n,-}(t)|^2\right] = 1.$$

The evolution of the quantum state $|\psi(t)\rangle$ is described by the following algorithms:

1. Initial state: The particle is initialized at some node on the graph with a specific quantum state $|\psi(0)\rangle$:

$$|\psi(0)\rangle = \sum_{n, s_n = \pm} \psi_{n, s_n}(0)|n, s_n\rangle.$$

For example, $\psi_{0,0}(0) = 1$, otherwise 0.

2. Quantum coin operation: The particle's state is modified by a quantum coin operator $C \in U(2)$, which is a 2-dimensional unitary operator that acts on a coin state $|c\rangle = \alpha_+|+\rangle + \alpha_-|-\rangle \in \mathcal{H}_C$. Explicitly, $|c'\rangle = C|c\rangle$ can be described by

$$C = \begin{pmatrix} a & b \\ c & d \end{pmatrix}, |c'\rangle = \begin{pmatrix} \alpha'_- \\ \alpha'_+ \end{pmatrix}, |c\rangle = \begin{pmatrix} \alpha_- \\ \alpha_+ \end{pmatrix}.$$

3. Conditional shift: The particle's position is then shifted according to the coin operation. For each node, there is a corresponding shift operator that acts on the state of the particle. The shift operator is often defined as

$$S = \sum_j |j\rangle\langle j - 1| \otimes |+\rangle\langle +| + |j\rangle\langle j + 1| \otimes |-\rangle\langle -| = S_+ \otimes P_+ + S_- \otimes P_-,$$

where $|j\rangle$ represents the state of the particle at node j, and $\otimes$ denotes the tensor product.

4. Total evolution: The total evolution of the quantum walk for each time step is given by the operator:

$$U = S(I \otimes C),$$

where C is the coin operator in $\mathcal{H}_C$, and I is the identity operator in $\mathcal{H}_P$. The total evolution of the quantum walk over t time steps is given by the product of U taken over t steps, U^t. The final form of the quantum state $|\psi(t)\rangle$ is obtained as follows:

$$|\psi(t)\rangle = U^t|\psi(0)\rangle.$$

These equations describe the basic dynamics of a quantum walk on a line. By choosing appropriate initial states, coin operators, and graph structures, quantum walks can be used to solve various problems in quantum computing, such as search and sampling.

Explicit form of the U operator is given by

$$U = S(I \otimes C) = \begin{pmatrix} & \vdots & \vdots & \vdots & \\ \cdots & 0 & P_- & 0 & \cdots \\ \cdots & P_+ & 0 & P_- & \cdots \\ \cdots & 0 & P_+ & 0 & \cdots \\ & \vdots & \vdots & \vdots & \end{pmatrix} \begin{pmatrix} & \vdots & \vdots & \vdots & \\ \cdots & C & 0 & 0 & \cdots \\ \cdots & 0 & C & 0 & \cdots \\ \cdots & 0 & 0 & C & \cdots \\ & \vdots & \vdots & \vdots & \end{pmatrix} = \begin{pmatrix} & \vdots & \vdots & \vdots & \\ \cdots & 0 & P & 0 & \cdots \\ \cdots & Q & 0 & P & \cdots \\ \cdots & 0 & Q & 0 & \cdots \\ & \vdots & \vdots & \vdots & \end{pmatrix},$$

where

$$P = P_- C = \begin{pmatrix} a & b \\ 0 & 0 \end{pmatrix}, Q = P_+ C = \begin{pmatrix} 0 & 0 \\ c & d \end{pmatrix}, C = P + Q.$$

Acting U once provides the probability of finding the walker at $(1, \pm 1)$ as $|P\psi_0|^2$ and $|Q\psi_0|^2$, respectively, so $|a|^2 + |c|^2 = 1$ gives the similar relation of the classical random walk system. However, if we consider more than two steps, U^t essentially provide $\psi_n(t)$ as a coherent sum of the amplitudes corresponding to the possible paths to reach the point (t, n) from $(0, 0)$ as in Fig. 1. It is conceptually happening for the quantum system when we don't observe the intermediate states and only observe the final wave function at time t.

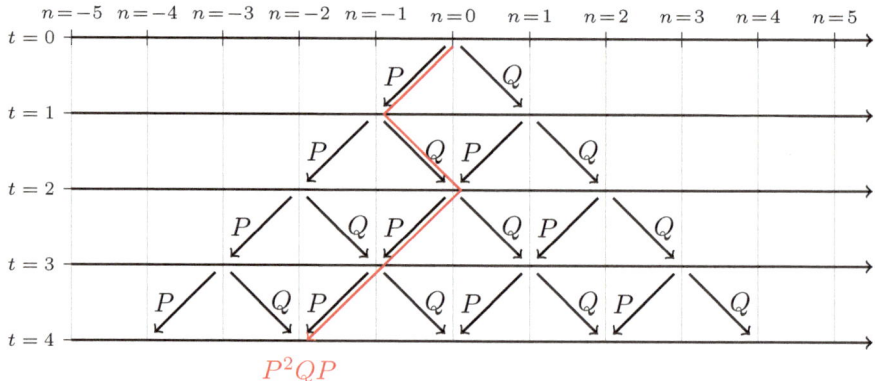

Fig. 1 Quantum walk paths. An example path is denoted in red line and the corresponding amplitude is obtained by the products of P and Q acting on the initial state $\psi_0(0)$

For an example discussion, the coin operator is often chosen to be a Hadamard coin operator H,

$$H = \frac{1}{\sqrt{2}} \begin{pmatrix} 1 & 1 \\ 1 & -1 \end{pmatrix},$$

which puts the coin into an equal superposition of $|+\rangle$ and $|-\rangle$ states. For example, the coin state obtained by H acting on $|+\rangle$ is given by $H|+\rangle = \frac{1}{\sqrt{2}}(|+\rangle + |-\rangle)$.

Here, in this setup, for the coin operator in general we can consider only the element of $SU(2)$, since the overall phase is not relevant in quantum computation. Thus, the variety of coin operator is parameterized with 3-dimensional real parameters, $a, b \in \mathbb{C}$ satisfying $|a|^2 + |b|^2 = 1$, and

$$C = \begin{pmatrix} a & b \\ -b^* & a^* \end{pmatrix} = \begin{pmatrix} \sqrt{1 - |b|^2} & b \\ -b & \sqrt{1 - |b|^2} \end{pmatrix}.$$

The last line can be obtained when a and b are restricted being real.

With this parameterization, it is known that the asymptotic probability distribution of X_t (the position of the walker at time t) in the Quantum Walk with the initial state $\psi_0(0) = \begin{pmatrix} \alpha \\ \beta \end{pmatrix}$, and $\psi_n(0) = \begin{pmatrix} 0 \\ 0 \end{pmatrix}$ ($n \neq 0$) is given by the following [1, 2]:

$$\lim_{t \to \infty} P(X_t/t \leq x) = \int_{\infty}^{x} f(y) I_{(-|a|, |a|)}(y) \mathrm{d}y,$$

$$f(x) = \frac{\sqrt{1 - |a|^2}}{\pi(1 - x^2)\sqrt{|a|^2 - x^2}} \left[1 - (|\alpha|^2 - |\beta|^2 + \frac{2Re[a\alpha b^* \beta^*]}{|a|^2})x \right],$$

where $I_A(y)$ is the compact support function giving $I_A(y) = 1$ for $y \in A$, and otherwise 0. Especially, most of the distributions accumulate around $x \sim \pm|a|$. The important fact for the Quantum Walk is that the standard deviation σ scales as $O(t)$ not $O(\sqrt{t})$, which would be advantageous for faster search algorithm and for generating samples far from initial states (Fig. 2).

1.3 Quantum Walk on FRET Networks

In the previous chapter, a mathematical model of the FRET network is introduced, where the reactions among the excited states and the ground states in an array of Quantum Dots (QDs) are considered. As a physical system, it should be more appropriate to treat it as a quantum system as a whole. We here introduce a way to include parts of the quantum effects, the interference effects, to the mathematical model of the FRET network.

Fig. 2 An example probability distribution of the walker at time $t = 100$ in Quantum Walk (blue). The corresponding asymptotic probability of the Quantum Walk (blue-dashed) and the Classical Random Walk (red) are also shown for comparison

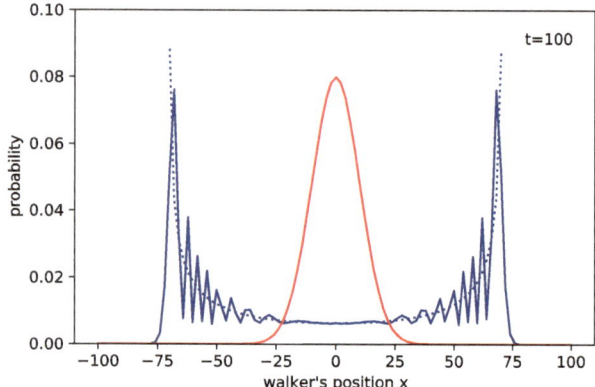

For simplicity, we consider the case where the interactions occur only between the QDs which are next to each other in the 1-dimensional array of QDs. That means, we consider the case $k_{nm} \neq 0$ only when $m = n + 1$, and otherwise $k_{nm} = 0$. We also assume the spontaneous decay process is negligible, i.e. $k_n = 0$. The resulting master formula is given by

$$\frac{\mathrm{d}}{\mathrm{d}t} P_n(t) = -k_{n,n+1} P_n(t) + k_{n-1,n} P_{n-1}(t).$$

Changing the continuous time to discrete time Δt,

$$P_n(t + \Delta t) = \left(1 - k_{n,n+1}^{\mathrm{FRET}} \Delta t\right) P_n(t) + k_{n-1,n}^{\mathrm{FRET}} \Delta t P_{n-1}(t). \tag{2}$$

Since the sum of the probability of all the possible configurations is conserved, we can describe this system as a unitary transformation acting on the vector in the Hilbert space that consists of the direct product of the Hilbert space representing n QDs $\{|n\rangle\}$ and that representing excited/non-excited states $\{|+\rangle, |-\rangle\}$ for each QD, which is originally in 2^N dimensions but restricted to the $2N$ dimensions since the number of excited states is restricted to one or zero. Thus, hot vector representation can be represented as $\psi_{0_1,\dots,0_{n-1},1_n,0_{n+1},\dots,0_N} |0_1, \dots, 0_{n-1}, 1_n, 0_{n+1}, \dots, 0_N\rangle = \psi_{n,+}(|n\rangle \otimes |+\rangle)$. For each step, the FRET interaction is acting as the transition from $|n\rangle \otimes |+\rangle \rightarrow |n+1\rangle \otimes |+\rangle$. Thus, to reproduce the correct transition probability using the coin operator in the QW algorithm, we can take $b = \sqrt{k^{\mathrm{FRET}} \Delta t}$. The explicit form of C is given as

$$C = \begin{pmatrix} \sqrt{1 - k^{\mathrm{FRET}} \Delta t} & \sqrt{k^{\mathrm{FRET}} \Delta t} \\ -\sqrt{k^{\mathrm{FRET}} \Delta t} & \sqrt{1 - k^{\mathrm{FRET}} \Delta t} \end{pmatrix},$$

and for this case we can take the shift operator $S' = S_+ \otimes |+\rangle\langle+| + S_0 \otimes |-\rangle\langle-|$. The corresponding U operator is given by

$$U' = S'(I \otimes C) = \begin{pmatrix} \vdots & \vdots & \vdots \\ \cdots P_- & 0 & 0 & \cdots \\ \cdots P_+ & P_- & 0 & \cdots \\ \cdots 0 & P_+ & P_- & \cdots \\ \vdots & \vdots & \vdots \end{pmatrix} \begin{pmatrix} \vdots & \vdots & \vdots \\ \cdots C & 0 & 0 & \cdots \\ \cdots 0 & C & 0 & \cdots \\ \cdots 0 & 0 & C & \cdots \\ \vdots & \vdots & \vdots \end{pmatrix} = \begin{pmatrix} \vdots & \vdots & \vdots \\ \cdots P & 0 & 0 & \cdots \\ \cdots Q & P & 0 & \cdots \\ \cdots 0 & Q & P & \cdots \\ \vdots & \vdots & \vdots \end{pmatrix}.$$

With this unitary operator, we can reproduce the relationship among the amplitude and the probability for one time step, which we assume to be the case when at each step we observe the configuration. For more time steps, we should get

$$\psi(t) = (U')^t \psi(0) = \begin{pmatrix} \vdots & \vdots & \vdots \\ \cdots & P^t & 0 & 0 & \cdots \\ \cdots \sum_{i=0}^{t-1} P^i Q P^{t-i-1} & P^t & 0 & \cdots \\ \cdots \sum_{path} Q^2 P^{t-2} & \sum_{i=0}^{t-1} P^i Q P^{t-i-1} & P^t & \cdots \\ \vdots & \vdots & \vdots \end{pmatrix} \psi(0).$$

The amplitude $\psi_n(t)$ is the coherent sum of all the amplitudes corresponding to the possible paths from $\psi_0(0)$. Note that the coin operator to reproduce the same Eq. (2) for one step is not unique. In this way, the dynamics of the FRET networks can be embedded in the Quantum Walk framework.

2 Application of Quantum Walk

One of the famous applications of Quantum Walk algorithm is search algorithm. Most of the cases are based on Grover's algorithm [3], and there are several examples including maze solving [4].

In this article, instead of considering the search algorithm application, we will introduce an application in particle physics. There is a well-studied phenomenon called "jet" which is originated by a quark production, and is well described by a parton shower algorithm [5–11] based on the Quantum Chromodynamics (QCD) theory. It is essentially a probabilistic process with the emission probabilities. Since processes in a microscopic world, such as this process observed in particle physics, are intrinsically described by quantum physics, the proper simulation requires a quantum computation or quantum simulation [12]. In particular, some properties in parton shower could be more efficiently implemented and described using the Quantum walk algorithm.

2.1 Classical Parton Shower

First, we review the parton shower algorithm to describe jets. At high-energy particle collider experiments, we expect quarks are produced. However, it is known that a bare quark is never observed because of the color confinements. Instead, due to the color charges, quark and gluon can emit a gluon, or split into quark and gluon, and we can compute the emission probability by the QCD theory. There are three types of splitting, $q \to qg$, $g \to gg$, $g \to q\bar{q}$, as depicted in Fig. 3.

It is known that the splitting probability is enhanced when the splitting occurs in a collinear way. For each step of splitting, $k \to ij$ the kinematics of the splitting is described by the 3-dimensional parameters, (θ, z, ϕ), where θ is the angle between i and j, z $(0 \le z \le 1)$ is the fraction of the momentum carried by i, that is $p_i = z p_k$, $p_j = (1 - z) p_k$, and ϕ is the azimuthal angle, which is just integrable to give 2π for simplicity. The differential cross sections between the split/non-split processes, corresponding to n-final states and $(n + 1)$-final states, are related as follows:

$$d\sigma_{n+1} = d\sigma_n \frac{\alpha_s}{2\pi} \frac{d\theta}{\theta} P(z) dz.$$

The QCD theory predicts the probability of the splitting with a parameter z as

$$P_{q \to qg}(z) = C_F \frac{1 + (1 - z)^2}{z}, \tag{3}$$

$$P_{g \to gg}(z) = C_A \left[\frac{2(1 - z)}{z} + z(1 - z) \right], \tag{4}$$

$$P_{g \to q\bar{q}}(z) = n_f T_R (z^2 + (1 - z)^2), \tag{5}$$

where $C_F = 4/3$, $C_A = 3$, $T_R = 1/2$ based on the color algebra, and n_f is the number of the massless quark flavors.

The above expression suggests that for all cases, the enhanced region is described by $P(z) \sim 1/z$. It is known that we can assume $\theta_1 > \theta_2 > \cdots > \theta_n$ due to the interference effects; thus, having an ensemble of the events with the variety of $\{\theta\}$ is interpreted as a time evolution process by considering $1/\theta$ as time t. In this interpretation, a certain time duration Δt corresponds to $\Delta \theta$.

Following those information, once a quark exists, it will evolve based on the Poisson process with those split/non-split probabilities. At each time, splitting/non-splitting is determined by these probability functions and the final set of the tree structure is obtained, which we call a shower history. We can consider the one-to-one correspondence to the "path" of the random walk and the shower history. Note

Fig. 3 QCD splitting patterns

that even for a classical parton shower algorithm, a part of the quantum interference effects is already taken into account during the computation of the splitting functions through the quantum corrections but not full.

In practice, to obtain the ensemble of the events from the parton shower algorithm, introducing the Sudakov factor is convenient, which is the non-splitting probability between the angle scale θ_i to θ,

$$\Delta(\theta_i, \theta) = \exp\left(-\frac{\alpha_s}{2\pi} \int_{\theta_i}^{\theta} \frac{d\theta}{\theta} \int dz\, P(z)\right). \tag{6}$$

Using the Monte Carlo method, based on the Sudakov factor, the next branching scale θ is determined by equating the random number sampled from uniform distribution $r \in [0, 1)$ as $r = \Delta(\theta_i, \theta)$. Note that $\Delta(\theta_i, \theta) \leq 1$. Alternatively, we can discretize the relevant range of the evolution between θ_i to θ_f into N steps, and introduce $\Delta\theta = (\theta_i - \theta_f)/N$. At step m, we obtain the non-splitting probability as

$$\Delta(\theta_m) = \Delta(\theta_m, \theta_{m+1}) = \exp\left(-\frac{\alpha_s}{2\pi} \frac{\Delta\theta}{\theta_m} \int dz\, P(z)\right). \tag{7}$$

As long as $\Delta\theta$ is small enough, the case with more than one splitting happening at step m is negligible, therefore the splitting probability is $1 - \Delta(\theta_m)$. We need to repeat this probabilistic process N-times. Thus, it reduces to the random walk system with the probability $\Delta(\theta_m)$. With the probability we can determine the N-set of non-splitting/splitting possibilities, which provide a "path". In the end, usually an order of 10–30 partons are generated by the splitting process.

2.2 Quantum Parton Shower Algorithm

Since the splitting history can be identified as the path, we can consider the superposition of the splitting history and the interference effects. The attempt implementing this system in Quantum Walk is discussed in Ref. [13–15]. We can identify the event of non-split/split in the parton shower as the shift to the left/right in the Quantum Walk. Explicitly, the coin operator for this problem can be taken as

$$C = \begin{pmatrix} \sqrt{\Delta(\theta_m)} & -\sqrt{1 - \Delta(\theta_m)} \\ \sqrt{1 - \Delta(\theta_m)} & \sqrt{\Delta(\theta_m)} \end{pmatrix},$$

where $\Delta(\theta_m)$ is the non-splitting probability of a particle at step m.

We consider here a simple shower, with only one particle species that exists. The operator C is acting on the coin space $\mathcal{H}_C = \{|0\rangle, |1\rangle\}$. The $|0\rangle$ state is identified as the "no emission" state, and the $|1\rangle$ state is identified as the "emission" state. The

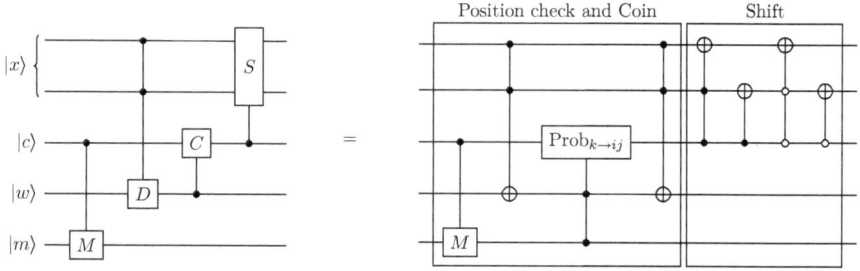

Fig. 4 Schematic quantum circuit to implement the quantum walk algorithm for parton shower. The figure is taken from Ref. [13]

position space $\mathcal{H}_P = \{|i\rangle | i \in \mathbb{N}_0\}$ represents the number of particles present in the shower and include only zero and positive integers as the parton shower cannot have a negative number of particles. The shift operation is taken as the S' in the previous section. In this way, the number of particles present in the shower is encoded in the position of the walker, with the initial state of the walker being at the $|0\rangle$ position.

It is possible to implement the Quantum Walk in the Quantum Circuit. The operator $U = S(I \otimes C)$ consists of the C acting on the coin space $\mathcal{H}_C$, which can be implemented in one qubit, and of the S, which is the conditional shift operator, which can be described by the CCNOT operator in Quantum Circuit. Figure 4 shows the schematic quantum circuit describing a single step of a quantum walk algorithm-based parton shower. In this simple shower, the number of particles present is encoded in the position of the walker, which is encoded in $|x\rangle$ in the figure. It shows a 2-qubit case, which can describe up to 4 shower particles with the initial state of the walker being at the zero position. The number of particles that the algorithm can simulate increases exponentially with the number of position qubits, x as 2^x. D describes the position check scheme, which is controlled from the position of the walker and applies the correct splitting probability accordingly in the coin operation C. The scheme is constructed from a series of CCNOT gates, thus the operation is entirely unitary. Furthermore, the position check scheme ensures that the coin operation is always applied to the $|0\rangle$ state on the coin qubit to recover the correct parton shower distribution. The subsequent shift operation then adjusts the number of particles present in the shower, depending on the outcome of the coin operation. If the coin qubit is in the $|1\rangle$ state after the coin operation, the splitting has occurred and the position of the walker is increased by one, otherwise the walker does not move. The shift operation is constructed from a series of Toffoli gates and thus is unitary. This step can be repeated for the number of discrete shower steps N in the parton shower, resembling the quantum random walk. Finally, we obtain the amplitude describing the superposition of the amplitudes with $1 - N$ shower particles. By measuring the amplitude, we can sample the "paths" with the appropriate probability.

3 Conclusion

We have reviewed the quantum walk algorithm, which can introduce the quantum interference effects to the system described by the classical random walk. From the physical setup of the FRET network, if all the quantum correlation is preserved, or the decoherence effects are negligible, the FRET network would provide a quantum device to simulate a quantum walk process. Although we need to consider the decoherence effects in a real device, it would be interesting to see what can be done in an ideal case. The real system would be modeled by the mixture of the classical random walk and the quantum walk, which would require further study. Although one of the famous applications of the quantum walk algorithm is the searching algorithm using the Grover algorithm, we have introduced an application in the parton shower algorithm in particle physics in this article. We explicitly show how to implement the quantum parton shower algorithm in the quantum walk approach. We hope a real device can help to simulate this system in future.

References

1. N. Konno, Quantum random walks in one dimension. Quantum Inf. Process. **1**(5), 345–354 (2002). https://doi.org/10.1023/A:1023413713008
2. N. Konno, A new type of limit theorems for the one-dimensional quantum random walk. J. Math. Soc. Jpn. **57**(4), 1179–1195 (2005). https://doi.org/10.2969/jmsj/1150287309
3. L.K. Grover, *A Fast Quantum Mechanical Algorithm for Database Search* (1996). arXiv:quant-ph/9605043
4. L. Matsuoka, K. Yuki, H. Lavička, E. Segawa, Maze solving by a quantum walk with sinks and self-loops: numerical analysis. Symmetry **13**(12) (2021). https://doi.org/10.3390/sym13122263, https://www.mdpi.com/2073-8994/13/12/2263
5. M. Tanabashi et al., Review of particle physics. Phys. Rev. D **98**(3), 030001 (2018). https://doi.org/10.1103/PhysRevD.98.030001
6. G.F. Sterman, S. Weinberg, Jets from quantum chromodynamics. Phys. Rev. Lett. **39**, 1436 (1977). https://doi.org/10.1103/PhysRevLett.39.1436, https://doi.org/10.1103/PhysRevLett.39.1436
7. L.G. Almeida, S.D. Ellis, C. Lee, G. Sterman, I. Sung, J.R. Walsh, Comparing and counting logs in direct and effective methods of QCD resummation. JHEP **04**, 174 (2014). https://doi.org/10.1007/JHEP04(2014)174, arXiv:1401.4460
8. T. Sjostrand, S. Mrenna, P.Z. Skands, PYTHIA 6.4 physics and manual. JHEP **05**, 026 (2006). arXiv:hep-ph/0603175, https://doi.org/10.1088/1126-6708/2006/05/026
9. M. Bahr et al., Herwig++ Physics and manual. Eur. Phys. J. C **58**, 639–707 (2008). arXiv:0803.0883, https://doi.org/10.1140/epjc/s10052-008-0798-9
10. T. Gleisberg, S. Hoeche, F. Krauss, M. Schonherr, S. Schumann, F. Siegert, J. Winter, Event generation with SHERPA 1.1. JHEP **02**, 007 (2009). arXiv:0811.4622, https://doi.org/10.1088/1126-6708/2009/02/007
11. A. Buckley et al., General-purpose event generators for LHC physics. Phys. Rept. **504**, 145–233 (2011). arXiv:1101.2599, https://doi.org/10.1016/j.physrep.2011.03.005
12. R.P. Feynman, Simulating physics with computers. Int. J. Theor. Phys. **21**, 467–488 (1982). https://doi.org/10.1007/BF02650179, https://doi.org/10.1007/BF02650179

13. K. Bepari, S. Malik, M. Spannowsky, S. Williams, Quantum walk approach to simulating parton showers. Phys. Rev. D **106**(5), 056002 (2022). arXiv:2109.13975, https://doi.org/10.1103/PhysRevD.106.056002
14. C.W. Bauer, W.A. de Jong, B. Nachman, D. Provasoli, Quantum algorithm for high energy physics simulations. Phys. Rev. Lett. **126**(6), 062001 (2021). arXiv:1904.03196, https://doi.org/10.1103/PhysRevLett.126.062001
15. K. Bepari, S. Malik, M. Spannowsky, S. Williams, Towards a quantum computing algorithm for helicity amplitudes and parton showers. Phys. Rev. D **103**(7), 076020 (2021). arXiv:2010.00046, https://doi.org/10.1103/PhysRevD.103.076020

Spatial-Photonic Spin System

Spatial Photonic Ising Machine with Time/Space Division Multiplexing

Yusuke Ogura

Abstract The spatial photonic Ising machine (SPIM) is an unconventional computing architecture based on parallel propagation/processing with spatial light modulation. SPIM enables the handling of an Ising model using light as a pseudospin. This chapter presents SPIMs with multiplexing to enhance their functionality. Handling a fully connected Ising model with a rank-2 or higher spin-interaction matrix becomes possible with multiplexing, drastically improving its applicability in practical applications. We constructed and examined systems based on time- and space-division multiplexing to handle Ising models with ranks of no less than one while maintaining high scalability owing to the features of spatial light modulation. Experimental results with knapsack problems demonstrate that these methods can compute the Hamiltonian consisting of objective and constraint terms, which require multiplexing, and can determine the ground-state spin configuration. In particular, in space-division multiplexing SPIM, the characteristics of the solution search vary based on the physical parameters of the optical system. A numerical study also suggested the effectiveness of the dynamic parameter settings in improving the Ising machine performance. These results demonstrate the high capability of SPIMs with multiplexing.

1 Introduction

Technologies for efficiently acquiring, processing, and utilizing a large amount of diverse information are becoming more important with the recent progress in data science, machine learning, and mathematical methods. Moreover, there is an increase in the computation needs for addressing social issues and scaling up computer simulation in various academic and industrial fields. Aiming to contribute to the remarkably advanced information society, research on optical/photonic computing is becoming more active. Light has a high potential for creating new computing architectures owing to its broadband processing capabilities, low energy consumption, interaction with various objects, multiplexing, and fast propagation. Novel optical/photonic

Y. Ogura (✉)
Osaka University, Osaka, Japan
e-mail: ogura@ist.osaka-u.ac.jp

© The Author(s) 2024

H. Suzuki et al. (eds.), *Photonic Neural Networks with Spatiotemporal Dynamics*,
https://doi.org/10.1007/978-981-99-5072-0_8

computing systems have been recently proposed, including computing based on integrated optical circuits [1, 2], optical reservoir computing [3], brain-morphic computing [4], optics-based deep leaning [5, 6], and photonic accelerator [7].

Combinatorial optimization addresses important problems in daily life, including the optimization of communication network routing and scheduling of apparatus usage. Metaheuristic algorithms, such as simulated annealing (SA) [8] and evolutionary computation [9] are often applied to these problems because they provide approximately optimal solutions that are sufficient for practical use. However, most combinatorial optimization problems are NP-hard, and unconventional architectures, such as physical and optical/photonic computing, are attracting significant attention for effectively solving large-scale problems.

Several combinatorial optimization problems can be mapped to the Ising model [10]. The Ising model is a mathematical model introduced to represent the ferromagnetic behavior. The system is expressed using spins with two states and the interaction between spins. Solving a combinatorial optimization problem is equivalent to determining the energy ground state of the Ising model with suitably determined interaction matrix.

Ising machines are dedicated computing systems where Ising models are implemented using pseudospins. Computations are carried out by developing a spin configuration toward the energy ground state of the Hamiltonian. Ising machines are realized using a variety of physical phenomena [11] and are expected to be fast solvers of optimization problems. For example, Ising machines based on the quantum-mechanics effect have been implemented using superconducting quantum circuits [12] and trapped ions [13]. Based on quantum fluctuations, these methods execute a solution search using quantum annealing [14]. CMOS annealing machines [15] and digital annealers [16] are other examples of SA using semiconductor integrated circuits. These machines can handle fully connected Ising models using suitable software.

Photonics-based Ising machines are also promising because they provide computing architectures capable of parallel data processing and high scalability. Good examples include the integrated nanophotonic recurrent Ising sampler (INPRIS) [17], the coherent Ising machine [18, 19], and the spatial photonic Ising machine (SPIM) [20]. In INPRIS, spin is realized by a coherent optical amplitude. The optical signal is passed through an optical matrix multiplication unit using a Mach-Zehnder interferometer, and the next spin configuration is created through noise addition to improve computing speed and thresholding. In a coherent Ising machine, spins are imitated using optical pulses generated by a degenerate optical parametric oscillator [21]. The phase and amplitude of the pulse in an optical fiber ring were measured. The interaction was realized by injecting optical pulses for modulation into the ring based on the feedback signal obtained through a matrix operation circuit. To date, the Ising machine consisting of 100,000 spins has been realized [19].

On the other hand, there are many research examples of computing by spatial light modulation as a method enjoying the parallel propagation property of light [6, 22, 23]. Based on this concept, SPIM [20] represents spin variables as the modulation of light using a spatial light modulator (SLM) and executes spin interaction by overlapping optical waves by free-space propagation. The SPIM system can be simpler than other methods, and the scalability of the spins is high because it uses the parallelism of light propagation based on Fourier optics. Moreover, fully connected Ising models can be handled using free-space optics. Owing to these characteristics, the SPIM has received considerable attention, and many derivative systems and methods have been proposed [24–26].

An issue with the primitive version of SPIM [20] by Pierangeli et al. is the low freedom to express the interaction coefficients. The light propagation model used in the computation can handle only a rank-1 interaction matrix. Because this is a major limitation in practical use, an extension of the computing model is required to apply SPIM to a wider range of problems. A few research examples can be found, including a quadrature SPIM that introduces quadrature phase modulation and an external magnetic field [27] and the implementation of a new computing model using gauge transformation by wavelength-division multiplexing (WDM) [28]. However, these methods deteriorate scalability because of the decrease in the number of spin variables owing to SLM segmentation for encoding spins. We investigated methods for increasing the interaction matrix rank without deteriorating scalability using multiplexing. Accordingly, in Sect. 2, the basic principle of the primitive SPIM is introduced and the concept of SPIM with multiplexing is explained. The procedure and experimental results for time-division multiplexing SPIM (TDM-SPIM) are presented in Sect. 3 and those of space-division multiplexing SPIM (SDM-SPIM) are presented in Sect. 4. Finally, the conclusions are presented in Sect. 5.

2   Spatial Photonic Ising Machine with Multiplexing

2.1   Basic Scheme of SPIM

The Ising model can be expressed using spins and their interactions. Let $\sigma = (\sigma_1, \ldots, \sigma_N) \in \{-1, 1\}^N$ be the spin variables and $\mathbf{J} = \{J_{jh}\}$ be the interaction coefficients between spins σ_j and σ_h, where j and h are the spin numbers and N is the total number of spins. When the external magnetic field is negligible, the Ising Hamiltonian $\mathcal{H}$ is represented as

$$\mathcal{H} = -\sum_{j,h} J_{jh}\sigma_j\sigma_h. \tag{1}$$

The concept of SPIM proposed by Pierangeli et al. in 2019 [20] is shown in Fig. 1. The optical hardware consists of an SLM, a lens, and an image sensor. An optical wave with a spatial amplitude distribution (uniform phase) is incident on the SLM. The amplitude distribution is determined based on the spin interaction $\mathbf{J}$ in the Ising model. The light modulated by the SLM, which encodes a spin configuration σ, is Fourier-transformed using the lens, and the intensity distribution $I(\mathbf{x})$ is acquired

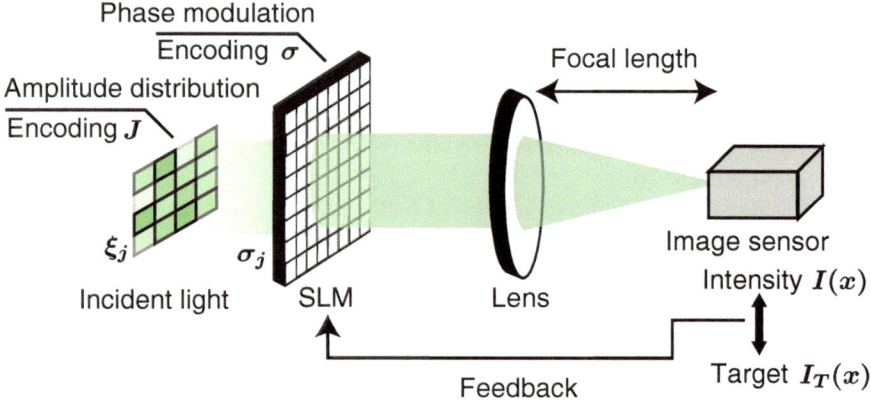

Fig. 1 Concept of the primitive SPIM

using the image sensor. The value of the Ising Hamiltonian is calculated from $I(\mathbf{x})$. The ground-state search is based on SA. The phase modulation of the SLM is updated for every calculation of the Hamiltonian. Repeating these operations provides a spin configuration with the minimum energy. In the primitive SPIM, the amplitude distribution that shines the SLM is fixed during the iterations.

The computation using SPIM is formulated as follows [20]: For simplicity, we consider a one-dimensional case. We assume that the amplitude distribution $\boldsymbol{\xi} = (\xi_1, \ldots, \xi_N)$ entering the system has a pixel structure similar to that of SLM. Each spin σ_j is encoded with binary phase modulation $\phi_j \in \{0, \pi\}$ using an SLM and is connected to $\sigma_j = \exp(i\phi_j) = \pm 1$. The width of a single SLM pixel is $2W$, the aperture is expressed as $\tilde{\delta}_W(k) = \text{rect}\left(\frac{k}{W}\right)$, and the optical field $\tilde{E}(k)$ immediately after the SLM is

$$\tilde{E}(k) = \sum_j \xi_j \sigma_j \tilde{\delta}_W(k - k_j), \qquad (2)$$

where $k_j = 2Wj$. The optical field $E(x)$ on the image sensor plane is obtained as a Fourier transform of $\tilde{E}(k)$, and the intensity distribution $I(x)$ is represented by

$$I(x) = |E(x)|^2 = \sum_{j,h} \xi_j \xi_h \sigma_j \sigma_h \delta_W^2(x) e^{2\iota W(h-j)x}, \qquad (3)$$

where $\delta_W(x) = \sin(Wx)/(Wx)$ denotes the inverse Fourier transform of $\tilde{\delta}_W(k)$.

Let $I_T(x)$ be an arbitrary target image. The minimization of $\|I_T(x) - I(x)\|$ corresponds to the minimization of the Ising Hamiltonian with interaction J_{jh} in Eq. (4):

$$J_{jh} = 2\xi_j\xi_h \int I_T(x)\delta_W^2(x)e^{2\iota W(h-j)x}dx. \tag{4}$$

If $2W$ is sufficiently small and $\delta_W(x) \sim 1$ is sufficient, J_{jh} can be approximated as

$$J_{jh} = 2\pi\xi_j\xi_h\tilde{I}_T[2W(j-h)], \tag{5}$$

where $\tilde{I}_T(k)$ denotes the Fourier transform of $I_T(x)$. In addition, when $I_T(x) = \delta(x)$ in Eq. (4), the interaction becomes simple: $J_{jh} \propto \xi_j\xi_h$. In this case, neglecting the constant of proportionality, the Ising Hamiltonian in Eq. (1) can be rewritten as

$$\mathcal{H} = -\sum_{jh}\xi_j\xi_h\sigma_j\sigma_h. \tag{6}$$

As seen from Eq. (6), the SPIM can handle fully connected Ising models using optical computation based on spatial light propagation. However, Eq. (6) is an Ising model with a special format known as the Mattis model. A pair of interactions between two spins is the product of two independent variables, and the interaction matrix is limited to a symmetric rank-1 matrix.

2.2 Concept of SPIMs with Multiplexing

As described above, the interaction matrix $\mathbf{J}$ has a restriction specific to SPIM. Thus, the computational model of SPIM should be improved to handle interaction matrices with a higher rank for application to diverse, practically useful optimization problems. A promising approach to address this issue is effectively utilizing multiplexing capabilities. Multiplexing is a well-known method for improving the performance and functionality of photonic information systems. Multiplexing strategies are used in methods using spatial light modulation, including holographic data storage [29] and computing [30], and would be effective for improving SPIM.

Consider the Hamiltonian configured using the linear sum of Eq. (6) [31]:

$$\mathcal{H} = -\sum_{l=1}^{L}\alpha^{(l)}\sum_{jh}\xi_j^{(l)}\xi_h^{(l)}\sigma_j\sigma_h. \tag{7}$$

Here, $l = 1, 2, \ldots, L$ is the multiplexing number, L is the total number of multiplexed components, $\alpha^{(l)}$ is an arbitrary constant, and $\xi_j^{(l)}$ is the amplitude. This extension enables the representation of an interaction matrix with a rank L or less in the Ising model. From Eq. (7), σ is common for all multiplexed terms; therefore, the hardware (SLM) for manipulating it can be shared among the multiplexed lights. In contrast, the amplitude distributions ξ must be treated independently by assigning individual multiplexed components to different amplitude distributions. Possible methods for multiplexing include time-division, space-division, angle-division, and wavelength-division. Figure 2 shows configuration examples of SPIMs with multiplexing.

TDM-SPIM (Fig. 2a) can be realized using a hardware configuration similar to that of the primitive SPIM. However, the amplitude distribution of the light incident on an SLM for encoding spins must change over time. This is achieved using, for example, an amplitude-type SLM. The intensity distribution was acquired for individual amplitude distributions while maintaining the spin configuration during each iteration. The system energy was calculated by summing L intensity distributions on a computer. When $\alpha^{(l)}$ has the same sign, the energy can be calculated optically by switching the amplitude distribution L times during the exposure of the image sensor. This method enables multiplexing without sacrificing the number of expressible spin variables, and the number of multiplexing channels can be easily increased while maintaining a simple hardware configuration. However, the computation time increases linearly with the number of multiplexing channels. The switching rate of the amplitude modulation can be a factor that restricts the computation speed.

In SDM-SPIM (Fig. 2b), mutually incoherent light waves with different amplitude distributions overlap and shine an SLM for encoding spins. Different amplitude distributions can be generated simultaneously using multiple amplitude modulation devices or by dividing the modulating area of a single device depending on the number of manipulated spins. When positive and negative signs are mixed in $\alpha^{(l)}$, switching between the amplitude distributions corresponding to the set of the positive and negative signs is necessary for calculating the system energy. However, the intensity acquisition required for each iteration is performed once or twice, independent of the number of multiplexing channels; thus, the time cost is low. The total number of pixels used for the amplitude modulation is divided according to the number of multiplexing channels, and the number of spin variables is determined as the number of pixels after division. However, as described above, introducing multiple devices can easily extend the total number of pixels for amplitude modulation.

In angle-division multiplexing SPIM (Fig. 2c), different amplitude distributions can be generated, for example, using a volume hologram with angle-multiplexed recording [32]. A light-wave readout with angle multiplexing leads to a single SLM for encoding spins. The computation of the Hamiltonian can be executed simultaneously and independently by reading the angle-multiplexed light using mutually incoherent light. In addition to the SDM, the acquisition of the intensity distribution required for every iteration is twice the maximum; thus, the computation time is independent of the multiplexing number. In addition, sharing pixels of amplitude distributions and phase-modulation SLM among multiplexing lights is not necessary, and this method is considered superior to TDM-SPIM and SDM-SPIM in terms of the scalability of the spin variables. However, introducing angle-multiplexing optics is necessary, and the system tends to be complicated. Moreover, the crosstalk of an angle-multiplexing device affects the Ising machine's performance.

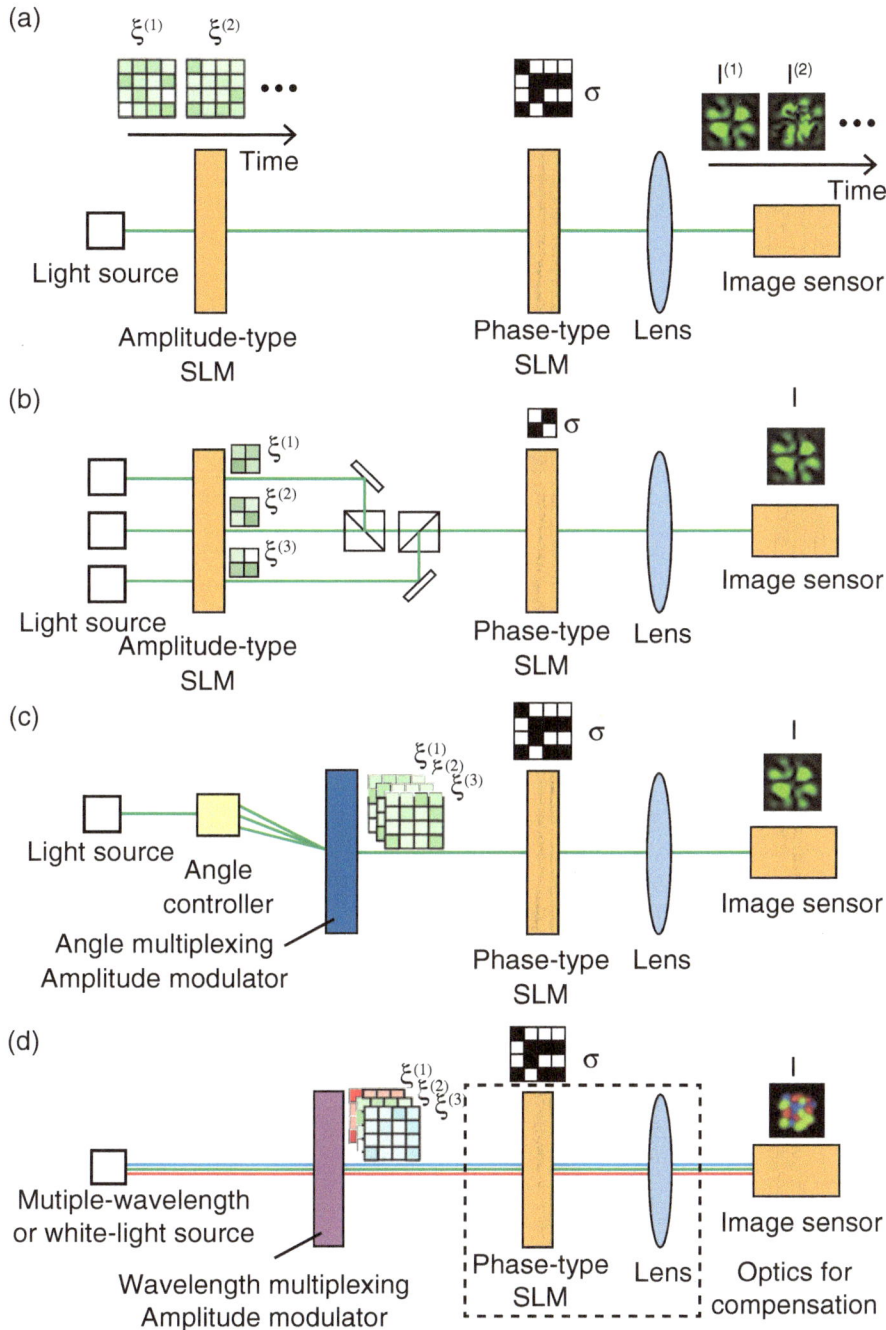

Fig. 2 Possible schemes of SPIM with multiplexing. **a** Time-division multiplexing SPIM, **b** space-division multiplexing SPIM, **c** angle-division multiplexing SPIM, and **d** wavelength-division multiplexing SPIM

In WDM-SPIM (Fig. 2d), we generate different amplitude distributions for multiple wavelengths while acquiring the sum of the energies (Eq. (7)), optically calculated for individual wavelengths. Volume holograms or other devices can generate wavelength-dependent amplitude distributions. Luo et al. recently proposed a system using the dispersion of supercontinuum light as an example of WDM [28]. The computational model was modified using gauge transformation, and the energy computation was executed by leading uniform-distribution multiple-wavelength optical waves to the phase-only SLM. In WDM methods, the computation time efficiency is high owing to the simultaneous calculation of multiplexed terms in the Hamiltonian. However, compensation is required for the wavelength dependence of the system behavior, such as the dependence of the intensity distribution scale after optical Fourier transformation on wavelengths. Moreover, it is difficult to satisfy the phase distribution for different wavelengths simultaneously; hence, some ingenuity is required.

This study investigated TDM- and SDM-SPIM, which provide relatively easy implementation. The two methods are discussed in the following two sections, along with the experimental results.

3 Time Division Multiplexed (TDM)-SPIM

We confirm that SPIM with multiplexing can handle a wider range of Ising models by applying it to a 0–1 knapsack problem with integer weights, which is a combinatorial optimization problem in the NP-hard class [10]. The primitive SPIM cannot be applied to this problem because the rank of the interaction matrix is greater than 1. A knapsack problem involves finding a set of items that maximizes the total value when a knapsack with a weight limit and items with predefined values and weights are given. This problem is related to several real-world decision-making processes.

Let us assume that there are n items and that the weights and values of the ith $(i = 1, 2, \ldots, n)$ item are w_i and v_i, respectively. $x_i \in \{0, 1\}$ $(i = 1, 2, \ldots, n)$ is a decision variable representing whether the ith item is selected $(x_i = 1)$ or not $(x_i = 0)$. The knapsack problem is then formulated as follows:

$$\text{maximize} \quad \sum_{i=1}^{n} v_i x_i, \tag{8}$$

$$\text{subject to} \quad \sum_{i=1}^{n} w_i x_i \leq W, \tag{9}$$

where W denotes the weight limit of the knapsack. The corresponding Ising Hamiltonian $\mathcal{H}$ is formulated using the log trick [10] as

$$\mathcal{H} = A\mathcal{H}_A - B\mathcal{H}_B, \tag{10}$$

$$\mathcal{H}_A = \left(W - \sum_{i=1}^{n} w_i x_i - \sum_{i=1}^{m} 2^{i-1} y_i \right)^2, \tag{11}$$

$$\mathcal{H}_B = \left(\sum_{i=1}^{n} v_i x_i \right)^2, \tag{12}$$

where $y_i \in \{0, 1\}$ denotes the auxiliary variables. The number of auxiliary variables is set to $m = \lceil \log_2 \max_i w_i \rceil$. A and B are constants. To find the optimal solution, the penalty for constraint violation must be greater than the gain from adding an item, and A and B must satisfy $0 < B \left[2 \sum_{i=1}^{n} v_i - \max_i v_i \right] \times \max_i v_i < A$. $\mathcal{H}_A$ is the constraint term and $\mathcal{H}_B$ is the objective term.

SPIM cannot handle Eqs. (11) and (12) directly; therefore, the Hamiltonian is transformed into a linear sum of the Mattis model with a variable transformation. Neglecting the constant term that does not affect the optimization, we obtain

$$\mathcal{H}(\sigma) = A\sigma^T \xi^{(1)} \xi^{(1)\, T} \sigma - B\sigma^T \xi^{(2)} \xi^{(2)\, T} \sigma, \tag{13}$$

where

$$\sigma = (2x_1 - 1, \ldots, 2x_n - 1, 2y_1 - 1, \ldots, 2y_m - 1, 1)^T, \tag{14}$$

$$\xi^{(1)} = (w_1, \ldots, w_n, 2^0, \ldots, 2^{m-1}, \sum_{i=1}^{n} w_i + 2^m - 1 - 2W)^T, \tag{15}$$

$$\xi^{(2)} = (v_1, \ldots, v_n, 0, \ldots, 0, \sum_{i=1}^{n} v_i)^T. \tag{16}$$

Equation (13) can be solved using SPIM with multiplexing. In this section, Eq. (13) was computed using TDM-SPIM [31]. By switching the amplitude distribution between $\xi^{(1)}$ and $\xi^{(2)}$, the intensity distributions with individual amplitude distributions were acquired sequentially, and the total energy was calculated using a computer. This method enables handling the same number of spins as in the primitive SPIM by securing the number of pixels for amplitude distributions equivalent to that of the SLM for encoding spins.

The optical setup of the TDM-SPIM is shown in Fig. 3. A plane-wave ray from a laser source (Shanghai Sanctity Laser, wavelength: 532 nm) was incident on SLM1 (Santec, SLM-200; pixel number: 1920 × 1080, pixel pitch: 8 µm) to spatially modulate the amplitudes and encode the problem to be solved. The light immediately after SLM1 was imaged on SLM2 (Hamamatsu Photonics, X15213-01; pixel number: 1272 × 1024; pixel pitch: 12.5 µm), where spatial phase modulation was applied to incorporate the spin configuration. The light was then Fourier-transformed by lens L3, and the intensity distribution was acquired using an image sensor (PixeLink, PL-B953U; pixel pitch: 4.65 µm). To eliminate the mismatch between the pixel

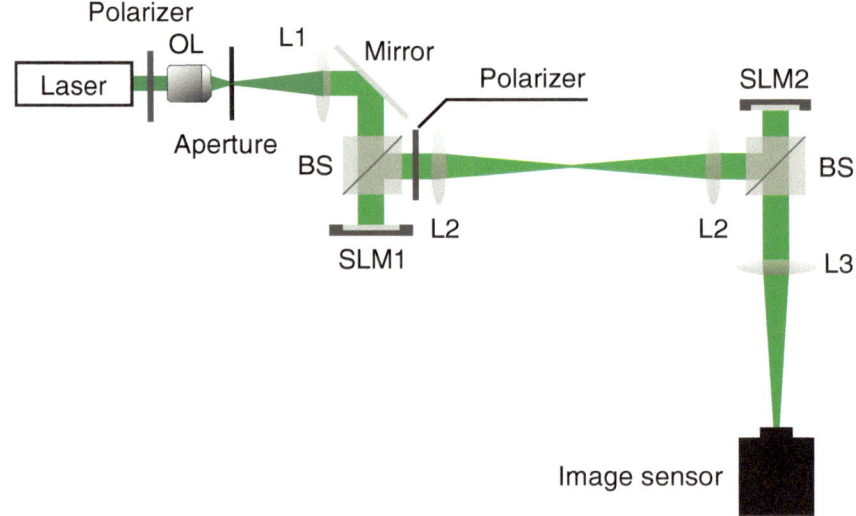

Fig. 3 Optical setup of TDM-SPIM. OL: objective lens (40×, NA: 0.6); BS: beam splitter; L1, L2, L3: lens (focal length: 150, 200, 300 mm)

sizes of SLM1 and SLM2, an area of $600 \times 600 \ \mu m^2$ (75 × 75 pixels for SLM1 and 48 × 48 pixels for SLM2) was considered the minimum modulation size for each spin. Because the amplitude range is limited from zero to one, ξ is normalized to $\max_i \xi_i$.

In the primitive SPIM, a target image I_T, associated with the Hamiltonian using Eq. (4), is employed to calculate the energy from the acquired intensity distribution. In our TDM-SPIM, the energy was calculated directly from Eq. (3) without using $I_T(x)$. By substituting $x = 0$ into Eq. (3), we obtain

$$I(0) = \sum_{j,h} \xi_j \xi_h \sigma_j \sigma_h, \tag{17}$$

and find

$$\mathcal{H} = -I(0). \tag{18}$$

The Hamiltonian value was obtained as the intensity at the center position. This method eliminates the cost of calculating $\|I_T(x) - I(x)\|$ from the intensity distribution and enables to employ a single sensor instead of an image sensor. In the experiments, we set the intensity within a single pixel at the center as $I(0)$.

The spin configuration was updated for each acquisition of a pair of constraint and objective terms. The next candidate of the spin configuration σ' is made by flipping individual spins except the last one, whose spin is fixed to "1," of the current spin configuration σ with the probability $3/(n + m)$. By simultaneously flipping multiple spins, overcoming a higher energy barrier becomes easier. The transition probability

P in the SA is determined by

$$P = \exp\left(-\frac{\mathcal{H}(\sigma') - \mathcal{H}(\sigma)}{T}\right), \tag{19}$$

where T is the temperature. We adopted a sample with the maximum total value among the feasible solutions obtained in the iterations as the final solution. This is because the Hamiltonian can be inconsistent with the total value because of the dependence in Eq. (10) on coefficients A and B, and to exclude samples that fail to satisfy the weight limit. In the experiments, $A = (\max_i v_i) \times (2 \sum v_i - \max_i v_i) + 1 = 2633$, $B = 1$, and the temperature was constant at $T = 10A = 26330$. The solved knapsack problem is as follows:

$$n = 13, \quad W = 80,$$
$$\mathbf{v} = (6, 7, 1, 15, 14, 8, 5, 6, 4, 7, 5, 12, 10),$$
$$\mathbf{w} = (7, 7, 8, 8, 2, 7, 12, 4, 0, 14, 2, 7, 14). \tag{20}$$

The total value and weight of the optimal solution are 95 and 80, respectively. The total number of spin variables, including auxiliary variables, is 17.

First, the accuracy of the Hamiltonian obtained using this system was investigated. We compared the energy values for the $8192(=2^{13})$ possible spin configurations between the theory and experiment. Figure 4 presents an almost linear relationship for the weight and total value terms with coefficients of determination of 0.8304 and 0.9726, respectively. In the weight calculation, we exclude the data saturated in the experiment. The results show that the matrix operations for calculating different terms are executed using a single system. A part of the Hamiltonian values for the weight in the experiment is measured with saturation owing to the limitation of the dynamic range of the image sensor. However, this does not hinder the system behavior because the values important for finding the ground state are those on the low-energy side. Nevertheless, it is necessary to suitably set the saturation threshold by considering A and B to effectively utilize the limited dynamic range.

An example of the system evolution during the search for solutions is shown in Fig. 5. Figure 5a shows the change in energy for each iteration. The total number of iterations was 3000. Although the energy did not converge because the temperature was set constant, searching was performed mainly in the low-energy area. Figures 5b and c show the transitions in the weight and total values for the spin configuration sampled at every iteration number. The spin configurations with high total values are broadly searched under the weight constraint. We confirm that the TDM-SPIM can deal with the Hamiltonian consisting of two terms; in particular, the constraint term, which is not dealt with in the primitive SPIM, works well.

We executed the TDM-SPIM 50 times and the characteristics of the generated samples were examined. Figure 6a presents a histogram of the feasible solutions, taking the maximal total value for every execution. The optimal solution was determined to be 48%. In addition, approximate solutions with high total values were

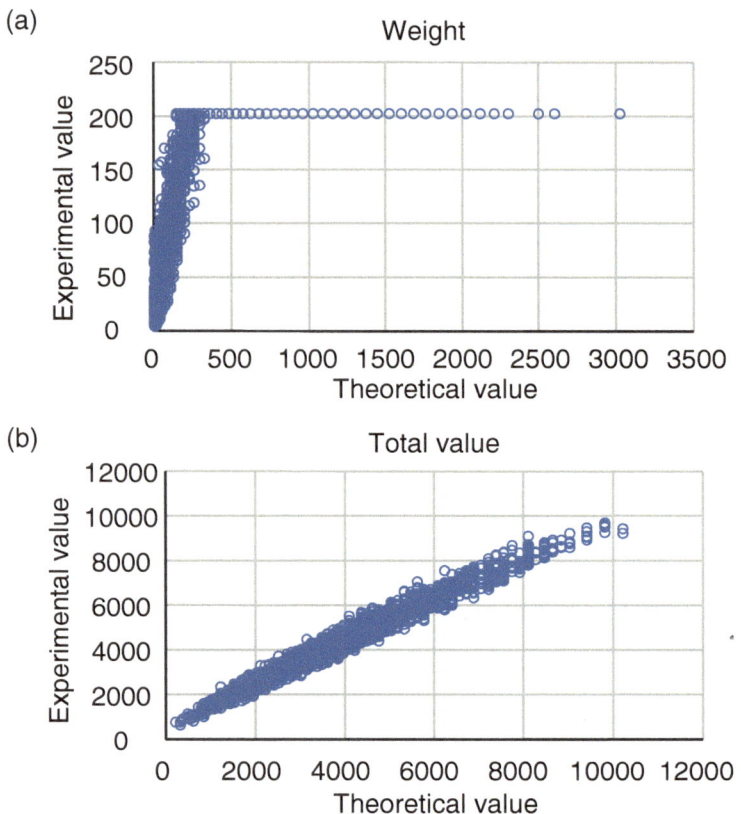

Fig. 4 Comparison of the Hamiltonian values between the theory and the experiment for **a** the weight term and **b** the total value term

found even if the optimal solution could not be found, demonstrating the system's capability as an Ising machine. Figure 6b shows a histogram of the energy values of 150,000 samples generated during the iteration for all executions. These statistical data show that the system generated many low-energy samples. Furthermore, an exponential decrease was observed within the area where the energy value was not too low. This is similar to the Boltzmann distribution, and the system has characteristics expected to be sufficient for determining the ground-state solution.

4   Space Division Multiplexing (SDM)-SPIM

TDM-SPIM can manage interaction matrices with a rank of two or more, but the computation time increases as the number of multiplexing channels increases. As an approach that provides other features, an SDM-SPIM system was constructed and

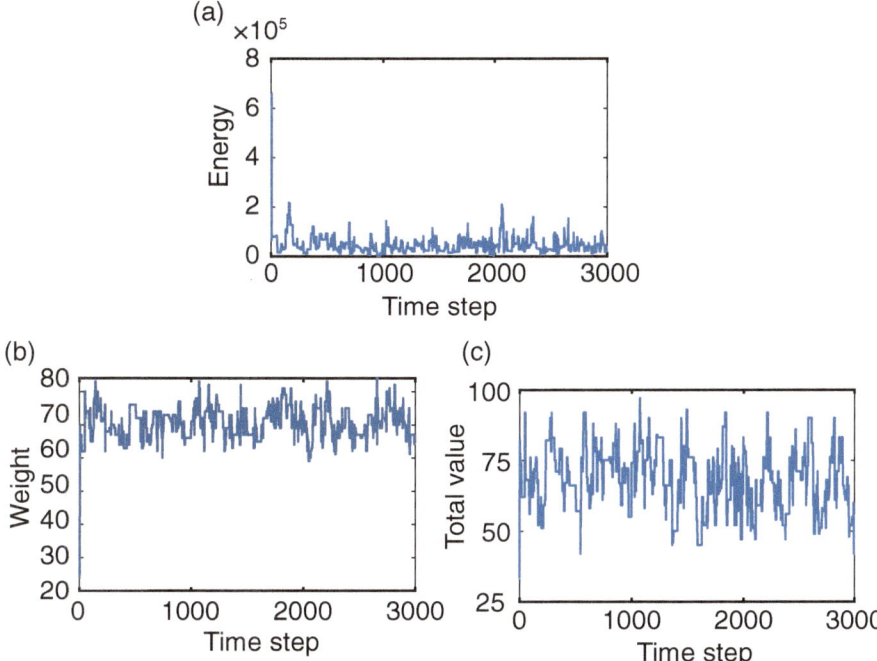

Fig. 5 Example of time evolution of **a** Hamiltonian, **b** weight, and **c** total value

demonstrated. The optical setup of the SDM-SPIM is shown in Fig. 7. We assume that two independent and mutually incoherent intensity distributions are created simultaneously in this setup. For this, we use two He-Ne laser sources with the same wavelength of 632.8 nm (LASOS, LGK7654-8; Melles Griot, 05-LHP-171). The individual beams from the sources shine in different areas of SLM1 (amplitude type, HOLOEYE, LC2012; pixel number: 1024×768; pixel pitch: 36 μm), and their amplitudes were modulated to independent distributions ξ. Beams 1 and 2 correspond to the objective and constraint terms, respectively. To control the degree of contribution of both terms in calculating the Hamiltonian, a neutral-density (ND) filter was inserted into the pass of beam 1 before SLM1 to adjust the intensity ratio between the two beams. The beams modulated by SLM1 were then coaxially combined and directed on the phase-only SLM2 (HOLOEYE, PURUTO-2; pixel number: 1920×1080; pixel pitch: 8.0 μm) for encoding spins. After receiving the same phase modulation, beams 1 and 2 were Fourier-transformed using lens $L3$. The CCD (PointGray Research, Grasshopper GS3-U3-32S4: pixel pitch: 3.45 μm) then captures the intensity images. The intensity ratio between the objective and constraint terms was $\beta = \frac{B}{A} = 4$ without the ND filter in the setup in Fig. 7. An area of 360×360 μm (10×10 pixels for SLM1 and 45×45 pixels for SLM2) is considered the minimum modulation size to eliminate the mismatch between the pixel sizes of SLM1 and SLM2.

(a)

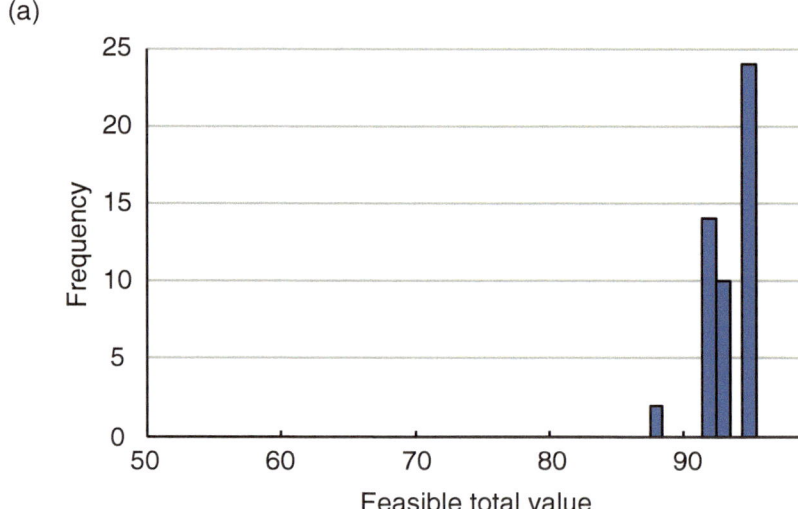

(b)

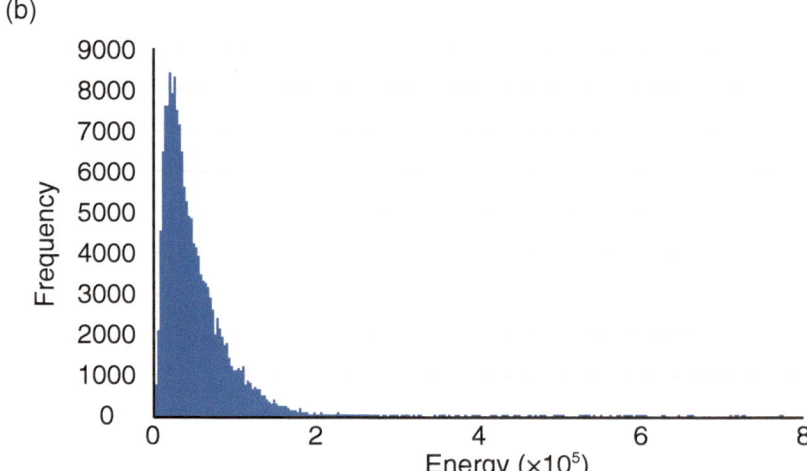

Fig. 6 Sampling behavior of TM-SPIM. **a** Histogram of the total value for obtained solutions. **b** Histogram of Hamiltonian values during all iterations

Knapsack problems were examined as well as the experiments described in the previous section. Here, the contribution of the total value term in the Hamiltonian is changed to linear such that $\mathcal{H}'_B$ is used instead of $\mathcal{H}_B$ in Eq. (12):

$$\mathcal{H}'_B = -B \sum_{i=1}^{n} v_i x_i. \tag{21}$$

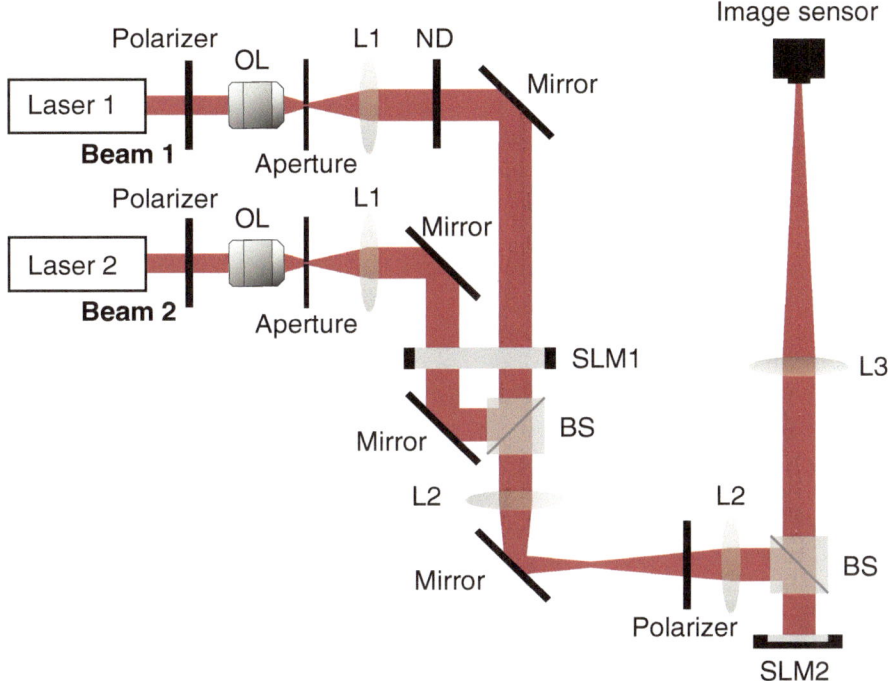

Fig. 7 Optical setup of the SDM-SPIM. The number of multiplexing is two. OL: Objective lens (10×, NA 0.25); ND: neutral-density filter; BS: beam splitter; L1, L2, L3: lens (focal length: 60, 150, 300 mm)

The Hamiltonian is represented as follows:

$$\mathcal{H}(\sigma) = A\sigma^T \xi^{(1)} \xi^{(1)\,T} \sigma - B\sigma^T \xi^{(2)} \xi^{(2)\,T} \sigma + B\sigma^T \xi^{(3)} \xi^{(3)\,T} \sigma, \tag{22}$$

$$\xi^{(1)} = (w_0, \dots, w_{N-1}, 2^0, \dots, 2^M, \sum_{i=0}^{N-1} w_i + \sum_{i=0}^{M} 2^i - 2W), \tag{23}$$

$$\xi^{(2)} = (v_0, \dots, v_{N-1}, v_N, \dots, v_{N+M}, 1), \tag{24}$$

$$\xi^{(3)} = (v_0, \dots, v_{N-1}, v_N, \dots, v_{N+M}, 0). \tag{25}$$

The image captured by this system is the sum of the intensity distributions of the individual amplitude distributions of the beams. In Eq. (22), the sign of the coefficient of the second term is different from that of the other terms, and it is not possible to obtain the sum of all Hamiltonians simultaneously. Therefore, TDM was utilized. Terms with the same sign were optically calculated simultaneously, and the energy was obtained separately for each sign. The separation of processing into two parts is sufficient, and the time cost for TDM is constant, regardless of the number of terms in the Hamiltonian. The spin configuration was updated based on SA. The initial temperature was 3000 and the cooling rate was 0.96. The next candidate of the spin configuration σ' is made by flipping individual spins except the last one, whose spin is fixed to "1," of the current spin configuration σ with the probability $3/(n + m)$. The spin configuration is updated according to Eq. (19). The delta function was employed as the target image $I_T(x)$. To represent the delta function in the experiments, we set 3×3 pixels around the center to 1, and the others to 0.

Proof-of-concept experiments were performed using the knapsack problem as follows:

$$n = 4, \; W = 11, \; \mathbf{v} = (6, 10, 12, 13), \; \mathbf{w} = (2, 4, 6, 7). \tag{26}$$

The total value of the optimal solution is 23 and the weight is 11. The total number of spin variables, including the auxiliary variables, is 8. Figure 8a presents a histogram of the total values of the final solutions obtained over 100 iterations. The total number of iterations was 300 and $\beta = 0.01$. The rate of execution in which the solution search converges to the optimal solution was 52%. The rate of execution in which the optimal solution is never sampled during iterations was 27%. The rate of convergence to reach the optimal solution out of the executions in which the optimal solution is sampled once or more was 71%. No solution significantly exceeded the weight constraint, and the constraint term was confirmed to work sufficiently. Figure 8b shows an example of the time evolution of a Hamiltonian during the iterations. The SDM-SPIM provides sufficient opportunities for convergence to the optimal solution.

It is necessary to set the ratio (β) of the constraint and objective terms suitably to determine the ground state in the Ising model. In the SDM-SPIM optical system, the ratio $\beta = \frac{B}{A}$ can be controlled by the light wave intensities related to the individual terms. We investigated the characteristics of the solution search when different ND filter transmittances were applied. Figure 9a shows the number of samples that exceed the weight limit during iteration, and (b) the histogram of the weights for the final solutions when the transmittance of the ND filter is 10% or 0.25% in 50 executions. The number of iterations was set to 300. When the intensity of light related to the objective term decreases (the transmittance of the ND filter decreases), the constraint is easily maintained in searches. In contrast, when the intensity increases, the constraint easily exceeds. No final solution exceeds the weight limit when $\beta = 0.01$ (ND:0.25%). However, many solutions violate this constraint when $\beta = 0.4$ (ND:10%). These experimental results demonstrate that the distributions of the samples during the iterations and the final solutions change depending on the transmittance of the ND filter or β. This indicates the manipulability of the space for solution search by controlling the optical parameters. In addition, for the constraint term to work effectively when using the Hamiltonian in Eq. (22), the following condition must be satisfied:

$$A > B \max_i v_i. \tag{27}$$

$\max_i v_i = 13$ for the examined problem, and the condition becomes $\beta = <1/13 \approx 0.077$. This is consistent with the results presented in Fig. 9.

(a)

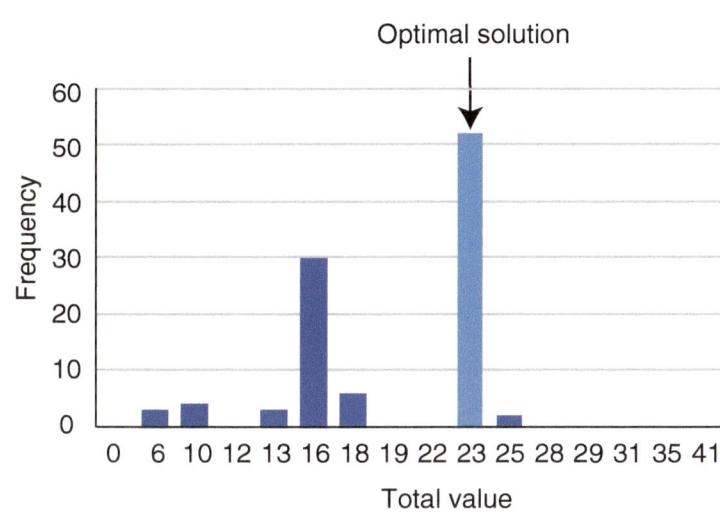

(b)

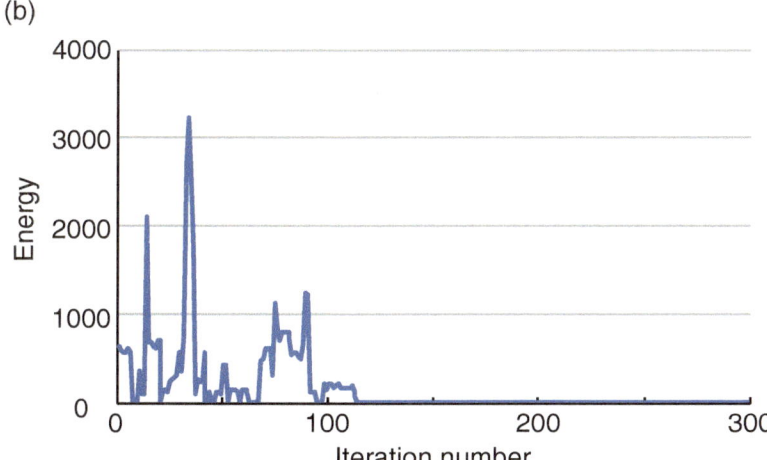

Fig. 8 **a** Histogram of the total value for the obtained solution. **b** Example of time evolution of Hamiltonian values

In the previous experiment, the ND filter's transmittance was fixed during iterations. The search characteristics can be improved by changing the optical parameters during the iterations. Thus, we investigated a method in which the iteration proceeded by changing the coefficient ratio β step-by-step. This method is referred to as the dynamic coefficient search in this study. The change in the coefficient can be realized by replacing the ND filter or controlling the light source emission intensity. SA with

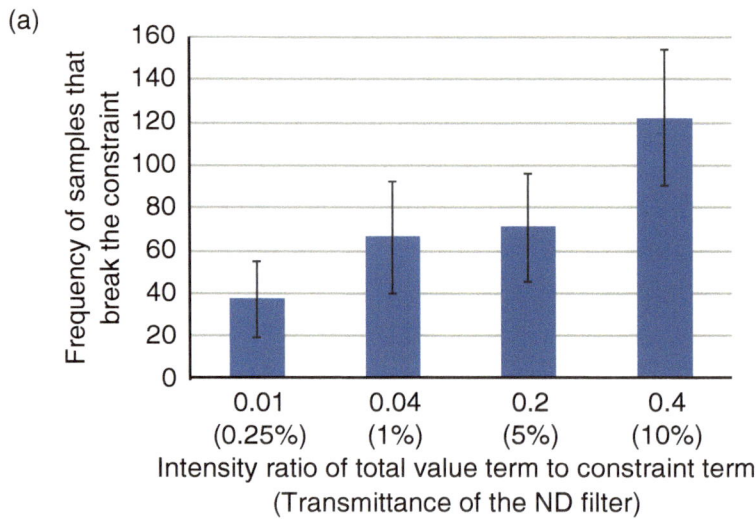

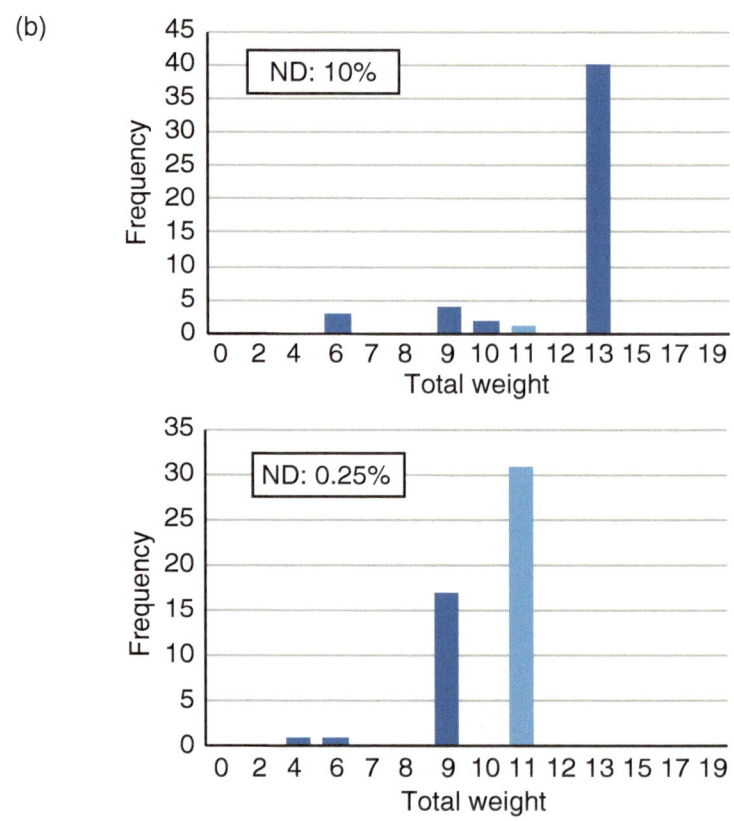

Fig. 9 Searching characteristics on the ND filter's transmittance for samples **a** during iterations and **b** final solutions

fixed coefficients and dynamic coefficient searches were compared using numerical experiments. The knapsack problem used is as follows:

$$n = 10, \ W = 60,$$
$$\mathbf{v} = (20, 18, 17, 15, 15, 10, 5, 3, 1, 1),$$
$$\mathbf{w} = (30, 25, 20, 18, 17, 11, 5, 2, 1, 1). \tag{28}$$

The total value of the optimal solution is 52 and the weight is 57–60. The total number of spin variables, including the auxiliary variables, is 16. The ratio $\beta = 0.05$. In the SA, the initial temperature was 300, 000, and the cooling rate was 0.96. In the dynamic coefficient search, the ratio was changed, $\beta = 2, \ 1, \ 0.8, \ 0.5, \ 0.1, \ 0.05$ for every 100 iterations. The annealing temperature was fixed at $T = 30$. The spin configuration with the minimum energy in iterations with the same β is used as the initial spin configuration in iterations with the next β. The total number of iterations was set to 600.

Figure 10 shows the histogram of the total values for 1000 executions. The dynamic coefficient search provides improved optimal or approximate solutions compared to SA with fixed coefficients. This tendency is also observed when the total number of iterations varies. A dynamic coefficient search has good potential. A possible reason for this is the difference in the search route leading to the optimal solution. In SA with fixed coefficients, the constraint term is strong from the beginning of the iteration, and solutions satisfying the constraint are preferentially searched. In contrast, in the dynamic coefficient search, the constraint term is weak at the beginning of the iterations, and the search proceeds from solutions with high total values. This suggests the possibility of the SDM-SPIM performance improvement by dynamic optical parameter tuning.

5 Conclusion

This study presents SPIMs with multiplexing to solve combinatorial optimization problems. An interaction coefficient matrix with a rank of two or more can be managed, and the applicability of SPIMs to practical applications is enhanced. We constructed TDM-SPIM and SDM SPIM systems among the possible multiplexing schemes and verified their performance using knapsack problems. In the TDM-SPIM experiments, the constraint and objective terms work well and the ground state of the system can be searched efficiently by considering the two terms. In the SDM-SPIM experiments, the search characteristics varied depending on the coefficient ratio, which can change with the transmittance of the ND filter, between the constraint and objective terms in the Hamiltonian. Furthermore, the numerical results suggest that dynamically decreasing the coefficient ratio during the iteration can enhance the performance of an Ising machine.

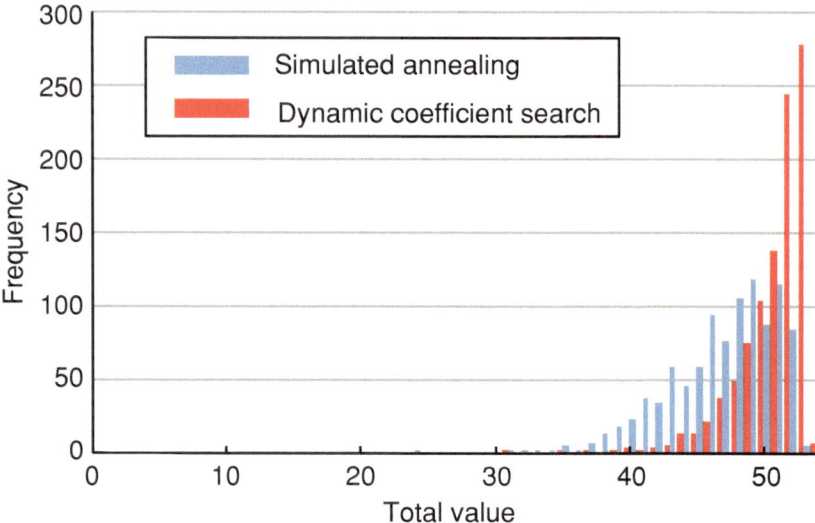

Fig. 10 Performance comparison between SA with fixed coefficients and dynamic coefficient search in SDM-SPIM

With support from the performance and functionality improvements of SLMs and the progress of mathematical methods, computing based on spatial light modulation and free-space propagation provides advantages in terms of scalability, controllability, and simplicity [6, 23]. The number of spin variables handled in the SPIM depends on the number of SLM's pixels. These pixels can be manipulated in parallel, and the time required to calculate the energy is independent of the number of spins and is constant. The degrees of freedom of the models that can be handled are determined by the number of multiplexing. The number of spins and multiplexing can be changed independently, thereby providing flexibility in the design of optical systems. Furthermore, physical operations are possible in simple energy calculations and when setting parameters related to annealing characteristics. These are significant features of SPIM with multiplexing, and they are expected to contribute to creating optics-based unconventional computing architectures in the future.

References

1. M. Ahmed, Y. Al-Hadeethi, A. Bakry, H. Dalir, V.J. Sorger, Integrated photonic FFT for photonic tensor operations towards efficient and high-speed neural networks. Nanophotonics **9**(13), 4097–4108 (2020). https://doi.org/10.1515/nanoph-2020-0055
2. J.S. Lee, N. Farmakidis, C.D. Wright, H. Bhaskaran, Polarization-selective reconfigurability in hybridized-active-dielectric nanowires. Sci. Adv. **8**(24), eabn9459 (2022)

3. K. Takano, C. Sugano, M. Inubushi, K. Yoshimura, S. Sunada, K. Kanno, A. Uchida, Compact reservoir computing with a photonic integrated circuit. Opt. Express **26**(22), 29424–29439 (2018)
4. B.J. Shastri, A.N. Tait, T. Ferreira de Lima, W.H.P. Pernice, H. Bhaskaran, C.D. Wright, P.R. Prucnal, Photonics for artificial intelligence and neuromorphic computing. Nat. Photonics **15**(2), 102–114 (2021)
5. Y. Shen, N.C. Harris, S. Skirlo, M. Prabhu, T. Baehr-Jones, M. Hochberg, X. Sun, S. Zhao, H. Larochelle, D. Englund, M. Soljačić, Deep learning with coherent nanophotonic circuits. Nat. Photonics **11**(7), 441–446 (2017)
6. X. Lin, Y. Rivenson, N.T. Yardimci, M. Veli, Y. Luo, M. Jarrahi, A. Ozcan, All-optical machine learning using diffractive deep neural networks. Science **361**(6406), 1004–1008 (2018). https://doi.org/10.1126/science.aat8084
7. K. Kitayama, M. Notomi, M. Naruse, K. Inoue, S. Kawakami, A. Uchida, Novel frontier of photonics for data processing–photonic accelerator. APL Photonics **4**(9), 090901 (2019)
8. S. Kirkpatrick, C.D. Gelatt, M.P. Vecchi, Optimization by simulated annealing. Science **220**(4598), 671–680 (1983)
9. H. Mühlenbein, M. Gorges-Schleuter, O. Krämer, Evolution algorithms in combinatorial optimization. Parallel Comput. **7**(1), 65–85 (1988)
10. A. Lucas, Ising formulations of many NP problems. Front. Phys. **2**(5) (2014)
11. N. Mohseni, P.L. McMahon, T. Byrnes, Ising machines as hardware solvers of combinatorial optimization problems. Nat. Rev. Phys. **4**(6), 363–379 (2022)
12. M.W. Johnson, M.H.S. Amin, S. Gildert, T. Lanting, F. Hamze, N. Dickson, R. Harris, A.J. Berkley, J. Johansson, P. Bunyk, E.M. Chapple, C. Enderud, J.P. Hilton, K. Karimi, E. Ladizinsky, N. Ladizinsky, T. Oh, I. Perminov, C. Rich, M.C. Thom, E. Tolkacheva, C.J.S. Truncik, S. Uchaikin, J. Wang, B. Wilson, G. Rose, Quantum annealing with manufactured spins. Nature **473**(7346), 194–198 (2011)
13. K. Kim, M.S. Chang, S. Korenblit, R. Islam, E.E. Edwards, J.K. Freericks, G.D. Lin, L.M. Duan, C. Monroe, Quantum simulation of frustrated ising spins with trapped ions. Nature **465**(7298), 590–593 (2010)
14. T. Kadowaki, H. Nishimori, Quantum annealing in the transverse Ising model. Phys. Rev. E **58**, 5355–5363 (1998). https://doi.org/10.1103/PhysRevE.58.5355
15. M. Yamaoka, C. Yoshimura, M. Hayashi, T. Okuyama, H. Aoki, H. Mizuno, A 20k-spin Ising chip to solve combinatorial optimization problems with CMOS annealing. IEEE J. Solid-State Circuits **51**(1), 303–309 (2016). https://doi.org/10.1109/JSSC.2015.2498601
16. M. Aramon, G. Rosenberg, E. Valiante, T. Miyazawa, H. Tamura, H.G. Katzgraber, Physics-inspired optimization for quadratic unconstrained problems using a digital annealer. Front. Phys. **7**(48) (2019)
17. M. Prabhu, C. Roques-Carmes, Y. Shen, N. Harris, L. Jing, J. Carolan, R. Hamerly, T. Baehr-Jones, M. Hochberg, V. Čeperić, J.D. Joannopoulos, D.R. Englund, M. Soljačić, Accelerating recurrent Ising machines in photonic integrated circuits. Optica **7**(5), 551–558 (2020). https://doi.org/10.1364/OPTICA.386623
18. T. Inagaki, Y. Haribara, K. Igarashi, T. Sonobe, S. Tamate, T. Honjo, A. Marandi, P.L. McMahon, T. Umeki, K. Enbutsu, O. Tadanaga, H. Takenouchi, K. Aihara, K. Kawarabayashi, K. Inoue, S. Utsunomiya, H. Takesue, A coherent Ising machine for 2000-node optimization problems. Science **354**(6312), 603–606 (2016)
19. T. Honjo, T. Sonobe, K. Inaba, T. Inagaki, T. Ikuta, Y. Yamada, T. Kazama, K. Enbutsu, T. Umeki, R. Kasahara, K. Kawarabayashi, H. Takesue, 100,000-spin coherent Ising machine. Sci. Adv. **7**(40), eabh0952 (2021). https://doi.org/10.1126/sciadv.abh0952
20. D. Pierangeli, G. Marcucci, C. Conti, Large-scale photonic Ising machine by spatial light modulation. Phys. Rev. Lett. **122**, 213902 (2019). https://doi.org/10.1103/PhysRevLett.122.213902
21. A. Marandi, Z. Wang, K. Takata, R.L. Byer, Y. Yamamoto, Network of time-multiplexed optical parametric oscillators as a coherent Ising machine. Nat. Photonics **8**(12), 937–942 (2014)

22. J. Chang, V. Sitzmann, X. Dun, W. Heidrich, G. Wetzstein, Hybrid optical-electronic convolutional neural networks with optimized diffractive optics for image classification. Sci. Rep. **8**(1), 12324 (2018)
23. J. Bueno, S. Maktoobi, L. Froehly, I. Fischer, M. Jacquot, L. Larger, D. Brunner, Reinforcement learning in a large-scale photonic recurrent neural network. Optica **5**(6), 756–760 (2018)
24. D. Pierangeli, G. Marcucci, D. Brunner, C. Conti, Noise-enhanced spatial-photonic Ising machine. Nanophotonics **9**(13), 4109–4116 (2020)
25. D. Pierangeli, G. Marcucci, C. Conti, Adiabatic evolution on a spatial-photonic ising machine. Optica **7**(11), 1535–1543 (2020)
26. J. Huang, Y. Fang, Z. Ruan, Antiferromagnetic spatial photonic Ising machine through optoelectronic correlation computing. Commun. Phys. **4**(1), 242 (2021)
27. W. Sun, W. Zhang, Y. Liu, Q. Liu, Z. He, Quadrature photonic spatial Ising machine. Opt. Lett. **47**(6), 1498–1501 (2022)
28. L. Luo, Z. Mi, J. Huang, Z. Ruan, Wavelength-division multiplexing optical Ising simulator enabling fully programmable spin couplings and external magnetic fields (2023). arXiv:2303.11565
29. L. Dhar, A. Hill, K. Curtis, W. Wilson, M. Ayres, *Holographic Data Storage: From Theory to Practical Systems* (Wiley, New York, 2011)
30. Y. Bai, X. Xu, M. Tan, Y. Sun, Y. Li, J. Wu, R. Morandotti, A. Mitchell, K. Xu, D.J. Moss, Photonic multiplexing techniques for neuromorphic computing. Nanophotonics **12**(5), 795–817 (2023). https://doi.org/10.1515/nanoph-2022-0485
31. H. Yamashita, K. ichi Okubo, S. Shimomura, Y. Ogura, J. Tanida, H. Suzuki, Low-rank combinatorial optimization and statistical learning by spatial photonic Ising machine. Phys. Rev. Lett. **131**(6), 063801 (2023). https://doi.org/10.1103/PhysRevLett.131.063801
32. F.H. Mok, Angle-multiplexed storage of 5000 holograms in lithium niobate. Opt. Lett. **18**(11), 915–917 (1993)

Investigation on Oscillator-Based Ising Machines

Sho Shirasaka

Abstract Moore's law is slowing down and, as traditional von Neumann computers face challenges in efficiently handling increasingly important issues in a modern
information society, there is a growing desire to find alternative computing and device
technologies. Ising machines are non-von Neumann computing systems designed to
solve combinatorial optimization problems. To explore their efficient implementation, Ising machines have been developed using a variety of physical principles
such as optics, electronics, and quantum mechanics. Among them, oscillator-based
Ising machines (OIMs) utilize synchronization dynamics of network-coupled spontaneous nonlinear oscillators. In these OIMs, phases of the oscillators undergo bina-
rization through second-harmonic injection signals, which effectively transform the
broad class of network-coupled oscillator systems into Ising machines. This makes
their implementation versatile across a wide variety of physical phenomena. In this
Chapter, we discuss the fundamentals and working mechanisms of the OIMs. We
also numerically investigate the relationship between their performance and their
properties, including some unexplored effects regarding driving stochastic process
and higher harmonics, which have not been addressed in the existing literature.

1 Introduction

In today's society, we are increasingly reliant on information devices in every aspect
of our lives. The remarkable progress in information technology has been largely
driven by the advancements in semiconductor technology, specifically the scaling
law known as Moore's law [1]. However, the pace of Moore's law slows down due
to physical and economic limitations [2].

The modern era of information processing has been largely dominated by the
von Neumann architecture, a paradigm that has served as the backbone of general-
purpose computing for decades. However, von Neumann machines have inherent

S. Shirasaka (✉)
Graduate School of Information Science and Technology, Osaka University, 1–5 Yamadaoka,
Suita, Osaka 565–0871, Japan
e-mail: shirasaka@ist.osaka-u.ac.jp

© The Author(s) 2024   175
H. Suzuki et al. (eds.), *Photonic Neural Networks with Spatiotemporal Dynamics*,
https://doi.org/10.1007/978-981-99-5072-0_9

limitations when solving certain types of problems, such as those involving combinatorial optimization. These problems have a wide spectrum of applications in the real world, including machine learning, computer vision, circuit wiring, route planning, and resource allocation [3–6].

Recognizing these challenges, there has been a growing interest in non-silicon-based, non-von Neumann architectures. These architectures, which admit massively parallel, asynchronous, and in-memory operations, are different from traditional general-purpose computing devices/models and are designed specifically to tackle these complex problems more effectively. These architectures, which include quantum computers, neuromorphic computers, and Ising machines, among others, offer promising alternatives for the advancement of next-generation information processing [7, 8].

Many combinatorial optimization problems can be translated into a problem of physics: finding the ground state of an Ising model, a system of interacting binary spins. Ising machines are physical systems specifically designed to find the ground states of Ising models [9]. Ising machines have been implemented using various physical systems, such as superconducting qubits, optical parametric oscillators, dedicated digital CMOS devices, memristors, and photonic simulators [10–15].

Oscillations are ubiquitous phenomena observed across the fields of natural science and engineering [16, 17]. Coupled oscillator systems, which can be realized through various physical phenomena, possess diverse information processing capacity and hold promise for building ultra energy efficient, high frequency and density scalable computing architecture [18, 19] (see [18, Table 1] for a comparison of several building block physical rhythmic elements). While the state of an oscillator is represented by a continuous phase value, sub-harmonic injection locking phenomena can be used to realize discrete states, as proposed since the time of von Neumann and Goto [20, 21]. These discrete states can be utilized to implement Ising spins, a principle that led to the foundation of oscillator-based Ising machines (OIMs) [22, 23]. OIMs have been experimentally demonstrated using various physical systems, such as analog electronic, insulator-to-metal phase transition, and spin oscillators [22, 24–26].

In this Chapter, we discuss the fundamentals and working mechanisms of the OIMs. We also numerically investigate the relationship between their performance and their properties, including some unexplored effects regarding driving stochastic process and higher harmonics, which have not been addressed in the existing literature.

2 Ising Model and Ising Machines

The Ising model, proposed by E. Ising in the early 20th century, is a theoretical model used to describe a system of interacting binary spins [27]. The model is specified by a collection of discrete variables, the "spins," $(s_i)_{i=1}^{N} \in \{-1, 1\}^N$, where N is the number of spins, and a cost function, or "Hamiltonian," $H : \{-1, 1\}^N \to \mathbb{R}$, which specifies how the spins interact. The Ising Hamiltonian is given by:

$$H = -\frac{1}{2} \sum_{i=1}^{N} \sum_{j=1}^{N} J_{ij} s_i s_j - \sum_{i}^{N} h_i s_i , \qquad (1)$$

where $J_{ij} \in \mathbb{R}$ is the interaction coefficient between the ith and jth spins, and $h_i \in \mathbb{R}$ is the external magnetic field for the ith spin. Many combinatorial optimization problems can be mapped onto the problem of finding the ground state of the Ising model, with instances of these problems specified by the symmetric adjacency matrix J and vector h. The problem is to find the spin configuration s that minimizes the above Hamiltonian. In this chapter, we limit to consider the Ising models with no external field:

$$H = -\frac{1}{2} \sum_{i=1}^{N} \sum_{j=1}^{N} J_{ij} s_i s_j . \qquad (2)$$

These models still encompass various important combinatorial optimization problems [28], called NP-complete problems, which can be computationally intractable for the traditional von Neumann architecture machines.

Ising machines are physical systems that are designed to efficiently explore the ground state of the Ising Hamiltonian. Various Ising machines have been proposed, using approaches including classical, quantum, classical-quantum hybrid, and quantum-inspired classical [9]. Also, these machines have been realized through various physical systems, such as superconducting qubits, optical parametric oscillators, dedicated digital CMOS devices, memristors, and photonic simulators [10–15]. Among these, the focus of this Chapter is on the classical ones. A classical physical system subjected to thermal fluctuation exhibits a stationary distribution p_s, known as the Boltzmann distribution, which takes the following form [29]:

$$p_s(x) = \mathcal{N} \exp\left(-V(x)/D\right) , \quad x \in \Omega , \qquad (3)$$

where x is the state of the physical system, Ω is the phase space, $\mathcal{N} \in \mathbb{R}$ is the normalization constant, $V(x) \in \mathbb{R}$ is the energy of the state x and $D \in \mathbb{R}$ is the strength of the thermal fluctuation. The Boltzmann distribution tells us that the lower energy states appear with higher probability, and the probability of obtaining the ground state increases as the fluctuation strength is lowered. While reducing the fluctuation strength can increase the probability of obtaining the ground state, it's not always advantageous to simply diminish the fluctuation. If the fluctuation is too weak, the system may become trapped in local minima of the potential and be unable to escape, which significantly increases the time it takes for the system to reach a stationary distribution. To address this challenge, a process known as annealing is often employed. In this process, the strength of the fluctuation is gradually reduced in order to achieve a balance between reaching a stationary distribution and enhancing the probability of finding the ground state. The discussion above leads to the following idea of a class of Ising machines: if we implement a physical gradient system with the potential V subject to the following conditions, and apply appropriate fluctuations to it, we can find the ground state of the Ising Hamiltonian and thus solve combinatorial optimization problems:

- There exists a set in the phase space Ω that can be regarded as spin configurations,
- The Ising Hamiltonian H is represented as the potential V evaluated at these spin configurations,
- The minimum value of V coincides with the minimum value of H.

Even continuous-state dynamical systems can be harnessed in the implementation of Ising machines. The Hopfield-Tank neural network [30] being a classical example, and coherent Ising machines [11] implemented using optical parametric oscillators and OIMs also belong to this group. Also, in addition to the method of utilizing thermal fluctuations as discussed above, other approaches utilizing deterministic chaotic fluctuations to implement Ising machines using classical continuous-state dynamical systems have been proposed [31, 32].

3   Oscillator-Based Ising Machines

Oscillations are ubiquitous phenomena observed across the fields of natural science and engineering [16, 17]. Coupled oscillator systems, which can be realized through various physical phenomena, possess diverse information processing capacity and hold promise for building ultra energy efficient, high frequency and density scalable computing architecture [18, 19].

In this section, we will discuss the background of the operating principle of oscillator-based Ising machines (OIMs) [22, 23]. This is summarized as follows: Under the assumption that the interaction and external forcing are sufficiently weak, network-coupled self-excited oscillators can universally be described using the Kuramoto model, which consists of network-coupled phase oscillators. Given certain symmetries in the topology and scheme of interaction, the Kuramoto model becomes a gradient system. Moreover, sub-harmonic injection allows for the introduction of spin configurations as a stable synchronized state within the phase space of the phase oscillator system.

These properties suggest that a broad class of network-coupled self-excited oscillator systems can be used to implement OIMs. OIMs have been experimentally demonstrated using various physical systems, such as analog electronic, insulator-to-metal phase transition, and spin oscillators [22, 24–26].

3.1 Phase Oscillators

In this Subsection, we introduce the notion of the phase for a stable self-excited oscillator and explain how its dynamics, when subjected to sufficiently weak fluctuation, can be reduced to a one-dimensional dynamics of a phase oscillator.

Consider a smooth autonomous dynamical system of N_d-dimensional state $x(t) \in \mathbb{R}^{N_d}$:

$$\dot{x}(t) = F(x(t)), \quad x(t) \in \mathbb{R}^{N_d}, \tag{4}$$

which has an exponentially stable limit-cycle $\chi : \tilde{x}_0(t)$ with a natural period T and frequency $\omega = 2\pi/T$, satisfying $\tilde{x}_0(t) = \tilde{x}_0(t + T)$. We first introduce a phase $\theta(x) \in [0, 2\pi)$ on χ, where 0 and 2π are considered identical. We can choose an arbitrary point $\tilde{x}_0(0)$ on χ as the origin of phase, i.e., $\theta(\tilde{x}_0(0)) = 0$, and define the phase of $\tilde{x}_0(t)$ as $\theta(\tilde{x}_0(t)) = \omega t \pmod{2\pi}$. In the following, we reparametrize a point on χ using θ instead of t. Specifically, we define $x_0(\theta) := \tilde{x}_0(t)$ for subsequent discussions. Apparently, $x_0(\theta) = x_0(\theta + 2\pi)$ holds.

To describe the dynamics when the system deviates from the periodic orbit χ due to perturbation, we extend the definition of the phase beyond χ. Here, it's important to note that $\dot{\theta} = \omega$ holds as long as x evolves on χ. Let us extend the definition of the phase such that $\dot{\theta} = \omega$ holds. With this extension, the phase difference between two solutions of (4) starting from different initial conditions should remain constant over time. The basin of attraction $\mathcal{B} \subset \mathbb{R}^{N_d}$ is the set of initial conditions that converge to χ. For smooth, exponentially stable limit-cycling system, the following holds [33]: For any point $x_* \in \mathcal{B}$, there exists a unique initial condition $x_0(\theta_*)$ on the periodic orbit, which yields a solution that maintains a constant phase difference of zero with the solution starting from x_*. Thus we can introduce a phase function $\theta(x) : \mathcal{B} \rightarrow [0, 2\pi)$ that maps the system state to a phase value as

$$\theta(x_*) = \theta(x_0(\theta_*)) = \theta_* . \tag{5}$$

For smooth systems, the phase function θ is also smooth, and thus

$$F(x) \cdot \nabla\theta(x) = \omega, \quad \forall x \in B, \tag{6}$$

holds due to the chain rule.

When an impulsive and sufficiently weak perturbation ϵk ($|\epsilon| \ll 1$) is given to the system at $x_0(\theta_*)$, the response of the phase can be linearly approximated by neglecting higher-order terms in ϵ as

$$\theta(x_0(\theta_*) + \epsilon k) - \theta_* = \nabla\theta(x_0(\theta_*)) \cdot \epsilon k . \tag{7}$$

Thus, the gradient $\nabla\theta(x_0(\theta_*))$ of θ, evaluated at $x = x_0(\theta_*)$ characterizes linear response property of the oscillator phase to weak perturbations. $\nabla\theta(x_0(\theta_*))$ is called the phase sensitivity function (a.k.a. infinitesimal phase resetting curve, perturbation projection vector) [34–37]. The phase sensitivity function plays central roles in analyzing synchronization dynamics of oscillatory systems. In the following, we denote the phase sensitivity function as $Z(\theta_*) := \nabla\theta(x_0(\theta_*))$.

Consider a limit-cycle oscillator subjected to a weak perturbation, described by the equation:

$$\dot{x}(t) = F(x(t)) + \epsilon p(x, t) , \quad p(x, t) \in \mathbb{R}^{N_d} , \tag{8}$$

where ϵp represents a small perturbation of magnitude ϵ, i.e., $|\epsilon| \ll 1$. The dynamics of the phase $\theta(x(t))$ can be obtained using the chain rule:

$$\dot{\theta}(x(t)) = \nabla\theta(x(t)) \cdot \{F(x(t)) + \epsilon p(x, t)\} = \omega + \epsilon\nabla\theta(x(t)) \cdot p(x, t) . \tag{9}$$

This equation is not yet closed in phase θ because $\nabla\theta(x)$ depends on x. In order to obtain an equation for θ, we used the fact that the perturbation is small and $O(\epsilon)$, implying that the deviation of the state x from χ is also small and $O(\epsilon)$, i.e., $x(t) = x_0(\theta_*) + O(\epsilon)$, where $\theta_* = \theta(x(t))$. The gradient $\nabla\theta$ at x can then be expressed as $\nabla\theta(x(t)) = \nabla\theta(x_0(\theta_*)) + O(\epsilon)$ and by substituting into (9), we obtain an approximate phase equation for θ,

$$\dot{\theta}(t) = \omega + \epsilon Z(\theta) \cdot p(x_0(\theta), t) , \tag{10}$$

by neglecting the terms of $O(\epsilon^2)$. This phase equation is now closed in θ and can be solved for θ when the phase sensitivity function Z and perturbation ϵp are given. Thus the N_d-dimensional nonlinear dynamics of the oscillator is successfully reduced the one-dimensional phase dynamics.

When the model of a dynamical system is known, a convenient method for calculating the phase sensitivity function is the adjoint method [38–40]. The adjoint method involves solving

$$\omega\frac{d}{d\theta}Y(\theta) = -DF^\top(x_0(\theta))Y(\theta) , \quad Y(\theta) \in \mathbb{R}^{N_d} , \tag{11}$$

where DF is the Jacobian matrix of F and $\top$ denotes the transposition. This equation is solved backward in time with an initial condition $Y(0)$ such that $F(x_0(0)) \cdot Y(0) \neq 0$. It then converges to a periodic solution. Normalizing this solution using the condition $F(x_0(\theta)) \cdot Y(\theta) = \omega$, which corresponds to (6), gives rise to the phase sensitivity function. While this is a simple method, it requires the calculation of the Jacobian matrix, which can often be challenging to use for high-dimensional oscillatory systems. Therefore, methods to avoid the calculation of the Jacobian matrix have also been proposed [41, 42].

The phase sensitivity functions can also be measured experimentally in model-free manners [43–47]. Furthermore, the phase function (and thus its gradient) can also be characterized by an eigenfunction of the associated Koopman operator for dynamical systems [48]. The Koopman operator allows for a data-driven spectral decomposition method, known as dynamic mode decomposition, which is a rapidly evolving field of study [49].

3.2 Second-Harmonic Injection Locking

An oscillator with a frequency ω can be entrained by an external periodic signal having a frequency close to $(l/k)\omega$, where k and l are natural numbers. This is phenomenon is known as synchronization of order $k : l$. If $k < l$ (resp. $k > l$), the locking is referred to as sub-harmonic (resp. superharmonic) [16, 17]. The term "second-harmonic injection locking" refers to $1 : 2$ sub-harmonic synchronization. Furthermore, when an oscillator is perturbed by an external second-harmonic injection signal, the phase difference between the oscillator and the injection signal settles down to one of two values, separated by π. This allows for the encoding of a spin state of an Ising model into the phase difference, using the two steady states to represent the spin down or up, respectively.

Let us provide a concrete discussion of this scenario. Consider a limit-cycle oscillator with the frequency ω subjected to a weak, almost second-harmonic perturbation of frequency $\omega_s \approx 2\omega$, described by the equation:

$$\dot{x}(t) = F(x(t)) + \epsilon p(t) . \tag{12}$$

We define $\omega - \omega_s/2 = \Delta\omega$ and the phase difference ψ between the oscillator and the forcing as

$$\psi := \theta - \frac{1}{2}\omega_s t . \tag{13}$$

The evolution of the phase difference is then given by

$$\frac{d\psi}{dt} = \Delta\omega + Z\left(\frac{1}{2}\omega_s t + \psi\right) \cdot \epsilon q\,(\omega_s t) , \tag{14}$$

where $q\,(\omega_s t) := p(t)$. From the assumptions $\omega_s \approx 2\omega$, $|\epsilon| \ll 1$, the right-hand side of (14) is very small, and ψ varies slowly. Hence, the averaging method [50] provides an approximate dynamics of (14) as

$$\frac{d\psi}{dt} = \Delta\omega + \epsilon\Gamma(\psi) , \tag{15}$$

$$\Gamma(\psi) = \frac{1}{2\pi} \int\limits_{0}^{2\pi} d\theta\, Z(\theta + \psi) \cdot q(2\theta) . \tag{16}$$

Consider the Fourier series expansions of Z and q:

$$\boldsymbol{Z}(\phi) = \left[\sum_{k=-\infty}^{\infty} Z_{1,k} e^{ik\phi}, \sum_{k=-\infty}^{\infty} Z_{2,k} e^{ik\phi}, \cdots, \sum_{k=-\infty}^{\infty} Z_{N_d,k} e^{ik\phi}, \right]^{\top}, \quad (17)$$

$$\boldsymbol{q}(\phi) = \left[\sum_{l=-\infty}^{\infty} q_{1,l} e^{il\phi}, \sum_{l=-\infty}^{\infty} q_{2,l} e^{il\phi}, \cdots, \sum_{l=-\infty}^{\infty} q_{N_d,l} e^{il\phi}, \right]^{\top}. \quad (18)$$

Then,

$$\Gamma(\psi) = \sum_{m=1}^{N_d} \sum_{l=-\infty}^{\infty} Z_{m,-2l} q_{m,l} e^{-2il\psi}. \quad (19)$$

If $\epsilon \boldsymbol{p}(t)$ is a second-harmonic injection, i.e., $q_{m,l} = 0$ for any $|l| \neq 1$, (19) simplifies to

$$\Gamma(\psi) = 2.0 \sum_{m=1}^{N_d} \mathrm{Re}\left(Z_{m,-2} q_{m,1} e^{-2i\psi} \right). \quad (20)$$

This equation represents a second-harmonic wave. Therefore, when the mismatch in the $1 : 2$ frequency relation $\Delta\omega$ is sufficiently small, the dynamics of the phase difference (15) exhibits two stable and two unstable equilibria, each of which are separated π. These stable equilibria can be utilized as the spin state of an Ising model.

Even if $\boldsymbol{q}$ is not a purely second-harmonic, as long as the frequency mismatch condition is met, the DC component of $(q_{m,0})_{m=1}^{N_d}$ is small, and $Z_{m,-2l} q_{m,l}$ ($|l| \geq 2$) do not create new equilibria, there continue to be only two stable equilibria separated by π. This separation can again be utilized to represent the spin states.

3.3 Kuramoto Model

In this Subsection, we derive a variant of the Kuramoto model from a general system of weakly coupled, weakly heterogeneous oscillators subjected to second-harmonic forcing.

Consider

$$\dot{\boldsymbol{x}}_i(t) = \boldsymbol{F}(\boldsymbol{x}_i(t)) + \tilde{\boldsymbol{f}}_i(\boldsymbol{x}_i(t)) + \sum_{j=1}^{N} J_{ij} \tilde{\boldsymbol{g}}_{ij}(\boldsymbol{x}_i(t), \boldsymbol{x}_j(t)) + \boldsymbol{p}(t). \quad (21)$$

Here, $\boldsymbol{F}$ has a "standard" oscillator with a periodic orbit $\tilde{\boldsymbol{x}}_0$ of frequency ω, while $\tilde{\boldsymbol{f}}_i$ characterizes the autonomous heterogeneity of the ith oscillator. $\boldsymbol{J} = (J_{ij})$ is the adjacency matrix of the coupling connectivity, and $\tilde{\boldsymbol{g}}_{ij}$ represents the interaction from oscillator j to i. $\boldsymbol{p}$ is the almost second-harmonic injection of frequency $\omega_s \approx 2\omega$. We assume that the magnitudes of $\tilde{\boldsymbol{f}}$, $\tilde{\boldsymbol{g}}_{ij}$, and $\boldsymbol{p}$ are sufficiently small. We introduce phase functions θ_i for the ith oscillator using the standard oscillator. We define $\Delta\omega$

and ψ_i as in (13). Then we obtain

$$
\dot{\psi}_i(t) = \Delta\omega + \mathbf{Z}\left(\frac{1}{2}\omega_s t + \psi_i\right) \cdot
$$

$$
\left[\mathbf{f}_i\left(\frac{1}{2}\omega_s t + \psi_i\right) + \sum_{j=1}^{N} J_{ij}\mathbf{g}_{ij}\left(\frac{1}{2}\omega_s t + \psi_i, \frac{1}{2}\omega_s t + \psi_j\right) + \mathbf{q}\left(\omega_s t\right)\right], \quad (22)
$$

where $\mathbf{x}_0(\theta) := \tilde{\mathbf{x}}_0(t)$, $\mathbf{f}_i(\theta_*) := \tilde{\mathbf{f}}_i(\mathbf{x}_0(\theta_*))$, $\mathbf{g}_{ij}(\theta_*, \theta_{**}) := \tilde{\mathbf{g}}_{ij}(\mathbf{x}_0(\theta_*), \mathbf{x}_0(\theta_{**}))$, $\mathbf{q}(\omega_s t) := \mathbf{p}(t)$. Given the assumptions above, the right-hand side of (22) is very small and the averaging approximation leads to

$$
\dot{\psi}_i(t) = \Delta\omega_i + \sum_{j=1}^{N} J_{ij}\Gamma_{ij}(\psi_i - \psi_j) + \Gamma(\psi_i), \quad (23)
$$

where $\Delta\omega_i := \Delta\omega + \delta\omega_i$ and

$$
\delta\omega_i = \frac{1}{2\pi}\int_0^{2\pi} d\theta\,\mathbf{Z}(\theta + \psi_i)\cdot\mathbf{f}_i(\theta + \psi_i), \quad (24)
$$

$$
\Gamma_{ij}(\psi_i - \psi_j) = \frac{1}{2\pi}\int_0^{2\pi} d\theta\,\mathbf{Z}(\theta + \psi_i)\cdot\mathbf{g}_{ij}(\theta + \psi_i, \theta + \psi_j), \quad (25)
$$

$$
\Gamma(\psi_i) = \frac{1}{2\pi}\int_0^{2\pi} d\theta\,\mathbf{Z}(\theta + \psi_i)\cdot\mathbf{q}(2\theta). \quad (26)
$$

Thus, the system of weakly coupled, weakly heterogeneous oscillators subjected to second-harmonic forcing (21) can be universally reduced to the variant of Kuramoto phase oscillator system (23). This type of Kuramoto model, which has an external field term Γ, is called the active rotator model [51–53].

3.4 Gradient Structure of the Kuramoto Model

In this Subsection, given certain assumptions about symmetries in the topology and the scheme of interaction, we show that the Kuramoto model (23) is a gradient system. Our discussion draws heavily on the material presented in Appendix C of [22], but we slightly relax the assumption therein and extend the result.

We assume that $J_{ij} = J_{ji}$, $\Gamma_{ij} = \Gamma_{ji}$ and its antisymmetricity:

$$\Gamma_{ij}(x) = -\Gamma_{ij}(-x) . \tag{27}$$

Such antisymmetry appears when the interaction g_{ij} is diffusive, i.e.,

$$g_{ij}(\psi_i, \psi_j) = -g_{ij}(\psi_j, \psi_i) . \tag{28}$$

Note that, in [22], it is assumed that $\Gamma_{ij} = \Gamma_{kl}$, i.e., the interaction scheme is uniform. We only assume its symmetricity instead.

Let us introduce a potential function L as

$$L(\boldsymbol{\psi}) := \frac{1}{2} \sum_{i=1}^{N} \sum_{j=1}^{N} \left\{ \frac{-1}{N} \left[\Delta\omega_i \psi_i + \Delta\omega_j \psi_j \right] \right.$$
$$\left. + \frac{1}{2N} \left[I_s(2\psi_i) + I_s(2\psi_j) \right] + J_{ij} I_{ij}(\psi_i - \psi_j) \right\} , \tag{29}$$

where

$$I_{ij}(x) := - \int_0^x \Gamma_{ij}(y) \mathrm{d}y + C_{ij} , \tag{30}$$

$$I_s(x) := - \int_0^x \Gamma(y) \mathrm{d}y . \tag{31}$$

Here $C_{ij} \in \mathbb{R}$ is a constant. Then we have

$$\frac{\partial}{\partial \psi_l} \left\{ \sum_{i=1}^{N} \sum_{j=1}^{N} \frac{-1}{N} \left[\Delta\omega_i \psi_i + \Delta\omega_j \psi_j \right] \right\} = -\frac{1}{N} \sum_{i=1}^{N} \sum_{j=1}^{N} \left[\Delta\omega_i \delta_{il} + \Delta\omega_j \delta_{jl} \right]$$
$$= -\frac{1}{N} \sum_{j=1}^{N} \sum_{i=1}^{N} \Delta\omega_i \delta_{il} - \frac{1}{N} \sum_{i=1}^{N} \sum_{j=1}^{N} \Delta\omega_j \delta_{jl}$$
$$= -\frac{1}{N} \sum_{j=1}^{N} \Delta\omega_l - \frac{1}{N} \sum_{i=1}^{N} \Delta\omega_l = -2\Delta\omega_l , \tag{32}$$

where δ_{ij} is the Kronecker delta. Also,

$$\frac{\partial}{\partial \psi_l} \left\{ \sum_{i=1}^{N} \sum_{j=1}^{N} \frac{1}{2N} \left[I_s \left(2\psi_i \right) + I_s \left(2\psi_j \right) \right] \right\}$$

$$= -\frac{1}{2N} \sum_{i=1}^{N} \sum_{j=1}^{N} \left[2\delta_{il} \Gamma \left(2\psi_i \right) + 2\delta_{jl} \Gamma \left(2\psi_j \right) \right]$$

$$= -\frac{1}{N} \sum_{j=1}^{N} \sum_{i=1}^{N} \delta_{il} \Gamma \left(2\psi_i \right) - \frac{1}{N} \sum_{i=1}^{N} \sum_{j=1}^{N} \delta_{jl} \Gamma \left(2\psi_j \right)$$

$$= -\frac{1}{N} \sum_{j=1}^{N} \Gamma \left(2\psi_l \right) - \frac{1}{N} \sum_{i=1}^{N} \Gamma \left(2\psi_l \right) = -2\Gamma \left(2\psi_l \right) , \tag{33}$$

$$\frac{\partial}{\partial \psi_l} \left\{ \sum_{i=1}^{N} \sum_{j=1}^{N} J_{ij} I_{ij} \left(\psi_i - \psi_j \right) \right\}$$

$$= -\sum_{i=1}^{N} \sum_{j=1}^{N} \left[J_{ij} \left(\delta_{il} - \delta_{jl} \right) \Gamma_{ij} \left(\psi_i - \psi_j \right) \right]$$

$$= -\left[\sum_{j=1}^{N} \sum_{i=1}^{N} \delta_{il} J_{ij} \Gamma_{ij} \left(\psi_i - \psi_j \right) - \sum_{i=1}^{N} \sum_{j=1}^{N} \delta_{jl} J_{ij} \Gamma_{ij} \left(\psi_i - \psi_j \right) \right]$$

$$= -\left[\sum_{j=1}^{N} J_{lj} \Gamma_{lj} \left(\psi_l - \psi_j \right) - \sum_{i=1}^{N} J_{il} \Gamma_{il} \left(\psi_i - \psi_l \right) \right]$$

$$= -\left[\sum_{j=1}^{N} J_{lj} \Gamma_{lj} \left(\psi_l - \psi_j \right) + \sum_{j=1}^{N} J_{lj} \Gamma_{lj} \left(\psi_l - \psi_j \right) \right]$$

$$(\because (27) \text{ and } J_{il} = J_{li}, \Gamma_{il} = \Gamma_{li})$$

$$= -2 \sum_{j=1}^{N} J_{lj} \Gamma_{lj} \left(\psi_l - \psi_j \right) . \tag{34}$$

Thus the Kuramoto model (23) is a gradient flow of L:

$$\dot{\boldsymbol{\psi}} = -\nabla L(\boldsymbol{\psi}) . \tag{35}$$

3.5   *Working Principle of OIMs*

In this Subsection, we explain how an OIM explores the ground state of the Ising Hamiltonian.

Consider a coupled oscillator system, where noisy fluctuation is introduced into (23), as follows:

$$\dot{\psi}_i(t) = \Delta\omega_i - K \sum_{j=1}^{N} J_{ij}\Gamma_{ij}(\psi_i - \psi_j) - K_s\Gamma(\psi_i) + K_n\xi_i(t) \,, \qquad (36)$$

where $\xi_i(t)$ is a white Gaussian noise with zero mean and unit variance, and K, K_s, K_n represent the strength of each term. We interpret the stochastic integral of the Langevin equation (36) in the Strantonovich sense, as we are considering real and physical noises [54]. We assume that the symmetry assumptions (27), $J_{ij} = J_{ji}, \Gamma_{ij} = \Gamma_{ji}$ so that the associated deterministic system has a potential function:

$$L(\boldsymbol{\psi}) := \frac{1}{2} \sum_{i=1}^{N} \sum_{j=1}^{N} \left\{ \frac{-1}{N} \left[\Delta\omega_i\psi_i + \Delta\omega_j\psi_j \right] \right.$$
$$\left. - \frac{K_s}{2N} \left[I_s(2\psi_i) + I_s(2\psi_j) \right] - K J_{ij} I_{ij}(\psi_i - \psi_j) \right\} \,. \qquad (37)$$

As discussed in Subsect. 3.2, the second-harmonic injection aids in creating two stable equilibria that are separated by π. Without loss of generality, we can consider these equilibria as 0 and π. This is because ψ represents the phase difference with respect to the second-harmonic injection, and the origin of the phase of the injection can be chosen arbitrarily. Assuming small phase mismatches $\Delta\omega_i$ and that all ψ_i have settled to either 0 or π, the potential energy can be approximated as

$$L(\boldsymbol{\psi}) \approx -\frac{N K_s}{2} I_s(0) - \frac{1}{2} \sum_{i=1}^{N} \sum_{j=1}^{N} K J_{ij} I_{ij}(\psi_i - \psi_j) \,. \qquad (38)$$

We used the fact that $I_s(0) = I_s(2\pi)$. As Γ_{ij} is antisymmetric, I_{ij} is symmetric, hence $I_{ij}(\pi) = I_{ij}(-\pi)$. Thus, if we can choose C_{ij} in (30) such that

$$I_{ij}(0) = -I_{ij}(\pi) = C \,, \qquad (39)$$

where $C \in \mathbb{R}_{\geq 0}$ is a constant independent of i, j, we obtain

$$L(\boldsymbol{\psi}) \approx -\frac{N K_s}{2} I_s(0) - \frac{KC}{2} \sum_{i=1}^{N} \sum_{j=1}^{N} J_{ij}\tilde{s}(\psi_i)\tilde{s}(\psi_j) \,, \qquad (40)$$

where $\tilde{s}(0) = 1, \tilde{s}(\pi) = \tilde{s}(-\pi) = -1$. Thus, the potential function evaluated at π-separated stable equilibria created by second-harmonic injection matches the Ising Hamiltonian (2), up to a constant offset and a constant scaling factor. Since the deterministic part of (36) has the gradient structure, the stationary distribution for $\boldsymbol{\psi}$ is given by the Boltzmann distribution:

$$p_s(\boldsymbol{\psi}) = \mathcal{N} \exp\left(-\frac{2}{K_n^2} L(\boldsymbol{\psi})\right). \tag{41}$$

In this stationary distribution, states with lower potential energy are more likely to occur. Therefore, when the second-harmonic injection establishes π-separated equilibria, the OIM effectively searches for the ground state of the Ising Hamiltonian.

4 Experiments

In this Section, we conduct numerical investigations to explore the relationship between the performance and properties of OIMs. Our focus is on the MAX-CUT problem, an important problem that is straightforwardly mapped onto the Ising model and is classified as NP-complete. Additionally, we delve into some aspects related to higher harmonics and time discretization, which have remained unexplored thoroughly in the existing literature.

4.1 MAX-CUT Problem

The MAX-CUT problem asks for the optimal decomposition of a graph's vertices into two groups, such that the number of edges between the two groups is maximized. The MAX-CUT problem is NP-complete for non-planar graphs [55].

In this Chapter, we consider the MAX-CUT problem for unweighted, undirected graphs. The problem can be formulated as follows: Given a simple graph $G = (V, E)$, where V is the set of vertices and E is the set of edges. Find a partition of V into two disjoint subsets V_1 and V_2 such that the number of edges between V_1 and V_2 is maximized. Let us assign each vertex $i \in V$ a binary variable $s_i \in \{-1, 1\}$. If $i \in V_1$ (resp. V_2), we set $s_i = 1$ (resp. $s_i = -1$). The term $1 - s_i s_j$ is 2 if vertices i and j are in different subsets (and thus contribute to the cut), and 0 otherwise. Then, the sum of $(1 - s_i s_j)/2$ over the graph provides the cut value:

$$c := \frac{1}{4} \sum_{i \in V} \sum_{j \in V} A_{ij}(1 - s_i s_j) = \frac{|E|}{2} - \frac{1}{4} \sum_{i \in V} \sum_{j \in V} A_{ij} s_i s_j. \tag{42}$$

Therefore, the MAX-CUT problem can be written as $\max_s\{-\frac{1}{2}\sum_{i \in V}\sum_{j \in V} A_{ij} s_i s_j\}$, which is eqivalent to

$$\min_s \left\{ -\frac{1}{2} \sum_{i \in V} \sum_{j \in V} J_{ij} s_i s_j \right\}, \tag{43}$$

by setting $J := -A$. Thus, by defining $J := -A$, the MAX-CUT problem is mapped to the Ising model (2).

In this work, we use MAX-CUT problems associated with Möbius ladder graphs in order to demonstrate the performance and characteristics of OIMs. Figure 1(a) depicts a Möbius ladder graph of size 8 along with its MAX-CUT solution. Möbius ladder graphs are non-planar and have been widely employed as benchmarking Ising machines [9, 11, 13, 14, 22, 24–26, 32]. Note that the weighted MAX-CUT problem on Möbius ladder graphs has recently been classified as "easy," falling into the complexity class P (NP-completeness does not imply all instances are hard) [56]. However, our focus is not on the qualitative side, such as the pursuit of polynomial scaling of required time to reach optimal or good solutions for NP-complete problems. Instead, by understanding the impact of the quantitative physical properties of OIMs, particularly the magnitudes and schemes of interaction and injection denoted by Γ_{ij} and Γ, we aim to lay the groundwork that could eventually lead to the derivation of effective design principles for physical rhythm elements, thus potentially enhancing the performance of OIMs.

The computational capability of OIMs for MAX-CUT problems, such as experimentally observed polynomial scaling, has been somewhat explained by exploring connections with rank-2 semidefinite programming (SDP) relaxation of MAX-CUT problems [57]. In this regard, the construction of physical coupled oscillator systems that could effectively integrate with randomized rounding [58] would be an intriguing research direction.

4.2 Experimental Setting and Evaluation Metrics

Unless otherwise specified, (36) is integrated over time using the Euler-Heun method [59] under the parameters listed in Table 1.

The waveforms of Γ_{ij} and Γ are normalized so that

$$\int_0^{2\pi} |\Gamma_{ij}(\psi)|d\psi = \int_0^{2\pi} |\sin\psi|d\psi , \quad \int_0^{2\pi} |\Gamma(\psi)|d\psi = \int_0^{2\pi} |\sin 2\psi|d\psi , \quad (44)$$

hold.

We define how to interpret the phase difference ψ as a spin state when it takes values other than 0, π. We extend the definition of $\tilde{s}(\psi)$ in (40) to $[0, 2\pi)$ as $\tilde{s}(\psi) := \text{sign}(\cos(\psi))$.

Figure 1(b–e) illustrates the time evolution of the state of OIMs solving the MAX-CUT problem and the corresponding cut values. (b) corresponds to weak coupling and weak noise. (c–f) are related to a moderate level of coupling, where (c) represents tiny noise, (d) somewhat weak noise, (e) moderate noise, and (f) excessively strong noise. In situations like (e), characterized by moderate coupling and noise intensity,

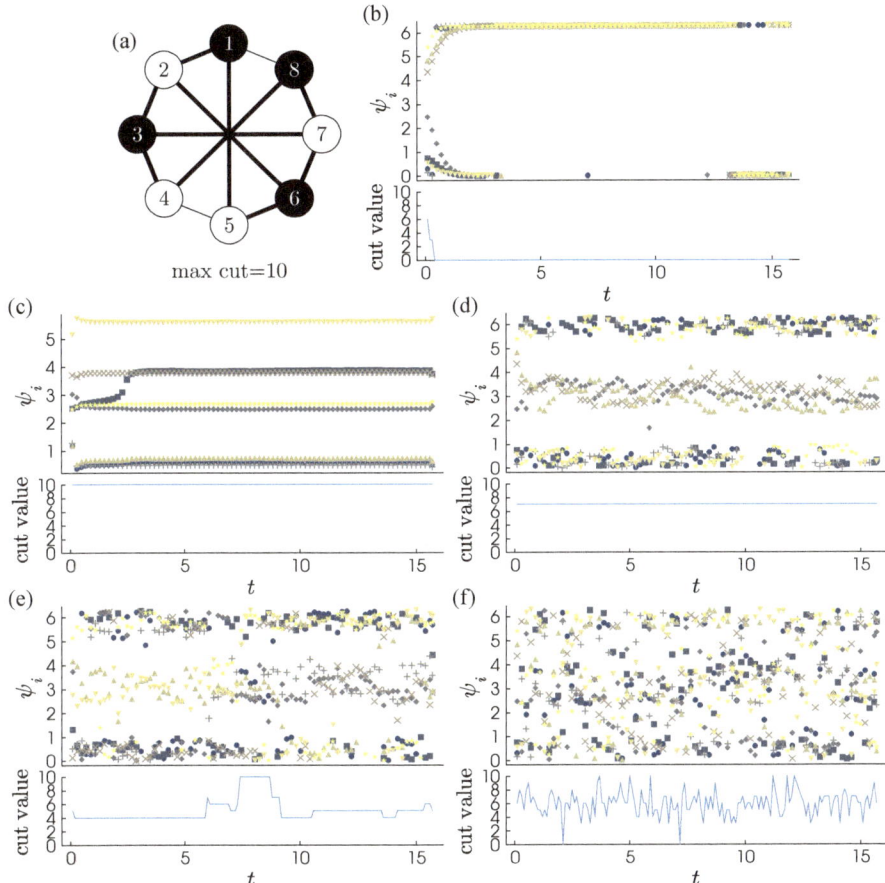

Fig. 1 (**a**) Möbius ladder graph of size 8 and its MAX-CUT solution. (**b–e**) Time evolution of the OIM state (top panel) and the corresponding cut value (bottom panel). The parameters for each plot are as follows: (**b**) $(K_s, K_n) = (0.1, 0.01)$; (**c**) $(K_s, K_n) = (13.5, 0.01)$; (**d**) $(K_s, K_n) = (13.5, 1.25)$; (**e**) $(K_s, K_n) = (13.5, 1.85)$; and (**f**) $(K_s, K_n) = (13.5, 5.27)$

it has been observed that even if the initial conditions do not lead to the spin configuration of the ground state (in this case, with a cut value of 10), the system can effectively navigate to the ground state due to the noise, subsequently sustaining this state for a certain time interval. Excessively strong noise (f) can also guide the OIM toward an instantaneous realization of the ground state. If we were to consider this as an achievement of the ground state, it would imply that the search performance could be infinitely improved by conducting exploration with pure white noise with unbounded strength and using unbounded frequency measurements, independent of the dynamical properties of OIMs. This is of course unphysical. In this study, we explore the performance of OIMs within the range where the dynamical characteristics of them matter. Specifically, when an OIM reaches a particular value of an

Table 1 Default parameters for the numerical integration of an OIM

Parameter	Setting
Number of oscillators (N)	8
Frequency mismatches ($\Delta\omega$)	**0**
Coupling strength (K)	1.0
Coupling matrix (J)	Adjacency matrix of a size 8 Möbius ladder graph negatively weighted with a value of -1
Time step size (dt)	0.1
Coupling function (Γ_{ij})	$\sin(\cdot)$
Injection coupling function (Γ)	$\sin(2\cdot)$
Integration time interval	$[0, 20\pi]$
Initial condition distribution	Uniform distribution over $[0, 2\pi)^N$
Number of initial conditions	100

Ising Hamiltonian and remains at that value for a time duration of τ_{duration} or more, we determine that this Ising Hamiltonian value has been achieved. We set $\tau_{\text{duration}} = \pi/4$ in this study.

We examine the performance of OIMs using Monte Carlo experiments with randomly generated initial conditions. We define the cut value for each trial as

$$\max \{\text{cut value} \mid \text{cut value persists for a time duration of } \tau_{\text{duration}} \text{ or more}\} . \quad (45)$$

Note that we do not merely use the cut value at the end point of the time integration interval.

Time-to-solution (TTS) metric is a standard quantitative measure of performance used for Ising machines [9]. TTS is introduced as follows: Consider a Monte Carlo experiment where the time taken for a single trial is denoted by τ. Assume that, after r trials, it is found that an acceptable performance can be achieved with probability p_{acc}. The probability that an acceptable performance is achieved at least once in r trials can be estimated as $1 - (1 - p_{\text{acc}})^r$. Let us denote the number of trials required to achieve a desired probability, typically set to 99%, as r_*. TTS refers to the time required to conduct r_* trials, represented as τr_*, and can be expressed as follows:

$$\text{TTS}(\tau) = \frac{\ln 0.01}{\ln (1 - p_{\text{acc}})} \tau . \quad (46)$$

TTS metric exhibits a nonlinear dependence on τ due to the nonlinear relationship of p_{acc} with τ. Therefore, in practice, TTS is defined to be

$$\text{TTS} = \min_{\tau} \text{TTS}(\tau) . \quad (47)$$

In this study, a trial is deemed to exhibit an acceptable performance if it yields a cut value greater than 0.9 times the best cut value discovered in the experiment. For the Möbius ladder graph of size 8, the maximum cut value is 10. Therefore, in this case, the acceptance probability corresponds to the ground state probability of the Ising model.

4.3   Effect of Properties of Noisy Fluctuation and Coarse Time Discretization

Figure 2(a) (resp. (d)) shows the color maps of the mean cut value of the OIM (resp. acceptance probability p_{acc}) when changing the ratio of injection strength to coupling strength K_s/K and noise intensity K_n, for a time discretization step of $dt = 0.01$. Figure 2(c, f) depicts those for a default time discretization step of $dt = 0.1$. It is observable that the mean cut values and acceptance probabilities have significantly improved by coarsening the time discretization step. Table 2 shows that by coarsening the time discretization step, the maximum of acceptance probabilities has improved by more than four times.

Discussing factors such as the coarsening of the time discretization step might initially appear artificial and tangential for physically implemented Ising machines. However, if we are able to identify a physical equivalent to this coarsening effect, these insights could serve as valuable guides to enhance the efficiency of OIMs.

Figure 2(b, e) plots the color maps for the OIM for a fine time discretization step of $dt = 0.01$, driven by random pulse inputs at coarse time intervals $d\tilde{t} = 0.1$ instead of the Wiener process. Here, each pulse follows an independent normal distribution, and its variance is taken to be $K_n^2 d\tilde{t}$, that is, identical to the quadratic variation of the Wiener process at the coarse time interval $d\tilde{t}$. Similar to the case where the time discretization step was coarsened, an improvement in performance can be observed.

In this way, it is conceivable that a solver with effects similar to coarsening the time discretization step can be physically implemented. Given its physical relevance, in this research, we have chosen to default to a coarser time discretization step.

Interestingly, performance improvements through coarsening the time discretization step have also been reported for Ising machines utilizing not stochastic fluctuation but deterministic chaotic fluctuations [60]. It is intriguing to explore effects equivalent to such coarse-graining from both deterministic and probabilistic perspectives.

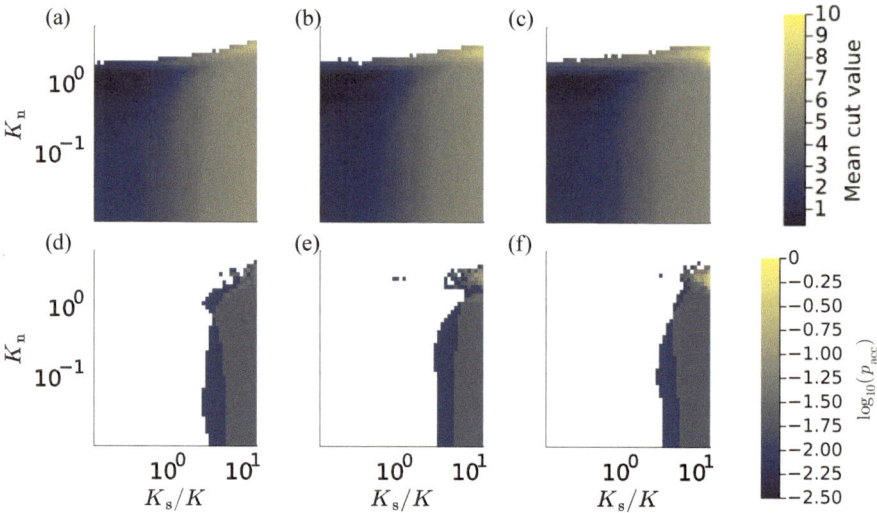

Fig. 2 (**a**) (resp. (d)) the mean cut value of the OIM (resp. acceptance probability p_{acc}) when varying the ratio of injection strength to coupling strength K_s/K and noise intensity K_n, for a time discretization step of d$t = 0.01$. (**b**) (resp. (e)) the same metrics for the OIM subjected to random pulse inputs at coarse time intervals d$\tilde{t} = 0.1$ instead of the Wiener process of the identical variance, with a fine time discretization step of d$t = 0.01$. (**c**) (resp. (f)) these measures for the OIM with a coarse time discretization step d$t = 0.1$. The horizontal and vertical axes of the color map are presented in a logarithmic scale

Table 2 Maximum mean cut value and maximum acceptance probability obtained from each simulation setting presented in Fig. 2

Timestep size	Random fluctuation	Maximum mean cut value	Maximum p_{acc}
d$t = 0.01$	Wiener process	7.8	0.11
d$t = 0.01$	Random pulses of d$\tilde{t} = 0.1$ Interval	8.7	0.24
d$t = 0.1$	Wiener process	9.3	0.48

4.4 Effect of Injection and Noise Strength

Figure 3 shows color maps of the mean cut value, the best cut value, the acceptance probability p_{acc}, and the time to solution when altering the ratio of injection strength to coupling strength K_s/K and noise intensity K_n, using the default parameter setting.

In situations where the relative injection strength is small, the best cut value does not reach the maximum cut value 10, indicating that the OIM does not converge to the ground state. Increasing the injection strength stabilizes the ground state, allowing the OIM to reach the ground state without requiring noise, if the initial condition is set within its basin of attraction (as shown in Fig. 1(c)).

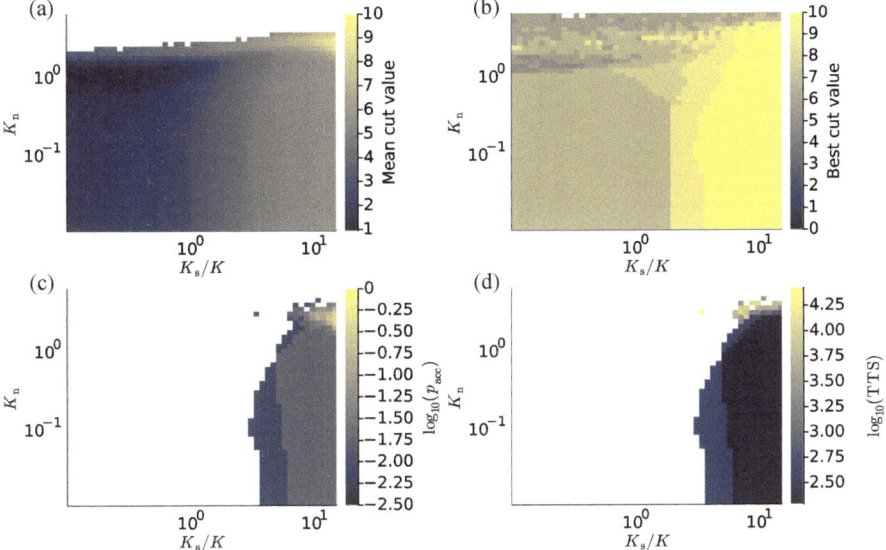

Fig. 3 Performance measures of the OIM when varying the ratio of injection strength to coupling strength K_s/K and noise intensity K_n using the default parameter setting. The horizontal and vertical axes of the color map are presented in a logarithmic scale. (**a**) Mean cut value. (**b**) Best cut value. (**c**) Acceptance probability. (**d**) Time to solution

However, once the ground state is stabilized, if the noise strength remains low, noise-driven exploration occurs infrequently (as depicted in Fig. 1(d)). Figure 3(c, d) shows that, within certain ranges of K_n, both p_{acc} and TTS show no significant improvement. There is an optimal level of noise magnitude that optimizes performance (as shown in Fig. 1(e)). Increasing the noise beyond this optimal level results in an inability to maintain the quasi-steady state, as observed in Fig. 1(f).

It should be noted that an excessively strong injection strength stabilizes all possible spin configurations [57], thereby degrading the performance of OIMs.

4.5 Effect of Higher Harmonics in Coupling and Injection Schemes

In [22], it was reported that the performance of the OIMs improves when a square wave type coupling function Γ_{ij} is used. However, there has been no comprehensive study investigating which types of coupling functions Γ_{ij} are effective, nor has there been research into the effectiveness of various injection schemes Γ.

Figure 4 shows performance metrics of the OIM when either the coupling scheme, the injection scheme, or both are implemented as square waves. The overall trend remains similar to that in Fig. 3. However, when the coupling scheme is implemented

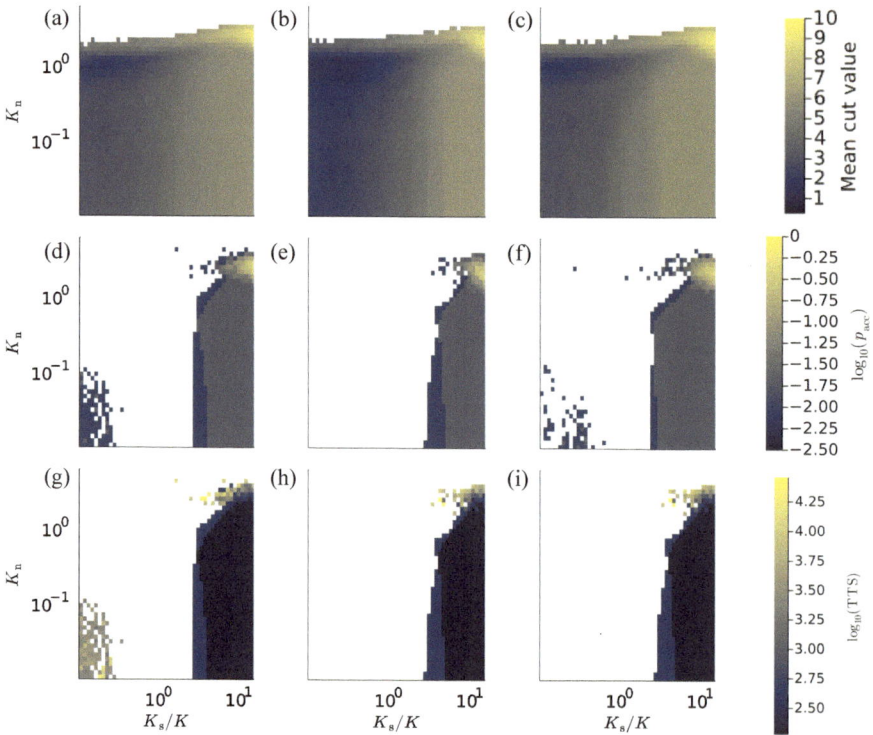

Fig. 4 (**a, d, g**) Performance metrics of the OIM with a square wave coupling scheme and second-harmonic sinusoidal injection scheme. (**b, e, h**) The same metrics for the OIM with a sinusoidal coupling scheme and square wave injection scheme. (**c, f, i**) The corresponding metrics for the OIM with both square wave coupling and injection schemes

as a square wave, it is observed that the ground state becomes stable even when the injection strength is small (as shown in Fig. 4(d, f)), and there is a significant improvement in performance metrics at optimal parameters, as shown in Table 3. In particular, TTS remains almost invariant over various magnitudes of noise intensity, and thus is largely dominated by whether the initial conditions belong to the attraction region of the ground state. However, when the coupling scheme is a square wave and the injection scheme is a sine wave, it can be observed that there is an improvement in TTS due to noise exploration, as shown in Table 3. This minimum TTS is attained at $K_s/K = 15$, $K_n = 1.43$.

To explore how optimal the square wave coupling is and what constitutes a good injection scheme, we conducted the following experiments using the parameter set $K_s/K = 15$, $K_n = 1.43$. We consider the following coupling and injection schemes:

Table 3 Optimal performance metrics obtained from each simulation setting presented in Figs. 3 and 4

Coupling scheme	Injection scheme	Maximum mean cut value	Maximum p_{acc}	Minimum TTS
Default	Default	9.33	0.48	196.5
Square wave	Default	9.48	0.61	188.5
Default	Square wave	9.2	0.44	196.5
Square wave	Square wave	9.28	0.5	196.5

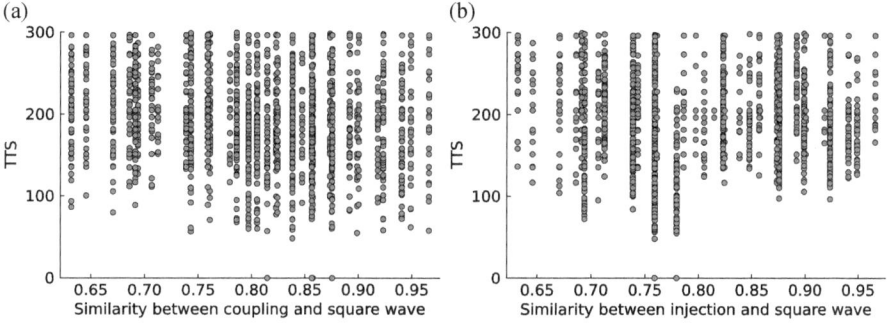

Fig. 5 TTS against the cosine similarity between a square wave and (**a**) the coupling scheme, and (**b**) the injection scheme

$$\Gamma_{ij}(\psi) = \mathcal{N}_c \left[\sin\psi + \sum_{k=2}^{5} \frac{l_k}{k} \sin k\psi \right], \quad l \in \{-1, 0, 1\}^4 , \qquad (48)$$

$$\Gamma(\psi) = \mathcal{N}_s \left[\sin 2\psi + \sum_{k=2}^{5} \frac{l_k}{2k} \sin 2k\psi \right], \quad l \in \{-1, 0, 1\}^4 , \qquad (49)$$

where $\mathcal{N}_c$, $\mathcal{N}_s$ are normalization constants to satisfy (44). Figure 5 shows TTS calculated for all combinations of the above coupling and injection schemes. The results are plotted against the cosine similarity between each scheme and a square wave. No clear correlation is observed between the similarity to a square wave and the performance of the coupling/injection scheme. Furthermore, a number of schemes demonstrate a TTS smaller than that achieved with a square wave coupling scheme, suggesting that the square wave scheme is not optimal. Most notably, we observed combinations of schemes that reached the ground state in every trial, resulting in a TTS of zero. Figure 6 shows the coupling and injection schemes that achieved a zero TTS. The results suggest that a sawtooth wave is more suitable as the coupling scheme than a square wave, and a triangular wave is effective for the injection scheme.

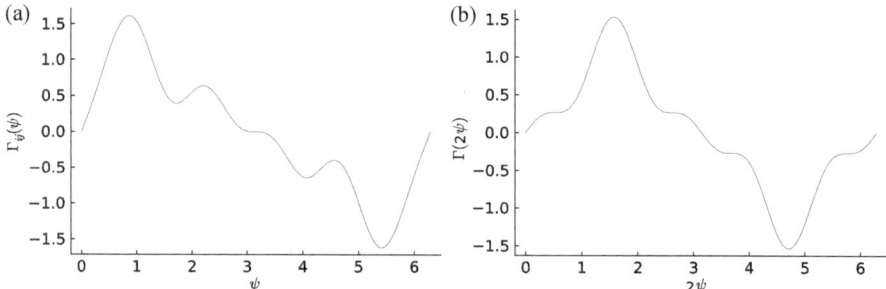

Fig. 6 (**a**) The coupling scheme, and (**b**) the injection scheme that achieved a zero TTS

5 Summary

Oscillations are ubiquitous phenomena observed across various fields of natural science and engineering. Coupled oscillator systems, manifested through diverse physical phenomena, exhibit significant information processing capabilities. These systems hold potential for the development of ultra energy efficient, high frequency, and density scalable computing architectures.

Oscillator-based Ising machines (OIMs) have shown great versatility, offering new paradigms for information processing. Although the inherent nonlinearity in spontaneously oscillating systems presents challenges in analysis and optimization, the application of phase reduction techniques can simplify the analysis and facilitate the optimization of the performance of the system.

The key to designing effective OIMs lies in several factors:

 (i) Tuning the strengths of coupling, injection, and noise.
 (ii) Designing good coupling and injection schemes: The choice of coupling and injection schemes, especially their higher harmonics, can greatly affect the performance of OIMs.
(iii) Properties of the driving stochastic process: The choice of the stochastic process, beyond the Wiener process, can have a significant impact on the performance of the OIM.

Related to (iii), exploring the physical implementations of performance-enhancing effects, which emerge from the coarse-graining of time discretization, in both deterministic and probabilistic aspects, presents an intriguing research direction. While not discussed in this Chapter, it is known that heterogeneity in the frequency of oscillators can degrade the performance of OIMs [22]. Additionally, performing appropriate annealing is also important [22, 24]. These factors highlight the complexity of designing effective OIMs and the need for a comprehensive approach that considers all these aspects.

These advancements together form the foundation for further improvements and innovations in the development of efficient computing architectures in a versatile manner using coupled oscillator systems.

Acknowledgements The author appreciates valuable comments from Dr. Ken-ichi Okubo.

References

1. G.E. Moore, Cramming more components onto integrated circuits. Proc. IEEE **86**(1), 82–85 (1998)
2. J. Shalf, The future of computing beyond Moore's law. Philos. Trans. R. Soc. A: Math., Phys. Eng. Sci. **378**(2166) (2020)
3. M.J. Schuetz, J.K. Brubaker, H.G. Katzgraber, Combinatorial optimization with physics-inspired graph neural networks. Nat. Mach. Intell. **4**(4), 367–377 (2022)
4. Y. Boykov, O. Veksler, *Graph Cuts in Vision and Graphics: Theories and Applications* (Springer, Berlin, 2006), pp. 79–96
5. F. Barahona, M. Grötschel, M. Jünger, G. Reinelt, An application of combinatorial optimization to statistical physics and circuit layout design. Oper. Res. **36**(3), 493–513 (1988)
6. B. Korte, J. Vygen, *Combinatorial Optimization: Theory and Algorithms* (Springer, Berlin, 2012)
7. S. Basu, R.E. Bryant, G. De Micheli, T. Theis, L. Whitman, Nonsilicon, non-von Neumann computing-part I [scanning the issue]. Proc. IEEE **107**(1), 11–18 (2019)
8. S. Basu, R.E. Bryant, G. De Micheli, T. Theis, L. Whitman, Nonsilicon, non-von Neumann computing-part II. Proc. IEEE **108**(8), 1211–1218 (2020)
9. N. Mohseni, P.L. McMahon, T. Byrnes, Ising machines as hardware solvers of combinatorial optimization problems. Nat. Rev. Phys. **4**(6), 363–379 (2022)
10. F. Arute, K. Arya, R. Babbush, D. Bacon, J.C. Bardin, R. Barends, R. Biswas, S. Boixo, F.G. Brandao, D.A. Buell et al., Quantum supremacy using a programmable superconducting processor. Nature **574**(7779), 505–510 (2019)
11. T. Inagaki, Y. Haribara, K. Igarashi, T. Sonobe, S. Tamate, T. Honjo, A. Marandi, P.L. McMahon, T. Umeki, K. Enbutsu et al., A coherent Ising machine for 2000-node optimization problems. Science **354**(6312), 603–606 (2016)
12. S. Tsukamoto, M. Takatsu, S. Matsubara, H. Tamura, An accelerator architecture for combinatorial optimization problems. Fujitsu Sci. Tech. J. **53**(5), 8–13 (2017)
13. F. Cai, S. Kumar, T. Van Vaerenbergh, X. Sheng, R. Liu, C. Li, Z. Liu, M. Foltin, S. Yu, Q. Xia et al., Power-efficient combinatorial optimization using intrinsic noise in memristor Hopfield neural networks. Nat. Electron. **3**(7), 409–418 (2020)
14. D. Pierangeli, G. Marcucci, C. Conti, Large-scale photonic Ising machine by spatial light modulation. Phys. Rev. Lett. **122**(21), 213902 (2019)
15. H. Yamashita, K. Okubo, S. Shimomura, Y. Ogura, J. Tanida, H. Suzuki, Low-rank combinatorial optimization and statistical learning by spatial photonic Ising machine. Phys. Rev. Lett. **131**(6), 063801 (2023)
16. A. Pikovsky, M. Rosenblum, J. Kurths, *Synchronization: A Universal Concept in Nonlinear Sciences, Cambridge Nonlinear Science Series* (Cambridge University Press, Cambridge, 2001)
17. Y. Kuramoto, Y. Kawamura, *Science of Synchronization: Phase Description Approach (in Japanese)* (Kyoto University Press, 2017)
18. G. Csaba, W. Porod, Coupled oscillators for computing: A review and perspective. Appl. Phys. Rev. **7**(1), 011302 (2020)
19. A. Raychowdhury, A. Parihar, G.H. Smith, V. Narayanan, G. Csaba, M. Jerry, W. Porod, S. Datta, Computing with networks of oscillatory dynamical systems. Proc. IEEE **107**(1), 73–89 (2019)

20. J. von Neumann, Non-linear capacitance or inductance switching, amplifying, and memory organs, US Patent 2,815,488 (1957)
21. E. Goto, The parametron, a digital computing element which utilizes parametric oscillation. Proc. IRE **47**(8), 1304–1316 (1959)
22. T. Wang, J. Roychowdhury, OIM: Oscillator-based Ising machines for solving combinatorial optimisation problems, in *Unconventional Computation and Natural Computation*. ed. by I. McQuillan, S. Seki (Springer, Berlin, 2019), pp.232–256
23. Y. Zhang, Y. Deng, Y. Lin, Y. Jiang, Y. Dong, X. Chen, G. Wang, D. Shang, Q. Wang, H. Yu, Z. Wang, Oscillator-network-based Ising machine. Micromachines **13**(7), 1016 (2022)
24. S. Dutta, A. Khanna, A. Assoa, H. Paik, D.G. Schlom, Z. Toroczkai, A. Raychowdhury, S. Datta, An Ising hamiltonian solver based on coupled stochastic phase-transition nano-oscillators. Nat. Electron. **4**(7), 502–512 (2021)
25. D.I. Albertsson, M. Zahedinejad, A. Houshang, R. Khymyn, J. Åkerman, A. Rusu, Ultrafast Ising machines using spin torque nano-oscillators. Appl. Phys. Lett. **118**(11), 112404 (2021)
26. B.C. McGoldrick, J.Z. Sun, L. Liu, Ising machine based on electrically coupled spin Hall nano-oscillators. Phys. Rev. Appl. **17**(1), 014006 (2022)
27. E. Ising, Beitrag zur theorie des ferromagnetismus. Zeitschrift für Physik **31**, 253–258 (1925)
28. A. Lucas, Ising formulations of many NP problems. Front. Phys. **2** (2014)
29. C. Gardiner, *Stochastic Methods* (Springer, Berlin, 2009)
30. J.J. Hopfield, D.W. Tank, "Neural" computation of decisions in optimization problems. Biol. Cybern. **52**(3), 141–152 (1985)
31. M. Ercsey-Ravasz, Z. Toroczkai, Optimization hardness as transient chaos in an analog approach to constraint satisfaction. Nat. Phys. **7**(12), 966–970 (2011)
32. H. Suzuki, J.-I. Imura, Y. Horio, K. Aihara, Chaotic Boltzmann machines. Sci. Rep. **3**(1), 1–5 (2013)
33. J. Guckenheimer, Isochrons and phaseless sets. J. Math. Biol. **1**, 259–273 (1975)
34. B. Ermentrout, D.H. Terman, *Mathematical Foundations of Neuroscience* (Springer, Berlin, 2010)
35. Y. Kuramoto, *Chemical Oscillations, Waves, and Turbulence* (Springer, Berlin, 2012)
36. A.T. Winfree, *The Geometry of Biological Time* (Springer, Berlin, 1980)
37. A. Demir, J. Roychowdhury, A reliable and efficient procedure for oscillator PPV computation, with phase noise macromodeling applications. IEEE Trans. Comput.-Aided Des. Integr. Circuits Syst. **22**(2), 188–197 (2003)
38. B. Ermentrout, Type I membranes, phase resetting curves, and synchrony. Neural Comput. **8**(5), 979–1001 (1996)
39. E. Brown, J. Moehlis, P. Holmes, On the phase reduction and response dynamics of neural oscillator populations. Neural Comput. **16**(4), 673–715 (2004)
40. H. Nakao, Phase reduction approach to synchronisation of nonlinear oscillators. Contemp. Phys. **57**(2), 188–214 (2016)
41. V. Novičenko, K. Pyragas, Computation of phase response curves via a direct method adapted to infinitesimal perturbations. Nonlinear Dyn. **67**, 517–526 (2012)
42. M. Iima, Jacobian-free algorithm to calculate the phase sensitivity function in the phase reduction theory and its applications to Kármán's vortex street. Phys. Rev. E **99**(6), 062203 (2019)
43. R.F. Galán, G.B. Ermentrout, N.N. Urban, Efficient estimation of phase-resetting curves in real neurons and its significance for neural-network modeling. Phys. Rev. Lett. **94**(15), 158101 (2005)
44. K. Ota, M. Nomura, T. Aoyagi, Weighted spike-triggered average of a fluctuating stimulus yielding the phase response curve. Phys. Rev. Lett. **103**(2), 024101 (2009)
45. K. Nakae, Y. Iba, Y. Tsubo, T. Fukai, T. Aoyagi, Bayesian estimation of phase response curves. Neural Netw. **23**(6), 752–763 (2010)
46. R. Cestnik, M. Rosenblum, Inferring the phase response curve from observation of a continuously perturbed oscillator. Sci. Rep. **8**(1), 13606 (2018)
47. N. Namura, S. Takata, K. Yamaguchi, R. Kobayashi, H. Nakao, Estimating asymptotic phase and amplitude functions of limit-cycle oscillators from time series data. Phys. Rev. E **106**(1), 014204 (2022)

48. A. Mauroy, I. Mezić, J. Moehlis, Isostables, isochrons, and Koopman spectrum for the action-angle representation of stable fixed point dynamics. Phys. D: Nonlinear Phenom. **261**, 19–30 (2013)
49. A. Mauroy, Y. Susuki, I. Mezić, *Koopman Operator in Systems and Control* (Springer, Berlin, 2020)
50. J.P. Keener, *Principles of Applied Mathematics: Transformation and Approximation* (CRC Press, 2019)
51. S. Shinomoto, Y. Kuramoto, Phase transitions in active rotator systems. Prog. Theor. Phys. **75**(5), 1105–1110 (1986)
52. S.H. Park, S. Kim, Noise-induced phase transitions in globally coupled active rotators. Phys. Rev. E **53**(4), 3425 (1996)
53. J.A. Acebrón, L.L. Bonilla, C.J.P. Vicente, F. Ritort, R. Spigler, The Kuramoto model: a simple paradigm for synchronization phenomena. Rev. Mod. Phys. **77**(1), 137 (2005)
54. E. Wong, M. Zakai, On the convergence of ordinary integrals to stochastic integrals. Ann. Math. Stat. **36**(5), 1560–1564 (1965)
55. M. Garey, D. Johnson, L. Stockmeyer, Some simplified NP-complete graph problems. Theor. Comput. Sci. **1**(3), 237–267 (1976)
56. K.P. Kalinin, N.G. Berloff, Computational complexity continuum within Ising formulation of NP problems. Commun. Phys. **5**(1), 20 (2022)
57. M. Erementchouk, A. Shukla, P. Mazumder, On computational capabilities of Ising machines based on nonlinear oscillators. Phys. D: Nonlinear Phenom. **437**, 133334 (2022)
58. S. Steinerberger, Max-Cut via Kuramoto-type oscillators. SIAM J. Appl. Dyn. Syst. **22**(2), 730–743 (2023)
59. P.E. Kloeden, E. Platen, H. Schurz, *Numerical Solution of Sde Through Computer Experiments* (Springer, Berlin, 2002)
60. H. Yamashita, K. Aihara, H. Suzuki, Accelerating numerical simulation of continuous-time Boolean satisfiability solver using discrete gradient. Commun. Nonlinear Sci. Numer. Simul. **102**, 105908 (2021)

Sampling-Like Dynamics of the Nonlinear Dynamical System Combined with Optimization

Hiroshi Yamashita

Abstract When considering computation using physical phenomena beyond established digital circuits, the variability of the device must be addressed. In this chapter, we will focus on random sampling for its algorithmic solution. In particular, we discuss the nonlinear dynamical system that achieves the sampling behavior. The system, called herding, is proposed as an algorithm that can be used in the same manner as Monte Carlo integration. The algorithm combines optimization methods in the system and does not depend on random number generators. In this chapter, we review this algorithm using nonlinear dynamics and related studies, including the author's previous results. Then we discuss the perspective of the application of herding in relation to the use of physical phenomena in computation.

As mentioned previously, the exploration of the use of physical phenomena in computation, not limited to digital electronics, is expected to achieve breakthroughs recently. The problem with computing with such physical phenomena is the variability of the device. In other words, physical devices generally have a limit to their accuracy in both fabrication and operation, and this is particularly problematic when the physical quantities are used as analog variables in the computation, as opposed to digital electronics. For example, physical reservoir computing, as discussed in earlier chapters, can be considered a method to deal with this problem. Owing to the physical limitations mentioned above, it may be difficult to tune the parameters of the physical reservoir for each; however, this problem can be addressed by training only a simple single-layer neural network inserted between the reservoir and the output. Similarly, in the case of the spatial photonic Ising machine (SPIM) [1], although the spatial light modulator (SLM) takes digital inputs, the computation is analog and the output can be degraded by the light propagation. When considering computations beyond the established digital circuits, this variability must be dealt with; in particular, an algorithmic solution is required.

H. Yamashita (✉)
Graduate School of Information Science and Technology, Osaka University, 1-5 Yamadaoka, Suita-shi, Osaka, Japan
e-mail: h.yamashita@ist.osaka-u.ac.jp

© The Author(s) 2024
H. Suzuki et al. (eds.), *Photonic Neural Networks with Spatiotemporal Dynamics*,
https://doi.org/10.1007/978-981-99-5072-0_10

In this chapter, we will focus on random sampling for this purpose. In statistics and machine learning, Monte Carlo (MC) integration is often used to compute expectations. The problem is to determine the expected value of a function $f(x)$ over a probability distribution p denoted by

$$\mathbb{E}_p[f(x)] = \int f(x)p(x)\mathrm{d}x .$$

(1)

In MC integration, a sequence of samples $x^{(1)}, x^{(2)}, \ldots, x^{(T)}$ that follow the probability distribution p is generated, and then the expectation is approximated as the sample average:

$$\mathbb{E}_p[f(x)] \simeq \frac{1}{T} \sum_{t=1}^{T} f(x^{(t)}) .$$

(2)

For example, MC integration is important for Bayesian inferences. When inferring the parameters θ of a probabilistic model $p(x; \theta)$ from the data $\mathcal{D}$, we often use the maximum likelihood method taking θ as an estimate that maximizes the likelihood $P(\mathcal{D}; \theta)$ of the data. Instead, Bayesian inference introduces a prior distribution $\pi(\theta)$ for the parameters and applies Bayes' rule to obtain the posterior distribution $P(\theta) \propto \pi(\theta)P(\mathcal{D}; \theta)$ as the inference result. The advantage of assuming distributions in the parameters is that estimates with uncertainty can be obtained.

Because the variability in physical phenomena is often understood as a stochastic behavior, this sampling and MC integration is a promising application of physical phenomena in computation. Usually, a random number generator (RNG) is used for sampling. Typically, a pseudo-RNG is used to generate a sequence of numbers using deterministic computations; however, physical random numbers can also be used. In either case, the RNG must be precisely designed to guarantee the quality of the output. Although the use of stochastic behavior observed in physical phenomena is promising, it is also difficult to precisely control its probabilistic properties.

Herding, proposed by [2, 3], is an algorithm that can be used in the same manner as MC integration. However, it does not use RNGs, so we can expect it to be a possible method for avoiding such difficulties. In this chapter, we review the herding algorithm and its related studies, including the author's previous results. Then we also discuss the prospects of studying herding from perspectives that include the use of physical phenomena in computation.

The remainder of this chapter is organized as follows. In Sects. 1 and 2, we introduce herding, focusing on its aspects as a method of MC integration. In particular, we introduce the herding algorithm in Sect. 1. In Sect. 2, the herded Gibbs algorithm is introduced as an application of herding. We discuss the improvement of the estimation error and its convergence and review the relevant literature. In Sect. 3, we consider herding from a different perspective, the maximum entropy principle. We explore the relationship between herding and entropy, including the role of high-dimensional nonlinear dynamics in herding. In Sect. 4, based on the discussion in the previous sections, we discuss the prospects of the study of herding, including the aspect of computation with the variability of physical phenomena, and finally, we provide concluding remarks in Sect. 5.

1 Herding: Sample Generation Using Nonlinear Dynamics and Monte Carlo Integration

In this section, we introduce the herding algorithm. We adopt a different approach to the introduction than that in the original work [2, 3].

1.1 Deterministic Tossup

We begin with a simple example. Consider a sequence of T random variables $x^{(1)}, \ldots, x^{(T)} \in \mathbb{R}$. Let $\mathbb{E}_p[\cdot]$ and $\mathrm{Var}_p[\cdot]$ denote the expectation and variance, respectively, of a function over a probability distribution p. Assume that each marginal distribution is equivalent to the distribution p whose expected value is $\mathbb{E}_p[x] = \mu$ and whose variance is $\mathrm{Var}_p[x] = V$. Then, the expected value of the sample mean

$$\hat{\mu} = \frac{1}{T} \sum_{t=1}^{T} x^{(t)} \tag{3}$$

satisfies $\mathbb{E}[\hat{\mu}] = \mu$ and the variance is calculated as

$$
\begin{aligned}
\mathrm{Var}\left[\hat{\mu}\right] &= \mathbb{E}\left[\left(\frac{1}{T} \sum_{t=1}^{T} (x^{(t)} - \mu) \right)^2 \right] \\
&= \frac{1}{T^2} \sum_{t=1}^{T} \sum_{t'=1}^{T} \mathbb{E}[(x^{(t)} - \mu)(x^{(t')} - \mu)] \\
&= \frac{1}{T^2} \left(\sum_{t=1}^{T} \mathbb{E}[(x^{(t)} - \mu)^2] + 2 \sum_{k=1}^{T-1} \sum_{t=1}^{T-k} \mathbb{E}[(x^{(t)} - \mu)(x^{(t+k)} - \mu)] \right) .
\end{aligned}
\tag{4}
$$

Assume that $x^{(1)}, \ldots, x^{(T)}$ are not independently and identically distributed (i.i.d.) and have a stationary autocorrelation as

$$\mathbb{E}[(x^{(t)} - \mu)(x^{(t+k)} - \mu)] = V \rho_k , \tag{5}$$

then the variance is calculated as

$$\text{Var}\left[\hat{\mu}\right] = \frac{V}{T}\left(1 + 2\sum_{k=1}^{T-1}\left(1 - \frac{k}{T}\right)\rho_k\right). \tag{6}$$

Considering these $x^{(t)}$ as the samples in MC integration, this suggests that the variance of the estimate is reduced when the number of samples T is large, and it can be further reduced when the autocorrelation ρ_k is small or especially negative.

Consider a distribution of the binary random variable $x \sim p_\pi$ with parameter π, representing a tossup that returns 1 with probability π and 0 with probability $1 - \pi$. This distribution is the well-known Bernoulli distribution with the parameter π. Instead of generating a sequence of i.i.d. samples from p_π, we consider generating a sequence that follows the deterministic process presented in Algorithm 1 and Fig. 1. According to the update rule (line 3) of the algorithm, $w^{(t)}$ is expected to be smaller after the step if $x^{(t)} = 1$, and $w^{(t)}$ is expected to be larger after the step if $x^{(t)} = 0$. Therefore, from the equation in line 2, we expect that this sequence will have negative autocorrelations.

Although the above arguments concern only ρ_1, in fact the sample mean $\hat{\mu}$ converges very quickly as $O(1/T)$. Let $C = [-1 + \pi, \pi)$ and assume that the initial value $w^{(0)}$ is taken from C. Then, one can easily check that $w^{(t)} \in C$ holds for all t, for example, from that we can change the variable $w^{(t)}$ to $w'^{(t)}$ to obtain an equivalent system including the rotation of the circle represented as

$$x^{(t)} = \begin{cases} 1 & (w'^{(t-1)} \le \pi) \\ 0 & (w'^{(t-1)} > \pi) \end{cases}, \tag{7}$$

$$w'^{(t)} = w'^{(t-1)} + \pi \bmod 1. \tag{8}$$

Adding the update formula (line 3) with respect to $t = 1, \ldots, T$, we obtain

$$w^{(T)} - w^{(0)} = T\pi - \sum_{t=1}^{T} x^{(t)}. \tag{9}$$

Then we evaluate the difference in average as

Algorithm 1 Deterministic tossup (herding for Bernoulli distribution)

1: **for** $t = 1, \ldots, T$ **do**

2:     $x^{(t)} = \begin{cases} 1 & (w^{(t-1)} \ge 0) \\ 0 & (w^{(t-1)} < 0) \end{cases}$

3:     $w^{(t)} = w^{(t-1)} + \pi - x^{(t)}$

4: **end for**

5: Output $x^{(1)}, \ldots, x^{(T)}$

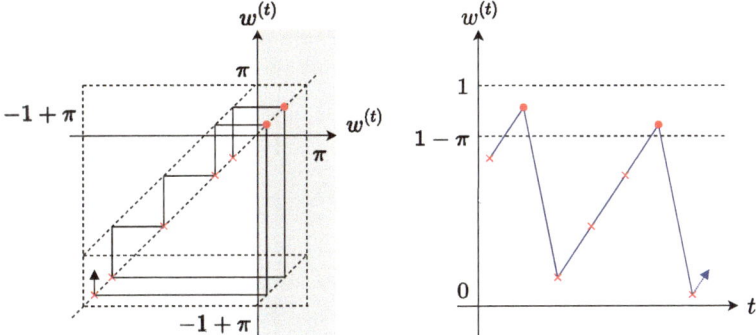

Fig. 1 An example trajectory of the deterministic tossup (herding for Bernoulli distribution). In the left panel, off-diagonal dashed lines represent the maps used to update $w^{(t)}$. The points on the diagonal dashed line indicate the $w^{(t)}$ for each t. The circles represent $x^{(t+1)} = 1$ and crosses represent $x^{(t+1)} = 0$. The gray area represents the condition for $w^{(t)}$ to output $x^{(t+1)} = 1$. The right panel represents the trajectory as the function of t

$$|\pi - \hat{\mu}| = \frac{1}{T}|w^{(T)} - w^{(0)}| \,, \tag{10}$$

which converges as $O(1/T)$ because $w^{(0)}, w^{(T)} \in C$.

1.2   Herding

We then generalize the deterministic tossup above to introduce herding. Let $\mathcal{X}$ be the sample space and $\varphi_m : \mathcal{X} \to \mathbb{R}$ for $m \in \mathcal{M}$ be the feature functions defined therein, where $\mathcal{M}$ is the set of their indices. Suppose that we have parameters $\mu_m \in \mathbb{R}$ that specify the target moment values of φ_m over output samples. The empirical moment values can be used as parameters if the dataset is available. We can also consider a situation in which only the aggregated observation is available, e.g., for privacy reasons or because the subject is microscopic so that individual observation is impossible. The problem considered here is the reconstruction of the distribution such that the expected value of the feature $\mathbb{E}_\pi[\varphi_m(x)]$ is equal to the given moment μ_m. Herding is an algorithm proposed by Welling [2], which generates a sequence of samples $x^{(1)}, x^{(2)}, \ldots$ that are diverse while satisfying the moment condition.

For ease of notation, we denote by $\varphi(x)$ and μ the vectors of features and moments, respectively. We denote by $\langle \cdot, \cdot \rangle$ the inner product of vectors and by $\|x\| \equiv \sqrt{\langle x, x \rangle}$ the square norm. The overall herding algorithm is presented in Algorithm 2. We refer to $w^{(t)}$ of the algorithm as the weight vector. As discussed below, herding is an extension of the above deterministic tossup where the sample mean of φ converges to μ. We can also consider the algorithm as a nonlinear dynamical system with the discrete-time variable t. As shown in Algorithm 2, the system includes optimization

---

Algorithm 2 Herding algorithm

1: **for** $t = 1, \ldots, T$ **do**
2: $\quad x^{(t)} = \arg \max_{x \in \mathcal{X}} \langle \boldsymbol{w}^{(t-1)}, \boldsymbol{\varphi}(x) \rangle$
3: $\quad \boldsymbol{w}^{(t)} = \boldsymbol{w}^{(t-1)} + \boldsymbol{\mu} - \boldsymbol{\varphi}(x^{(t)})$
4: **end for**
5: Output $x^{(1)}, \ldots, x^{(T)}$

---

problem as a component. If we obtain the unique solution to the optimization problem, then updating the weight vector is deterministic. However, the system behaves in a random or chaotic manner. This random-like behavior can generate a pseudo-random sequence of samples while satisfying the moment conditions. Thus, the generated sequence that reflects the given information can be used as samples drawn from the background distribution. We discuss the importance of the weight vector behavior later.

There are many studies related to herding; for example, it is extended to include hidden variables through Markov probability fields and is applied to data compression by learning the rate coefficients of the dynamical system of herding [4]. It is also combined with the kernel trick to sample from a continuous distribution [5]. However, these are not discussed in detail here.

1.3 Convergence of Herding

Let us consider the herding algorithm for the Bernoulli distribution. In this situation, the sample space is $\mathcal{X} = \{0, 1\}$ and the features are $\boldsymbol{\varphi}(x) = (\phi_0(x), \phi_1(x))^\top = (1 - x, x)^\top$, whose moments are $\boldsymbol{\mu} = (1 - \pi, \pi)^\top$. The herding algorithm in this situation is equivalent to

$$x^{(t)} = \begin{cases} 1 & (f^{(t-1)}(1) \geq f^{(t-1)}(0)) \\ 0 & (f^{(t-1)}(1) < f^{(t-1)}(0)) \end{cases}, \tag{11}$$

$$w_0^{(t)} = w_0^{(t-1)} + (1 - \pi) - (1 - x)$$
$$= w_0^{(t-1)} - \pi + x, \tag{12}$$

$$w_1^{(t)} = w_1^{(t-1)} + \pi - x, \tag{13}$$

where $f^{(t-1)}(x) \equiv \langle \boldsymbol{w}^{(t-1)}, \boldsymbol{\phi}(x) \rangle = w_0^{(t-1)} + (w_1^{(t-1)} - w_0^{(t-1)})x$. This updated formula corresponds to the deterministic tossup (Algorithm 1), with the relation $(w_0^{(t)}, w_1^{(t)}) = (-w^{(t)}, w^{(t)})$; the two algorithms become equivalent if the initial value satisfies $w_0^{(t)} = -w_1^{(t)}$.

Generalizing and using this relation conversely, we can perform an analysis similar to that of the deterministic tossup in the general case. Suppose $x^{(t)}$ is chosen such that $\varphi_m(x^{(t)})$ is small at step t and the target value is $\varphi_m(x^{(t)}) < \mu_m$. In this case, the corresponding weight $w_m^{(t)}$ increases as

$$w^{(t)} = w^{(t-1)} + \mu - \varphi(x^{(t)}) , \tag{14}$$

which corresponds to line 3 of Algorithm 2. Therefore, φ_m becomes more important in the optimization step and $\varphi_m(x^{(t+1)})$ is expected to be larger in the next step. Similarly, $\varphi_m(x^{(t+1)})$ is expected to be smaller when $\varphi_m(x^{(t)})$ is large. In other words, the sequence $\varphi_m(x^{(t)})$ is expected to have negative autocorrelation, which makes its average converge to μ_m faster.

In addition, we obtain a convergence result for herding similar to that of the deterministic tossup. In particular, the average of the features for the generated sequence converges as follows:

$$\frac{1}{T} \sum_{t=1}^{T} \varphi(x^{(t)}) \to \mu , \tag{15}$$

where the convergence rate is $O(1/T)$. This convergence is obtained by using the boundedness of $w^{(T)}$ and equation

$$w^{(T)} - w^{(0)} = T\mu - \sum_{t=1}^{T} \varphi(x^{(t)}) . \tag{16}$$

This equation is obtained by summing both sides of (14) for $t = 1, \ldots, T$.

Using the optimality for the optimization problem

$$\arg\max_{x \in \mathcal{X}} \langle w^{(t-1)}, \varphi(x) \rangle , \tag{17}$$

we can guarantee the boundedness of $w^{(T)}$ as the following summarized version of the proof in [2]: Let $R^{(t)} = \|w^{(t)}\|$. The change in the value of R is calculated as follows:

$$\begin{aligned} (R^{(t)})^2 &= \|w^{(t)}\|^2 \\ &= \|w^{(t-1)} - \varphi(x^{(t)}) + \mu\|^2 \\ &= (R^{(t-1)})^2 - 2\langle w^{(t-1)}, \varphi(x^{(t)}) \rangle + 2\langle w^{(t-1)}, \mu \rangle + \|\varphi(x^{(t)}) - \mu\|^2 . \end{aligned} \tag{18}$$

Note that $x^{(t)}$ is obtained by the optimization problem in (17); we define the following quantity

$$A = \min_{\|\tilde{w}\|=1} \left(\langle \tilde{w}, \varphi(\hat{x}) \rangle - \langle \tilde{w}, \mu \rangle \right) , \tag{19}$$

where $\hat{x}$ is defined as $\hat{x} = \arg\max_{x}\langle\tilde{w}, \varphi(x)\rangle$. From the optimality of $\hat{x}$, we obtain $A > 0$ under a mild assumption of μ. We also assume that the third term in (18) is bounded; that is, $\|\varphi(x^{(t)}) - \mu\|^2 \leq B$ holds for some $B \in \mathbb{R}$. This holds if the functions φ are bounded. Then, we obtain

$$(R^{(t)})^2 \leq (R^{(t-1)})^2 - 2R^{(t-1)}A + B . \tag{20}$$

In other words, if $R^{(t-1)} > R$ holds for $R \equiv B/2A$, then $R^{(t)}$ is decreasing; $R^{(t)} < R^{(t-1)}$. On the other hand, if $R^{(t-1)} \leq R$, then $R^{(t)}$ is bounded. Thus, there are $R' \in \mathbb{R}$ such that $R^{(t)} \leq R'$ always holds if $R^{(0)} \leq R'$.

2 Herded Gibbs: Model-Based Herding on Spin System

Herding typically requires the feature moments μ as the inputs. However, in sampling applications, the probability model π is often available instead. In this section, we describe herded Gibbs (HG), an extension of herding that can be applied to such model-based situations.

2.1 Markov Chain Monte Carlo Method

Sampling independently from a general probability distribution π is difficult in general; thus the Markov chain Monte Carlo (MCMC) method is often used. In MCMC, we consider the state variable $x^{(t)}$ that is repeatedly updated to be the next state $x^{(t+1)}$, where this transition is represented as a Markov chain. By appropriately designing this transition, the probability distribution of $x^{(t)}$ converges to π. In addition, the two samples $x^{(t)}$ and $x^{(t+T)}$ become nearly independent for sufficiently large T; the measure of such a time difference T is called the mixing time. If the transitions are designed such that the mixing time is short, then the MCMC becomes more effective for MC integration.

MCMC transitions are often designed to be local, i.e., the state $x^{(t)}$ moves only in its neighborhood at each step. In such a case, the mixing time strongly depends on the energy landscape of probability distribution π to be sampled. The energy landscape is represented as the graph of $E(x)$, where the probability distribution is represented by the Gibbs-Boltzmann distribution, $\pi(x) \propto \exp(-E(x))$. In particular, if there are large energy barriers in the phase space, the mixing time increases because it takes time for $x^{(t)}$ to pass over them.

A commonly used method for designing MCMC is the Metropolis-Hastings method [6, 7], and it is a good example of a local transition. In the Metropolis-Hastings method, given the current state x, the state y is generated according to the proposal distribution $q_x(y)$. Typically, it is designed so that y is in the neighborhood of x. The proposed state y is accepted as the next state with an acceptance probability $\alpha \in [0, 1]$ and is rejected with probability $1 - \alpha$. If rejected, state x is kept as the next state. If the proposal distribution is symmetric, i.e., $q_x(y) = q_y(x) \ \forall x, y$, and the acceptance probability is defined as $\alpha = \min(1, \exp(-E(y))/\exp(-E(x)))$, the chain has $\pi(x) \propto \exp(-E(x))$ as its stationary distribution. Transition probabilities are defined such that a transition to y with higher energy $E(y)$ has a lower acceptance probability.

Gibbs sampling [8] is another type of MCMC method. Let us denote by $p(\cdot \mid \cdot)$ the conditional distribution for a distribution p. Suppose π is a probability distribution of N variables $x_1, \ldots, x_N$, and we can compute the conditional distribution $\pi(x_i \mid \boldsymbol{x}_{-i})$ for each i, where $\boldsymbol{x}_{-i}$ is the vector of all variables except x_i. Then, we can construct a Markov chain as presented in Algorithm 3.

Algorithm 3 Gibbs sampling

1: **for** $t := 1, \ldots, T$ **do**
2:     **for** $i := 1, \ldots, N$ **do**
3:         Update x_i by sampling from $\pi(x_i \mid \boldsymbol{x}_{-i})$
4:         Maintain other variables x_j for $j \neq i$
5:     **end for**
6:             ▷ The above can be repeated as many times as the mixing time to reduce correlation between samples
7:     Output current values $(x_1, \ldots, x_N)$
8: **end for**

In particular, in this chapter, we consider the following spin system: Let us consider N binary random variables, $x_1, \ldots, x_N \in \{0, 1\}$. Suppose that the random variables $\mathcal{V} = \{x_1, \ldots, x_N\}$ form a network $G = (\mathcal{V}, \mathcal{E})$, where $\mathcal{E}$ denotes the set of edges, and the joint distribution π is represented as

$$\pi(\boldsymbol{x}) = \frac{1}{Z} \exp(-E(\boldsymbol{x})), \tag{21}$$

$$E(\boldsymbol{x}) = -\sum_{(i,j)\in\mathcal{E}} W_{ij} x_i x_j - \sum_{i=1}^{N} b_i x_i, \tag{22}$$

where Z is the normalization factor that makes the sum of the probabilities equal to one;

$$Z = \sum_{x} \exp(-E(\boldsymbol{x})). \tag{23}$$

This probability model is widely known as the Boltzmann machine (BM), which is also discussed in Chap. 2.

For simplicity, let us suppose that there is no bias term, such that

$$E(\boldsymbol{x}) = - \sum_{(i,j)\in\mathcal{E}} W_{ij} x_i x_j \,. \tag{24}$$

Let $\mathcal{N}(i)$ be the index set of neighboring variables of x_i on G and $\boldsymbol{x}_{\mathcal{N}(i)}$ be the vector of the corresponding variables. For the BM, the conditional distribution can be easily obtained as

$$\pi(x_i \mid \boldsymbol{x}_{-i}) = \pi(x_i \mid \boldsymbol{x}_{\mathcal{N}(i)}) = \frac{1}{Z_i} \exp\left(x_i \left(\sum_{j\in\mathcal{N}(i)} W_{ij} x_j \right) \right), \tag{25}$$

where Z_i is also the normalization factor. Therefore, sampling from this distribution can be easily performed using Gibbs sampling.

2.2 Herded Gibbs

Herding can be used as a deterministic sampling algorithm, but it cannot be applied directly to the BM because the input parameter $\boldsymbol{\mu}$ is typically unavailable. By combining Gibbs sampling with herding, a deterministic sampling method for BMs called the herded Gibbs (HG) algorithm [9] is obtained. The structure of the algorithm is the same as that of Gibbs sampling, but the random update step for each variable x_i is replaced by herding, as presented in Algorithm 4. The variable neighborhood $\boldsymbol{x}_{\mathcal{N}(i)}$ of x_i has $2^{|\mathcal{N}(i)|}$ configurations, so let us give indices for such configurations. HG uses weight variables, denoted by $w_{i,j}$ for the ith variable x_i and the j-th configuration of the variable neighborhood $\boldsymbol{x}_{\mathcal{N}(i)}$. We can compute the corresponding conditional probability given that $\boldsymbol{x}_{\mathcal{N}(i)}$ takes the jth state denoted by $\pi_{i,j} \equiv \pi(x_i = 1 \mid \boldsymbol{x}_{\mathcal{N}(i)} = j)$.

Algorithm 4 Herded Gibbs

1: **for** $t := 1, \ldots, T$ **do**
2:    **for** $i := 1, \ldots, N$ **do**
3:       Let j be the index of the configuration of $\boldsymbol{x}_{\mathcal{N}(i)}$
4:       Compute the conditional probability $\pi_{i,j} = \pi(x_i = 1 \mid \boldsymbol{x}_{\mathcal{N}(i)} = j)$
5:       Update $x_i \leftarrow \begin{cases} 1 & (w_{i,j} < \pi_{i,j}) \\ 0 & (w_{i,j} \geq \pi_{i,j}) \end{cases}$
6:       $w_{i,j} \leftarrow w_{i,j} + \pi_{i,j} \bmod 1$
7:                         ▷ applying the equivalent version of herding ((7), (8)) to x_i and $w_{i,j}$
8:       Maintain other variables $x_{i'}$ for $i' \neq i$ and weights $w_{i',j'}$ for $(i', j') \neq (i, j)$
9:    **end for**
10:   Output current values $(x_1, \ldots, x_N)$
11: **end for**

HG is not only a deterministic variant of Gibbs sampling but is reported to have better performance than Gibbs sampling [9]. For a function f of the spins x in the BM, let us consider estimating the expected value $\mathbb{E}_\pi[f(x)]$. Theoretically, for the BM on the complete graph, the error of the estimate by HG decreases at a rate of $O(1/T)$, whereas it decreases at a rate of $O(1/\sqrt{T})$ for random sampling. Experimental results show that HG outperforms Gibbs sampling for image processing and natural language processing tasks.

When a BM is used as a probabilistic model, its parameters must be learned from data. The learning procedure is expressed as

$$W_{ij} \leftarrow W_{ij} + \mathbb{E}_{\text{data}}[x_i x_j] - \mathbb{E}_{\text{model}}[x_i x_j] \, , \tag{26}$$

where $\mathbb{E}_{\text{data}}[\cdot]$ is the mean of the training data and $\mathbb{E}_{\text{model}}[\cdot] = \mathbb{E}_\pi[\cdot]$ is the expected value in the model (21) with the current parameters W_{ij}. The exact calculation of the third term is typically difficult owing to the exponentially increasing number of terms, but MCMC can be used to estimate it [10–12]. HG can also be applied to the learning process via its estimation; the variance reduction by HG can have a positive effect on learning, as demonstrated for BM learning on handwritten digit images [13].

2.3 Sharing Weight Variables

The original literature on HG [9] also introduces the idea of "weight sharing." For a variable x_i, let j, j' be indices of different configurations with equal conditional probabilities $\pi_{i,j} = \pi_{i,j'}$. For example, this can occur when the coupling coefficients are restricted to $W_{i,j} \in \{0, \pm c\}$ for $c \in \mathbb{R}$; thus the conditional probability is determined only by counting neighboring variable values. In weight sharing, such configurations are classified into groups, and those in the same group, indexed by y, share a weight variable $w_{i,y}$. We will also refer to these versions, which are based on general graphs and use weight sharing, as HG.

Eskelinen [14] proposed a variant of this algorithm called discretized herded Gibbs (DHG), which is presented in Algorithm 5. It uses B disjoint intervals of the conditional probability value dividing the unit interval $[0, 1]$. Let us call them bins and denote by C_y the y-th interval. At the time of the update, the conditional probability $\pi(x_i = 1 \mid x_{N(i)})$ is approximated by the representative value $\tilde{\pi}_{i,y} \in C_y$, where y is the index of the interval to which the configuration $x_{N(i)}$ belongs. The weight variable used for the update is shared among several spin configurations with similar conditional probability values. The probability value used for the update is replaced by the representative value, but this replacement introduces an error in the distribution of the output samples. A trade-off exists between the computational complexity proportional to B and the magnitude of the error.

These ideas aim to reduce computational complexity by reducing the number of weights. However, Bornn et al. [9] reported a performance improvement for image

---

Algorithm 5 Discretized herded Gibbs

1: **for** $t := 1, \ldots, T$ **do**
2:    **for** $i := 1, \ldots, N$ **do**
3:       Compute the conditional probability $\pi(x_i = 1 \mid \boldsymbol{x}_{\mathcal{N}(i)})$
4:       Quantize the conditional probability to obtain the discrete index y and the representative value of conditional probability $\tilde{\pi}_{i,y}$
5:       $x_i \leftarrow \begin{cases} 1 & (w_{i,y} < \tilde{\pi}_{i,y}) \\ 0 & (w_{i,y} \geq \tilde{\pi}_{i,y}) \end{cases}$
6:       $w_{i,y} = w_{i,y} + \tilde{\pi}_{i,y} \bmod 1$
7:                         ▷ applying the equivalent version of herding ((7), (8)) to x_i and $w_{i,y}$
8:       Maintain other variables $x_{i'}$ for $i' \neq i$ and weights $w_{i',y'}$ for $(i', y') \neq (i, y)$
9:    **end for**
10:   Output current values $(x_1, \ldots, x_N)$
11: **end for**

---

restoration using the BM, and Eskelinen [14] reported a reduction in the estimation error in the early iterations for the BM on a small complete graph.

2.4 Monte Carlo Integration Using HG

On complete graphs, HG has been shown to be consistent; the estimates converge to the true value associated with the target distribution π. On general graphs, however, this is not always the case. In particular, the error decay for HG in the general case has two characteristics: faster convergence in the early iterations and convergence bias in the later iterations. These are explained by the "weight sharing" and the temporal correlation of the weights.

Specifically, the accuracy of the sample approximation of the function f with HG was evaluated as follows [13]: For HG with a target BM π (21), let P be the empirical distribution of the obtained samples. Let the error be defined as

$$D \equiv \mathbb{E}_P[f(\boldsymbol{x})] - \mathbb{E}_\pi[f(\boldsymbol{x})] . \tag{27}$$

The magnitude of D is evaluated as

$$|D| \leq \lambda_{\text{cor}} D^{\text{cor}} + \lambda_{\text{herding}} D^{\text{herding}} + \lambda_{\text{approx}} D^{\text{approx}} + \lambda_z D^z . \tag{28}$$

This is obtained by decomposing the estimation error with respect to a variable, where each term has the following meaning: Let $x = x_i$ with fixing i, and let $z = \boldsymbol{x}_{-i}$ be all the variables except x_i. Let y be the index of the weight used to generate x_i (line 3 of Algorithm 4 or line 4 of Algorithm 5) at each step. D^{approx} is the error term resulting from the replacement of the conditional probability $\pi(x = 1 \mid z)$ by the representative value $\tilde{\pi}_{i,y}$, which occurs in the DHG. D^z is the error in the distribution of z, namely, the joint distribution of the variables other than x_i, and expected to depend largely on the mixing time of the original Gibbs sampling.

The other two, D^{cor} and D^{herding}, are the terms most relevant to the herding dynamics. D^{herding} is the term corresponding to the herding algorithm for the Bernoulli distribution or the deterministic tossup that decays by $O(1/T)$. D^{cor} is a non-vanishing term. Thus, the error decay is characterized by the fast decay in early iterations dominated by D^{herding} and its stagnation in later iterations dominated by D^{cor}. D^{cor} is the term owing to the temporal correlation of the weight variables in the HG and is represented by the following equation:

$$D^{\text{cor}} = \sum_y P(y) \sum_{x,z} |P(x, z \mid y) - P(x \mid y)P(z \mid y)| . \tag{29}$$

Let us consider the ideal case that the weight index y and the conditional distribution $\pi(x \mid z)$ have a one-to-one correspondence. For DHG, this means that there are enough many small bins. Then, for the target distribution, x and z conditioned on y are independent; $\pi(x, z \mid y) = \pi(x \mid y)\pi(z \mid y)$. This is because $\pi(x \mid y)$ becomes constant on z. For the output distribution, however, this independence does not hold even under this assumption because the internal state $w_{i,y}$ determines the value of x and also has the temporal correlation with z. D^{cor} evaluates the correlation between the conditional distributions of x and z.

This temporal correlation is mainly caused by the deterministic nature of the herding dynamics. To mitigate this, the algorithm can be modified by introducing some stochasticity in the transitions. This has been shown to solve the bias problem [13], but at the cost of a degradation in the estimation for small T.

2.5 Rao-Blackwellization

Let us further discuss this analysis with a more concrete example. Let us fix i as before. Let $f(x) = x_i$ and consider the accuracy of estimating the expected value $\mathbb{E}_\pi[f(x)] = \mathbb{E}_\pi[x_i]$. The estimation error is bounded as

$$
\begin{aligned}
|D| = |\mathbb{E}_P[x_i] - \mathbb{E}_\pi[x_i]| &= \left| \sum_z P(z)P(x = 1 \mid z) - \sum_z \pi(z)\pi(x = 1 \mid z) \right| \\
&\leq \sum_z P(z) |P(x = 1 \mid z) - \pi(x = 1 \mid z)| \\
&\quad + \sum_z |P(z) - \pi(z)| \pi(x = 1 \mid z) .
\end{aligned} \tag{30}
$$

Under some assumptions, the first term corresponds to D^{herding} and the second term corresponds to D^z. Therefore, we can evaluate the error decay as the first term decaying by $O(1/T)$ as in herding but is eventually dominated by the second term for a large T.

The results were compared with those of a similar algorithm as follows: First, an i.i.d. sample sequence $x^{(t)}$ is generated via random sampling. Subsequently, the expected value of $f(x) = x_i$ is approximated as the sample mean of the conditional probability $\pi(x = 1 \mid z)$. In this case, the approximation error is bounded above as

$$|D| \leq \sum_z |P(z) - \pi(z)| \, \pi(x = 1 \mid z) , \tag{31}$$

which is equal to the term that is dominant in the HG case. The substitution of the conditional mean (or distribution) into an estimator is sometimes referred to as Rao-Blackwellization [15, 16], after the Rao-Blackwell theorem that guarantees the improvement in estimation accuracy from this substitution. In other words, herding can be viewed as a sample-based estimation method that achieves error reduction through Rao-Blackwellization, although it is implemented implicitly.

However, HG is not equivalent to this; the evaluation of the decaying error simultaneously holds for any other x_j $(j \neq i)$ without changing the algorithm. If the conditional probability $\pi(x = 1 \mid z)$ is considered as a function of z, this function itself can have an approximation similar to that of (28). That is, we can decompose this function as

$$\pi(x = 1 \mid z) = \hat{f}(z) + \sum_{j \neq i} \beta_j x_j , \tag{32}$$

and evaluate each term in the summation by (28) for $f(x) = x_j$. Then we find that the first term in (32) is dominant, which leads to the improvement of the error evaluation in HG.

We can also use this formula for i.i.d. random sampling to improve the estimation. However, in this case, it is necessary to determine the appropriate β_j by estimating from the random samples obtained. However, in the case of herding, because this evaluation is valid for any β_j, selecting β_j is not necessary. Extending this discussion, we may be able to analyze the accuracy of estimation by herding in general; however, this is left open for future study.

3   Entropic Herding: Regularizer for Dynamical Sampling

In the previous section, we discussed the characteristics of the herding algorithm as a Monte Carlo numerical integration algorithm. In particular, there is a negative autocorrelation in the weight variables and samples, which is important for numerical integration. However, this does not cover all the characteristics of herding as a high-dimensional nonlinear dynamics. This section discusses the connection between herding and the maximum entropy principle.

3.1 Maximum Entropy Principle

The maximum entropy principle [17] is a common approach to statistical inference. It states that the distribution with the greatest uncertainty among those consistent with the information at hand should be used. As the name implies, entropy is often used to measure uncertainty. Specifically, for a distribution with a probability mass (or density) function p, the (differential) entropy is defined as

$$H(p) = \mathbb{E}_p[-\log p(x)] . \tag{33}$$

Furthermore, we assume that information is collected on features $\varphi_m \colon X \to \mathbb{R}$ indexed with $m \in M$, where X is the sample space and M is the set of feature indices. Assume further that the collected data is available as the mean μ_m of each feature. According to the maximum entropy principle, the estimated distribution p should satisfy the condition that the expected value $\mathbb{E}_p[\varphi_m(x)]$ of the feature values is equal to the given value μ_m. Therefore, the maximum entropy principle can be explicitly expressed as the following optimization problem:

$$\begin{aligned} \underset{p}{\text{maximize}} \quad & H(p) \\ \text{subject to} \quad & \mathbb{E}_p[\varphi_m(x)] = \mu_m \quad \forall m \in M . \end{aligned} \tag{34}$$

We obtain the condition that the solution should satisfy as follows: For simplicity, let $X = \{x_1, \ldots, x_N\}$ be a discrete set and, for each i, let p_i denote the probability of the i-th state x_i. The gradient of H is

$$\frac{\partial}{\partial p_i} H(p) = -\log p_i - 1 . \tag{35}$$

Using the Lagrange multiplier method, we obtain the condition

$$p_i \propto \exp\left(-\sum \theta_m \varphi_m(x_i)\right) , \tag{36}$$

where θ_m corresponds to the Lagrange multiplier for the moment condition $\mathbb{E}_p[\varphi_m(x)] = \mu_m$. For continuous X, we can make a similar argument using the functional derivative.

In general, a family of distributions of the form (36) where the parameters θ_m are unconstrained is called an exponential family, and is also called Gibbs-Boltzmann distribution. Optimizing the parameters θ_m requires a learning algorithm similar to that of the Boltzmann machine (21), which is generally computationally difficult.

For example, if $X = \mathbb{R}$ and $(\varphi_1(x), \varphi_2(x)) = (x, x^2)$, the maximum entropy principle gives the normal distribution as

$$p(x) \propto \exp\left(-\theta_1 x - \theta_2 x^2\right)$$

$$\propto \exp\left(-\frac{(x-m)^2}{2\sigma^2}\right), \tag{37}$$

where $(m, \sigma) = (-\theta_1/2\theta_2, \sqrt{1/2\theta_2})$.

If we use $\mathcal{X} \in \{0, 1\}^N$ and $\varphi_m(x) = x_i x_j$ for all pairs $i < j$ where $m = (i, j)$, we obtain

$$p(x) \propto \exp\left(-\sum_{i<j} \theta_{ij} x_i x_j\right), \tag{38}$$

which is identical to the Boltzmann machine (21) without bias on the fully connected graph.

The output of the herding algorithm is also such that the moment condition $\mathbb{E}_p[\varphi_m(x)] = \mu_m \ \forall m \in M$ is satisfied in the limit of $T \to +\infty$, and is expected to be diversified due to the complexity of the herding dynamics. Therefore, we can expect the output sequence to follow, at least partially, the maximum entropy principle. The original literature on herding [2, 3] also describes its motive in the context of the maximum entropy principle.

3.2 Entropic Herding

We observed that herding is an algorithm closely related to the maximum entropy principle. Entropic herding, which is described in this section, is an algorithm that incorporates this principle in a more explicit manner.

Let us consider the same features φ_m and target means μ_m as above. Pseudocode for entropic herding is provided in Algorithm 6. Here, we introduce scale parameters $\Lambda_m \geq 0$ for each condition $m \in M$, which are used to control the penalty for the condition $\mathbb{E}_p[\varphi_m] = \mu_m$ as described below. In addition, we introduce step-size parameters $\varepsilon^{(t)} \geq 0$ to the algorithm.

Similar to the original herding algorithm, entropic herding is an iterative process and the time-varying weight a_m for each feature is included in the system. Each

---

Algorithm 6 Entropic herding

1: Choose an initial distribution $r^{(0)}$ and let $\eta_m^{(0)} = \mathbb{E}_{r^{(0)}}[\varphi_m(x)]$ and $a_m^{(0)} = \Lambda_m(\eta_m^{(0)} - \mu_m)$
2: **for** $t = 1, \ldots, T$ **do**
3:     Solve $\min_{q \in Q}\left(\sum_{m \in M} a_m^{(t-1)} \mathbb{E}_q[\varphi_m(x)]\right) - H(q)$ (Equation (39))
4:     Let $r^{(t)}$ be the minimizer
5:     Update $a_m^{(t)} = a_m^{(t-1)} + \varepsilon^{(t)}\left(\Lambda_m(\mathbb{E}_{r^{(t)}}[\varphi_m(x)] - \mu_m) - a_m^{(t-1)}\right)$ (Equation (40))
6: **end for**

---

iteration of the algorithm, indexed by t, consists of two steps: the first step solves the optimization problem and the second step updates the parameters based on the solution. Unlike the original herding algorithm, the entropic herding algorithm outputs a sequence of distributions $r^{(1)}, r^{(2)}, \ldots$ instead of points. The two steps for each time step, derived later, are as follows:

$$r^{(t)} = \arg\min_{q \in Q} \left(\left(\sum_{m \in M} a_m^{(t-1)} \eta_m(q) \right) - H(q) \right), \tag{39}$$

$$a_m^{(t)} = a_m^{(t-1)} + \varepsilon^{(t)} \left(\Lambda_m(\eta_m(r^{(t)}) - \mu_m) - a_m^{(t-1)} \right), \tag{40}$$

where Q is the (sometimes parameterized) family of distributions and $\eta_m(q) \equiv \mathbb{E}_q[\varphi_m(x)]$ is the feature mean for the distribution q.

Obtaining the exact solution to the optimization problem (39) is often computationally intractable. We can restrict the candidate distributions Q or allow suboptimal solutions of (39) to reduce the computational cost.

3.3 Dynamical Entropic Maximization in Herding

Entropic herding is derived from the minimization problem $\min_p \mathcal{L}(p)$ with the following objective function:

$$\mathcal{L}(p) = \frac{1}{2} \left(\sum_{m \in M} \Lambda_m(\eta_m(p) - \mu_m)^2 \right) - H(p) . \tag{41}$$

The optimal solution of (41) is expressed with parameter θ^* as

$$p_{\theta^*}(x) \propto \exp\left(- \sum_{m \in M} \theta_m^* \varphi_m(x) \right) . \tag{42}$$

This has the same form as the distribution obtained using the maximum entropy principle (36). However, parameter θ^* does not coincide with the optimal solution θ expressed in (36). Specifically, it satisfies the following equation:

$$\theta_m^* = \Lambda_m(\eta_m(p_{\theta^*}) - \mu_m) . \tag{43}$$

It roughly implies that the moment error becomes smaller when Λ_m is large.

Entropic herding can be interpreted as an approximate optimization algorithm for this problem with a restricted search space. Suppose it only considers P that has the following form:

$$P(x) = \sum_{t=0}^{T} \rho_t r^{(t)}(x) , \tag{44}$$

where ρ_t are the fixed component weights and each component $r^{(t)}$ is determined sequentially using the results of the previous steps $r^{(0)}, \dots, r^{(t-1)}$.

Let $\tilde{\mathcal{L}}$ be the function obtained by replacing the entropy term $H(p)$ of $\mathcal{L}$ with the weighted average of the entropies of each component $\tilde{H} \equiv \sum_{t=0}^{T} \rho_t H(r^{(t)})$:

$$\tilde{\mathcal{L}} = \frac{1}{2} \left(\sum_{m \in \mathcal{M}} \Lambda_m \left(\eta_m(P) - \mu_m \right)^2 \right) - \sum_{t=0}^{T} \rho_t H(r^{(t)}) . \tag{45}$$

Since $H \geq \tilde{H}$ holds due to the convexity of the entropy, $\tilde{\mathcal{L}}$ is an upper bound on $\mathcal{L}$. Equation (39) is obtained by optimizing $r^{(t)}$ with fixing the components $r^{(0)}, \dots, r^{(t-1)}$ to minimize $\tilde{\mathcal{L}}(P^{(t)})$, where $P^{(t)} \propto \sum_{t'=0}^{t} \rho_{t'} r^{(t')}$ is the tentative solution. The optimal condition is equivalent to (39), where the parameters are represented using the tentative solution $P^{(t)}$ as

$$a_m^{(t-1)} = \Lambda_m(\eta_m(P^{(t-1)}) - \mu_m) . \tag{46}$$

If we choose ρ_t appropriately, the update formula (40) can be derived by (46) and a recursive relation between the tentative solutions for consecutive steps, $P^{(t-1)}$ and $P^{(t)}$.

We expect that the formalization of entropic herding will help theoretical studies of the herding algorithm. Equation (41) contains the entropy term $H(p)$, and we need to make it larger. Entropic herding does so through the following two mechanisms:

(a) Explicit optimization: Greedy minimization of $\tilde{\mathcal{L}}$, which is the upper bound of $\mathcal{L}$, by solving (39) including the entropy term.
(b) Implicit diversification: The additional reduction of $\mathcal{L}$ owing to the diversity of $r^{(t)}$ caused by the complicated joint dynamics of the optimization steps and the weight update steps.

We rely on entropy maximization to reconstruct the target distribution, especially to reproduce the distributional characteristics not directly included in the inputs $\varphi(x)$ and μ. In the case of the original non-entropic herding algorithm, this depends entirely on the complexity of the dynamics, which is difficult to analyze accurately. Entropic herding explicitly incorporates this concept.

The added entropy term regularizes the output sample distribution. From (41), the error of η_m is small when Λ_m is large; however, the absolute value of a_m defined in

(46) becomes large, and thus, the entropy values of the output components $r^{(t)}$ are expected to be small. Thus, a trade-off exists in the choice of Λ_m.

Additionally, setting Λ_m to $+\infty$ implies ignoring the entropy terms in (39). This is equivalent to the original herding algorithm. That is, it is an extremum of entropic herding such that only (b) implicit diversification is used for entropy maximization.

3.4 Other Notes on Entropic Herding

Entropic herding yields the mixture distribution expressed in (44). The required number of components T of the mixture distribution is large when the target distribution is complex. In this case, the usual parameter fitting requires the simultaneous optimization of $O(T)$ parameters, which causes computational difficulties. On the other hand, entropic herding can determine $r^{(t)}$ sequentially. Therefore, the number of parameters to be optimized in each step can be kept small, which simplifies the implementation and improves its numerical stability.

In general, the optimization (39) is non-convex. Therefore, a local improvement strategy alone may lead to a local optimum. One possible solution to this problem is to prepare several candidate initial values. Unlike normal situations in optimization, herding solves optimization problems that differ only in their weights repeatedly. As a result, a good initial value can again become a good initial value again in later steps. Furthermore, the candidate initial values themselves can be improved during the algorithm. This introduces dynamics into them, which has been shown to exhibit interesting behavior [18].

4 Discussion

We reviewed the herding algorithm that exploits negative autocorrelation and the complexity of nonlinear dynamical systems. In this section, we present two perspectives on herding with a review of related studies.

4.1 Application of Physical Combinatorial Optimization Using Herding

To use physical phenomena in devices, we must precisely control them, particularly for computation. However, a certain amount of variability usually remains in both fabrication and operation, which hinders their practical use in computation. For example, quantum mechanics inherently involves probabilistic behavior, and current devices for quantum computing are also susceptible to thermal noise. Given that information processing in the human brain occurs through neurons that are energy efficient but noisy, perfect control may not be an essential element of information processing.

Sampling is a promising direction for the application of physical systems in computation, because such variability can be used as a source of complexity. However, completely disorganized behavior is not useful, and the system should be balanced with appropriate control; herding can be the basis for such control. Specifically, this incompletely controlled noisy device can be applied to the optimization step of the herding algorithm. Although the output may deviate from the optimal solution owing to the variability, this is compatible with the formulation of entropic herding, in which a point distribution is not expected. The gap of the output from the ideal solution may affect R, the upper bound of the weight norm $\|w\|$ used in Sect. 1.3, but a larger R may be acceptable because it can be handled by increasing the number of samples with a trade-off in the overall cost of the algorithm as long as R does not diverge to infinity.

Conventional computer systems have important advantages, such as the versatility of the von Neumann architecture, established communication standards, and the efficiency from the long history of performance improvements. Therefore, even if a novel device that achieves a breakthrough in computation, it is expected to eventually be used as a hybrid system coupled to conventional computers, in which we expect that herding can serve as the interface. For example, herding and optimization with quantum computers such as quantum annealing (QA) [19, 20] or quantum approximate optimization algorithm (QAOA) [21] can be a good combination. As shown in Sect. 1, herding reduces the variance of the estimate through negative autocorrelation. However, this does not mean that the mixing time of MCMC is improved. For herding, when the objective function in the optimization step is multimodal and steep, finding the optimal solution should be difficult even for herding. When an optimization method that can overcome this difficulty, such as quantum annealing, is realized, its combination with herding is expected to lead to further improvements.

4.2 Generative Models and Herding

Among the latest machine learning technologies, deep generative models are one of the areas that have attracted the most attention in recent years. In machine learning, a generative model is a mathematical model that represents the process of generating points from the sample space, which represents the set of possible targets, such as images or sentences. For example, flow-based models [22] and diffusion models [23] are among the most famous deep generative models, and other popular methods such as variational autoencoders (VAE) [24] and generative adversarial networks (GAN) [25] use generative models as components. These models are known for their amazing performance in AI, but their behavior with a massive number of tunable parameters is generally a black-box.

One of the unique approaches used in machine learning and other new mathematical techniques that take full advantage of computer technology is the use of *implicit representation* of probability distributions through black-box generative processes. For example, the stochastic gradient method, which is the basis of deep learning, requires only its generative method as an algorithm, not a density function in analytic form. When there is a large amount of data, randomly selecting from it can be considered as generating a point from a true generative process. Therefore, it is possible to implement the minimization of the objective function defined using an inaccessible true probability distribution. This is also exemplified by the famous reparameterization trick for VAE [24], where the variational model to be optimized is represented as a generative model instead of an explicit density function. Deep reinforcement learning is another good example, where the probability distribution of the possible time series of agent and environment states is represented as records of past trials, and the agent is trained using randomly sampled records of past experience [26]. This way of using the models increases their affinity with evolving computing technologies and the big-data accumulated from the growing Internet.

On the other hand, one of the main purposes of using more traditional *explicit modeling* of the density functions using parameterized analytic forms is to correlate model parameters and knowledge. The simplest example of this is regression analysis, as expressed by $Y \sim \beta X$. Here, the probability density function representing the model is fitted to the data, and whether or not the model parameter β is 0 is interpreted as an indication of whether the variables X and Y are related. We can also use a mathematical model to represent knowledge or hypotheses about a subject and use it to mathematically formalize the arguments based on that knowledge or hypotheses.

If we look at herding from the above perspective, we can find a unique feature of herding. Herding is the process of generating a sequence of points and can be thought of as generative models that implicitly represent a probability distribution. In addition, another important feature is that the algorithmic parameters $\varphi(x)$ and μ can be directly associated with knowledge as parameters of a mathematical model. Herding is a combination of implicit representation and explicit modeling. Although random sampling methods such as MCMC can be used to generate a set of points in the same manner as herding, the process is derived from the explicit model of the distribution, as opposed to the implicit definition of probability distributions as in herding.

5 Conclusion

In this chapter, we reviewed the herding algorithm, which can be used in the same manner as random sampling, although it is implemented as a deterministic dynamical system. Herding is expected to play an important role in the application of physical phenomena in computation because of the two aspects discussed in this chapter: a Monte Carlo numerical integration algorithm in Sects. 1 and 2, which are related to the mathematical advantages of the herding algorithm, and the connection to the maximum entropy principle in Sect. 3, which is related to the application of physical phenomena with herding. In addition, as shown in Sect. 4, herding is an interesting algorithm not only from these perspectives. We expect that theoretical studies of herding and its applications will continue in both directions.

Acknowledgements The author would like to thank Hideyuki Suzuki and Naoki Watamura for their valuable discussions and comments on the manuscript.

References

1. D. Pierangeli, G. Marcucci, C. Conti, Large-scale photonic Ising machine by spatial light modulation. Phys. Rev. Lett. **122**(21), 213902 (2019)
2. M. Welling, Herding dynamical weights to learn, in *Proceedings of the 26th Annual International Conference on Machine Learning* (2009), pp. 1121–1128
3. M. Welling, Y. Chen, Statistical inference using weak chaos and infinite memory. J. Phys.: Conf. Ser. **233**, 012005 (2010)
4. M. Welling, Herding dynamic weights for partially observed random field models, in *Proceedings of the Twenty-Fifth Conference on Uncertainty in Artificial Intelligence* (2009), pp. 599–606
5. Y. Chen, M. Welling, A. Smola, Super-samples from kernel herding, in *Proceedings of the Twenty-Sixth Conference on Uncertainty in Artificial Intelligence* (2010), pp. 109–116
6. N. Metropolis, A.W. Rosenbluth, M.N. Rosenbluth, A.H. Teller, E. Teller, Equations of state calculations by fast computing machines. J. Chem. Phys. **21**(6), 1087–1092 (1953)
7. W. Hastings, Monte Carlo sampling methods using Markov chains and their application. Biometrika **57**, 97–109 (1970)
8. S. Geman, D. Geman, Stochastic relaxation, Gibbs distribution, and the Bayesian restoration of images. IEEE Trans. Pattern Anal. Mach. Intell. **6**(6), 721–741 (1984)
9. Y. Chen, L. Bornn, N.D. Freitas, M. Eskelinen, J. Fang, M. Welling, Herded Gibbs sampling. J. Mach. Learn. Res. **17**(10), 1–29 (2016)
10. G.E. Hinton, Training products of experts by minimizing contrastive divergence. Neural Comput. **14**(8), 1771–1800 (2002)
11. T. Tieleman, Training restricted Boltzmann machines using approximations to the likelihood gradient, in *Proceedings of the 25th International Conference on Machine Learning* (2008), pp. 1064–1071
12. T. Tieleman, G. Hinton, Using fast weights to improve persistent contrastive divergence, in *Proceedings of the 26th Annual International Conference on Machine Learning* (2009), pp. 1033–1040
13. H. Yamashita, H. Suzuki, Convergence analysis of herded-Gibbs-type sampling algorithms: effects of weight sharing. Stat. Comput. **29**(5), 1035–1053 (2019)
14. M. Eskelinen, Herded Gibbs and discretized herded Gibbs sampling, Master's thesis, The University of British Columbia (2013)
15. A.E. Gelfand, A.F.M. Smith, Sampling-based approaches to calculating marginal densities. J. Am. Stat. Assoc. **85**(410), 398–409 (1990)
16. C.P. Robert, G. Roberts, Rao-Blackwellisation in the Markov chain Monte Carlo era. Int. Stat. Rev. **89**(2), 237–249 (2021)
17. E.T. Jaynes, Information theory and statistical mechanics. Phys. Rev. **106**(4), 620–630 (1957)
18. H. Yamashita, H. Suzuki, K. Aihara, Herding with self-organizing multiple starting point optimization, in *Proceedings of the 2022 International Symposium on Nonlinear Theory and its Applications* (2022), pp. 33–36
19. T. Kadowaki, H. Nishimori, Quantum annealing in the transverse Ising model. Phys. Rev. E **58**, 5355–5363 (1998)

20. M.W. Johnson, M.H.S. Amin, S. Gildert, T. Lanting, F. Hamze, N. Dickson, R. Harris, A.J. Berkley, J. Johansson, P. Bunyk, E.M. Chapple, C. Enderud, J.P. Hilton, K. Karimi, E. Ladizinsky, N. Ladizinsky, T. Oh, I. Perminov, C. Rich, M.C. Thom, E. Tolkacheva, C.J.S. Truncik, S. Uchaikin, J. Wang, B. Wilson, G. Rose, Quantum annealing with manufactured spins. Nature **473**, 194–198 (2011)
21. E. Farhi, J. Goldstone, S. Gutmann, A quantum approximate optimization algorithm. arXiv:1411.4028 [quant-ph] (2014)
22. I. Kobyzev, S.J. Prince, M.A. Brubaker, Normalizing flows: an introduction and review of current methods. IEEE Trans. Pattern Anal. Mach. Intell. **43**(11), 3964–3979 (2021)
23. J. Sohl-Dickstein, E. Weiss, N. Maheswaranathan, S. Ganguli, Deep unsupervised learning using nonequilibrium thermodynamics, in *Proceedings of the 32nd International Conference on Machine Learning* (2015), pp. 2256–2265
24. D. P. Kingma, M. Welling, Auto-encoding variational Bayes, in *2nd International Conference on Learning Representations* (ICLR, 2014)
25. I. J. Goodfellow, J. Pouget-Abadie, M. Mirza, B. Xu, D. Warde-Farley, S. Ozair, A. Courville, Y. Bengio, Generative adversarial nets, in *Advances in Neural Information Processing Systems*, vol. 27 (NIPS 2014) (2014), pp. 2672–2680
26. V. Mnih, K. Kavukcuoglu, D. Silver, A.A. Rusu, J. Veness, M.G. Bellemare, A. Graves, M. Riedmiller, A.K. Fidjeland, G. Ostrovski, S. Petersen, C. Beattie, A. Sadik, I. Antonoglou, H. King, D. Kumaran, D. Wierstra, S. Legg, D. Hassabis, Human-level control through deep reinforcement learning. Nature **518**, 529–533 (2015)

Photonic Reservoir Computing

Reservoir Computing Based on Iterative Function Systems

Suguru Shimomura

Abstract Various approaches have been proposed to construct reservoir computing systems. However, the network structure and information processing capacity
of these systems are often tied to their individual implementations, which typically
become difficult to modify after physical setup. This limitation can hinder performance when the system is required to handle a wide spectrum of prediction tasks.
To address this limitation, it is crucial to develop tunable systems that can adapt to a
wide range of problem domains. This chapter presents a tunable optical computing
method based on the iterative function system (IFS). The tuning capability of IFS
provides adjustment of the network structure and optimizes the performance of the
optical system. Numerical and experimental results show the tuning capability of the
IFS reservoir computing. The relationship between tuning parameters and reservoir
properties is discussed. We further investigate the impact of optical feedback on the
reservoir properties and present the prediction results.

1 Introduction

An artificial neural network (ANN) is a brain-inspired computing model and contributes to a wide field of information processing including image classification and
speech recognition [1]. The ANN is represented by a network structure connected by
weighted links. By optimizing the weight of the connections, ANNs have capabilities
for desired information processing [2]. However, the optimization requires updating
all connections in ANNs, and it is difficult to realize large-scale ANNs. Reservoir
computing, which is a kind of recurrent neural network for processing time-series
data emerged [3]. Typical models of reservoir computing are divided into an echo
state network (ESN) and a liquid state machine [4, 5]. The idea of ESN has been
employed for hardware implementation of recurrent neural network, and various
architectures have been proposed (see Chap. 13). This chapter focuses on reservoir

S. Shimomura (✉)
Graduate School of Information Science and Technology, Osaka University, 1–5, Yamadaoka,
Suita, Osaka, Japan
e-mail: s-shimomura@ist.osaka-u.ac.jp

© The Author(s) 2024
H. Suzuki et al. (eds.), *Photonic Neural Networks with Spatiotemporal Dynamics*,
https://doi.org/10.1007/978-981-99-5072-0_11

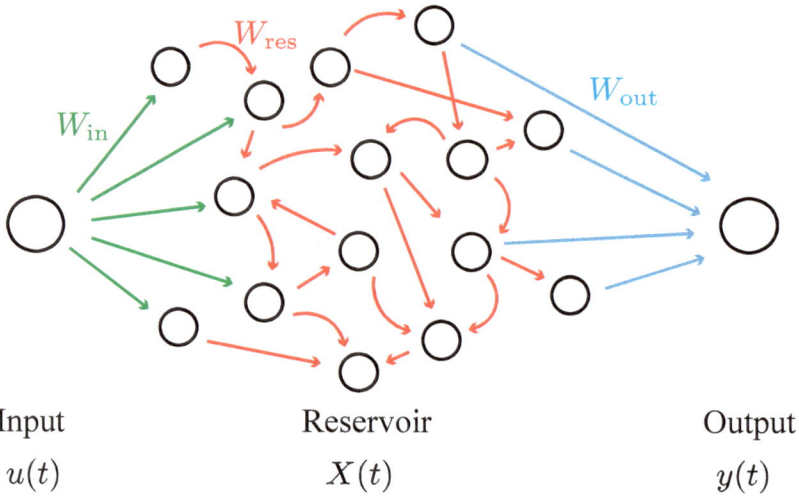

Fig. 1 Model of reservoir computing

computing based on the ESN. A model of reservoir computing is shown in Fig. 1. An echo state machine consists of three layers: input, reservoir, and output layers. Nodes in the reservoir layer are connected, and the structure is a recurrent network. A state of the nodes in the reservoir layer at time t, $X(t)$, is updated by

$$X(t + 1) = f[W_{res}X(t) + W_{in}u(t)], \tag{1}$$

where $u(t)$ is the input signal at time t, and f is a nonlinear function such as a hyperbolic tangent and a sigmoid function. Each component of $X(t)$ is transferred to the other nodes in the reservoir layer according to the connecting weight W_{res}. After adding with the weighted input signals $W_{in}u(t)$, the nonlinear function is applied, and the next state is updated as $X(t + 1)$. The connection weights W_{in} between the input and reservoir layers and W_{res} in the reservoir layer are fixed and not updated in learning process. The ESN is optimized by a linear regression of weights W_{out} between the reservoir and the output layers. Owing to simple structure of the echo state network and low computational processing in the learning process, reservoir computing can be implemented as hardware.

Reservoir computing is an intriguing and dynamic research field, offering a wide range of possibilities for hardware implementation by leveraging diverse types of materials and phenomena [6]. To construct a reservoir computing system, it is essential to design and implement a reservoir as the hardware component, and its response must satisfy Eq. 1. Thus far, various types of reservoir computing systems with individual properties of the utilized phenomenon have been proposed. For instance, the dynamic motion of soft materials has been used to determine the response of a reservoir [7]. The interaction of spin-torque oscillators based on spin waves provides small-scale reservoir devices [8]. In the field of optics, recurrent network circuits implemented on silicon chips and time-delayed feedback loop systems utilizing optical fibers have been successfully employed as reservoirs [9]. Individual reservoirs exhibit a specific reservoir property, which is related to the prediction performance of time-series data [10]. For example, the coupling matrix W_{res} shown in Eq. 1 is determined by the characteristics of utilized materials and phenomenon. Consequently, the performance may decrease depending on the prediction task. Tuning of the coupling matrix W_{res} is crucial in optimizing the prediction performance of a reservoir computing system, allowing it to effectively address a wide range of problems. To achieve this, the integration of the tuning function within an optical system is imperative. However, once an optical reservoir computing system is deployed, its physical configuration becomes fixed. This fixed configuration poses a challenge for any subsequent tuning. To predict various types of time-series data, it is necessary to tune systems' parameters after the construction of the system.

Free-space optics, which expresses the coupling matrix as a light-transfer matrix, is a promising solution for optimizing the performance of RC after the construction of the system. The use of a spatial light modulator provides a flexible adjustment of the transfer matrix by controlling the wavefront. In optical reservoir computing, transmission through scattering media is used to multiply the signal by W_{res} [11]. However, the controllability is limited by the SLM pixel size and pitch, which affect the representable coupling matrix.

In this chapter, we describe an optical RC approach using iterative function systems (IFSs) as a method to achieve optical tuning of the coupling matrix [12]. By employing optical affine transformation and video feedback, the coupling matrix can be flexibly tuned, allowing for the optimization of specific tasks.

2 Iterative Function Systems

For the adjustment of the coupling matrix W_{res}, an optical fractal synthesizer (OFS) was employed as the tuning function. The OFS utilizes an optical computing system to generate a fractal pattern using a pseudorandom signal, which can be applied to various applications, including steam ciphers [13, 14]. Pseudo-random signals are generated based on an IFS using a collection of deterministic contraction mappings [15]. Figure 2 shows the generation of pseudorandom signals by IFS. IFS mapping comprises affine transformations of signals, including rotation, scaling, and shifting, as follows:

$$\begin{bmatrix} x' \\ y' \end{bmatrix} = \begin{bmatrix} s & 0 \\ 0 & s \end{bmatrix} \begin{bmatrix} \cos\theta & -\sin\theta \\ \sin\theta & \cos\theta \end{bmatrix} \begin{bmatrix} x \\ y \end{bmatrix} + \begin{bmatrix} t_x \\ t_y \end{bmatrix}, \tag{2}$$

where x' and y' are the coordinates after translation, x and y are those before translation, s is the scaling factor, θ is the rotation angle, and t_x and t_y are the translation parameters. The OFS can generate pseudorandom signals when $s > 1$, or fractal patterns when $s < 1$. Owing to the simultaneous processing of input images, IFS

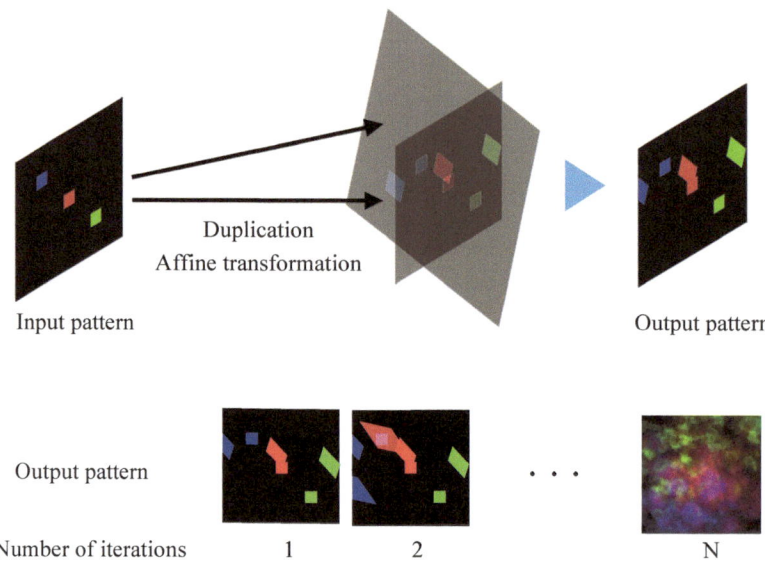

Duplication
Affine transformation

Input pattern   Output pattern

Output pattern

Number of iterations   1   2   N

Fig. 2 Pattern generation by iterative function systems

provides spatio-parallel processing. The operations in Eq. 2 can be implemented by using two dove prisms and lenses, and the parameters can be tuned by adjusting the optical components. Moreover, the IFS allows for the duplication and transformation of signals in a recursive manner, enabling more intricate and complex pattern generation. This results in an increased number of degrees of freedom in the IFS operation. In the proposed system, the IFS is utilized for the operation of the coupling matrix W_{res}, which can be tuned by controlling the optical setup including the rotation and tilt of the dove mirrors.

3   Iterative Function System-Based Reservoir Computing

Hardware can be implemented for tunable reservoir computing by utilizing IFS. We refer to reservoir computing using IFS as IFS reservoir computing.

Figure 3 shows the model of optical RC based on the IFS. The input signal $u(t)$ at time t is multiplied by the coupling matrix W_{in} and converted into a two-dimensional image. The reservoir state $X(t)$, which is generated from the input signal, is assigned as the input of the IFS after undergoing an electric-optic conversion and multiplication quantization operation B. The signal is duplicated and combined after individual optical affine transformations. This signal processing is represented by the multiplication with the matrix W_{res}, which can be adjusted using

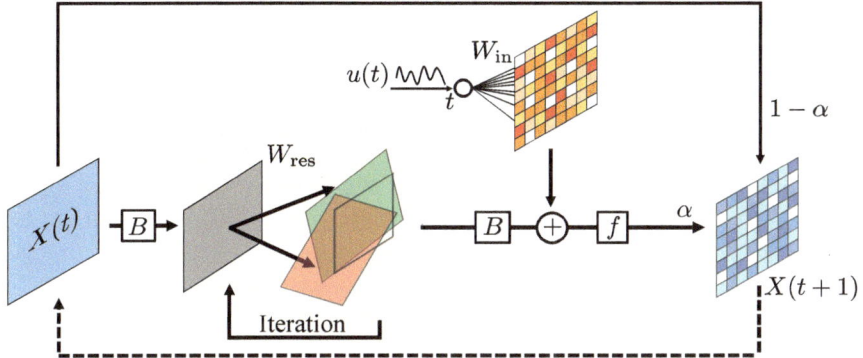

Fig. 3 Model of IFS reservoir computing [12]

the parameters of the optical affine transformation and the number of iterations. Following the addition of the input image $\boldsymbol{u}(t)$ to the transferred signal, the reservoir state $\boldsymbol{X}(t)$ is updated as

$$\boldsymbol{X}(t+1) = \alpha f[\boldsymbol{B}[\boldsymbol{W}_{\mathrm{res}}\boldsymbol{B}[\boldsymbol{X}(t)]] + \boldsymbol{W}_{\mathrm{in}}\boldsymbol{u}(t)] + (1-\alpha)\boldsymbol{X}(t), \quad 0 < \alpha < 1, \quad (3)$$

where α is a leaking rate, which determines the memory capacity of the signal in the reservoir layer. Thus, the sequence of reservoir states corresponding to a sequence of input signals is obtained. The output signal is generated by multiplying the reservoir state $\boldsymbol{X}(t)$ with a variable weight $\boldsymbol{W}_{\mathrm{out}}$, as follows:

$$y(t) = \boldsymbol{W}_{\mathrm{out}}\boldsymbol{X}'(t), \quad (4)$$

where $\boldsymbol{X}'(t)$ is a subset of pixels extracted from the reservoir state $\boldsymbol{X}(t)$. In reservoir computing, only the output connection weights $\boldsymbol{W}_{\mathrm{out}}$ is updated using a dataset with a sequence of input signals $\boldsymbol{u}(t)$ and the corresponding sequence of the reservoir state $\boldsymbol{X}(t)$. During the training phase, Ridge regression was adopted to optimize the output signal. The loss function E is expressed as

$$E = \frac{1}{n}\sum_{t=1}^{n}(\boldsymbol{y}(t) - \hat{\boldsymbol{y}}(t))^2 + \lambda\sum_{i=1}^{N}\omega_i^2, \quad (5)$$

where n is the number of training sets, $\hat{\boldsymbol{y}}(t)$ is the correct value, λ is the regulation parameter, and ω_i is the ith element of $\boldsymbol{W}_{\mathrm{out}}$. By minimizing the loss function E, $\boldsymbol{W}_{\mathrm{out}}$ is optimized.

4 Prediction Performance of IFSRC

Time-series data prediction is divided into two types: multistep and one-step-ahead predictions. The former is a task involving continuous predictions of the input signal by updating the input to the IFS reservoir computing based on the predicted value, and the latter involves predicting the input signal one-step-ahead based on the reservoir state at each time. To verify whether the proposed system can predict various types of tasks, the prediction performance was evaluated for both types of prediction tasks.

4.1 Multi-step Ahead Prediction

In the evaluation of the prediction performance for multistep ahead prediction, we employed the prediction of the Mackey-Glass equation which represents a chaotic signal and is used as a benchmark for time-series signal prediction [16]. The Mackey-Glass equation in this study is given by:

$$u(t+1) = au(t) + \frac{bu(t-\tau)}{c + u(t-\tau)^m} + 0.5, \tag{6}$$

where a, b, c and m are constants, and τ is the delay parameter. A dataset of 30,000 inputs and the next predicted values obtained from the equation were prepared, and W_{out} was optimized by using Ridge regression which is the optimization method using Eq. 5. After optimization of W_{out}, we assessed the system's ability to replicate a chaotic signal by inputting the predicted output into the system. The pixel size of reservoir state $X(t)$ was set to 64×64 pixels.

Figure 4 shows the predicted results. The best parameters of the IFS reservoir to predict the Mackey-Glass equation were in Table 1. The IFS reservoir parameters are listed in Table 1. The inital time step in prediction phase was 300. The chaotic behavior of the Mackey-Glass equation was reproduced even for prediction phase.

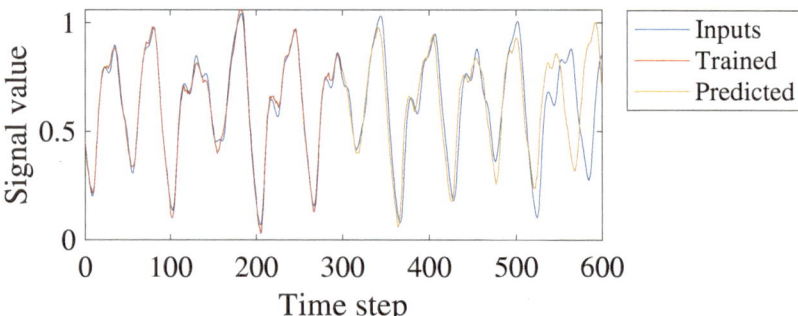

Fig. 4 Prediction of Mackey-Glass equation [12]

Table 1 Simulation parameters in IFS reservoir computing to predict Mackey-Glass equation [12]

Parameter	Value	
Number of iterations	3	
Leaking rate	0.1	
	Affine transformation 1	Affine transformation 2
Rotation angle θ (degree)	10	50
Scaling ratio s	1.0	1.2
Horizontal shift t_x (pixel)	−10	−10
Vertical shift t_y (pixel)	0	0

To evaluate the performance, the mean squared error (MSE) between the target signal and prediction output was estimated. The target signals were predicted for 261 time steps with a satisfactory MSE of <0.01. These results demonstrate the capability of IFS reservoir computing in predicting time-series data.

4.2 Single-Step Ahead Prediction of Santa Fe Time-Series Data

Single-step-ahead prediction is a task that predicts the next time signal from the input. To evaluate the performance of the system, we employed the Santa Fe time-series data, which requires memory to be predicted accurately. The Santa Fe time-series data, which models the behavior of a chaotic laser, is a widely recognized benchmark for evaluating reservoir computing systems [17]. The number of samples used for training and performing the test were 3,000, and 1,000, respectively. Figure 5a, b shows the targeted data and the prediction result. Table 2 presents the parameters of the IFS reservoir that exhibited the highest performance. Note that the best IFS parameter was different from that in case of the signal prediction of the Mackey-Glass equation. The system predicted signals similar to the label data. To evaluate the prediction performance, the normalized mean squared error was calculated between the predicted output and label. The definition of NMSE is described as follows:

$$\text{NMSE} = \frac{1}{n\sigma^2} \sum_{t=1}^{n} (y(t) - \hat{y}(t))^2, \tag{7}$$

where n is the number of dataset, σ is the standard deviation of the inputs, $y(t)$ is the prediction, and $\hat{y}(t)$ is the label value. As shown in Fig. 5c, the NMSE was 8.5×10^{-3}. These results demonstrate that prediction performance can be improved by adjusting the IFS reservoir.

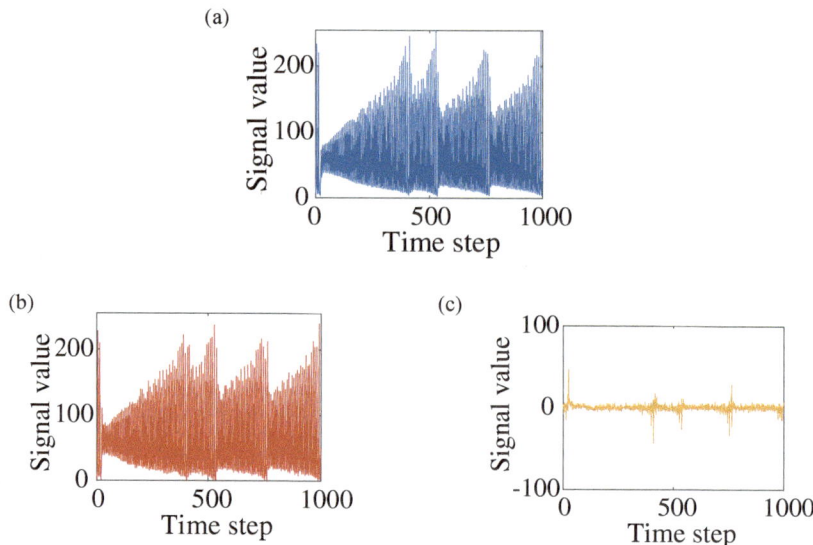

Fig. 5 Prediction result of Santa Fe time-series **a** target data, **b** predicted data, and **c** difference between (**a**) and (**b**) [12]

Table 2 Simulation parameters in IFS reservoir computing to predict Santa Fe time-series [12]

Parameter	Value	
Number of iterations	1	
Leaking rate	1.0	
	Affine transformation 1	Affine transformation 2
Rotation angle θ (degree)	10	80
Scaling ratio s	0.8	1.2
Horizontal shift t_x (pixel)	−10	10
Vertical shift t_y (pixel)	0	0

5 Experimental Performance of IFS Reservoir Computing

5.1 Optical Setup

To evaluate the hardware performance of the IFS reservoir computing system, an optical system featuring a video feedback system was constructed, as depicted in Fig. 6. First, the image representing the reservoir state $X(t)$ is projected into the display (MIP3508, Prament, number of pixels: 480 × 320), and replicated by a beam splitter (BS). By using Dove prisms, optical affine transformations through rotation and tilt are processed to the individual images. The scaling factor is determined by the difference between the focal lengths of lenses L2 and L3. Individual images are

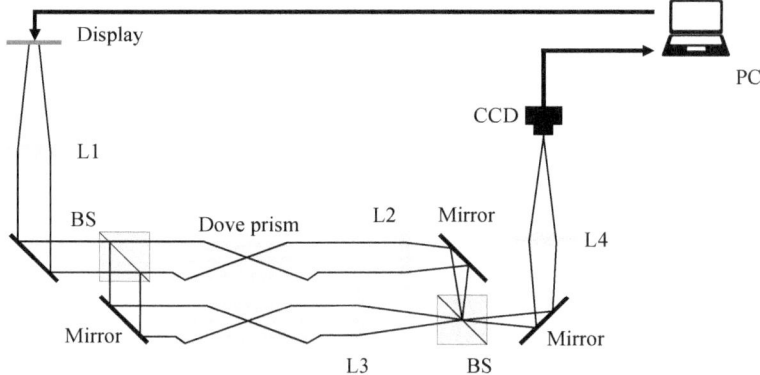

Fig. 6 Optical setup of IFS reservoir [12]

combined by using a BS and pass through lens L4. Finally, the image is captured by the image sensor (S3–U3–123S6, FLIR, number of pixels: 4096 × 3000). The captured image is resized to perform a predetermined number of iterations and is fed back to the display. The finally-obtained image is updated to next reservoir state $X(t + 1)$ by Eq. 3. After the processing, $X(t + 1)$ is fed back to the display as the next IFS reservoir state. The same process is repeated, and learning is performed with pairs of reservoir states and label data. In the experiment, a region of 37 × 30 pixels in the display was sampled and used as the signals of the IFS reservoir to decrease the computational cost in Ridge regression. Moreover, a hyperbolic tangent function was used as the nonlinear function.

5.2 Multi-step Ahead Prediction of Mackey-Glass Equation

Figure 7 shows the predicted results for the Mackey-Glass equation by using the optical system. The IFS parameters used in the experiment are listed in Table 3, and each value was estimated from the obtained images. The initial status of the reservoir was set to zero, and the number of training data points was 30,000. The prediction output is the chaotic signal similar to the Mackey-Glass equation. This result shows that the IFS reservoir can perform the prediction of time-series data. However, the prediction point with MSE <0.01 was 85 steps, which is lower than the time step in numerical simulation. The reason is that the iteration parameter was fixed to 1 in the optical setup, and the captured image was resized for feedback to the display. Increasing the output signal from the reservoir layer improves the performance in the physical reservoir computing [18, 19]. However, too large size of $X(t)$ takes computational cost in Ridge regression to optimize W_{out}. Therefore it is important to

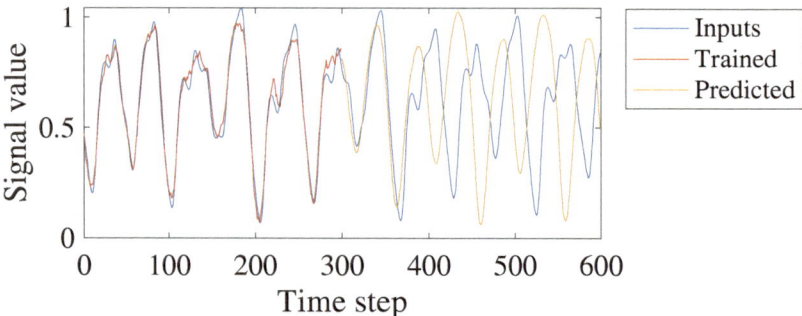

Fig. 7 Prediction result of Mackey-Glass equation **a** target data, **b** predicted data, and **c** difference between (**a**) and (**b**) [12]

Table 3 Experimental parameters in IFS reservoir computing [12]

Parameter	Value	
Number of iterations	1	
Leaking rate (Mackey-Glass equation)	0.1	
Leaking rate (Santa Fe time-series)	1.0	
	Affine transformation 1	Affine transformation 2
Rotation angle θ (degree)	–20	43
Scaling ratio s	1.0	1.0
Horizontal shift t_x (pixel)	50	460
Vertical shift t_y (pixel)	370	480

adjust the resolution of an image sensor approximately depending on the time-series data to be predicted.

5.3 Single-step Ahead Prediction of Santa Fe Time-Series Data

Next, the one-step-ahead prediction of the Santa Fe time-series was evaluated. The number of data points for training and prediction was set to 3,000 and 1,000, respectively. The parameters used are listed in Table 3. Figures 8 show the label data, predicted signal, and their difference. Similar to the prediction of the Mackey-Glass equation shown in Fig. 7, the IFS reservoir computing generates a signal waveform similar to the target signal. From the difference, the NMSE is estimated as 0.033. Although the IFS parameters are not fine-tuned, the performance of the experimental IFS reservoir system is higher than that of existing physical reservoir computers [20,

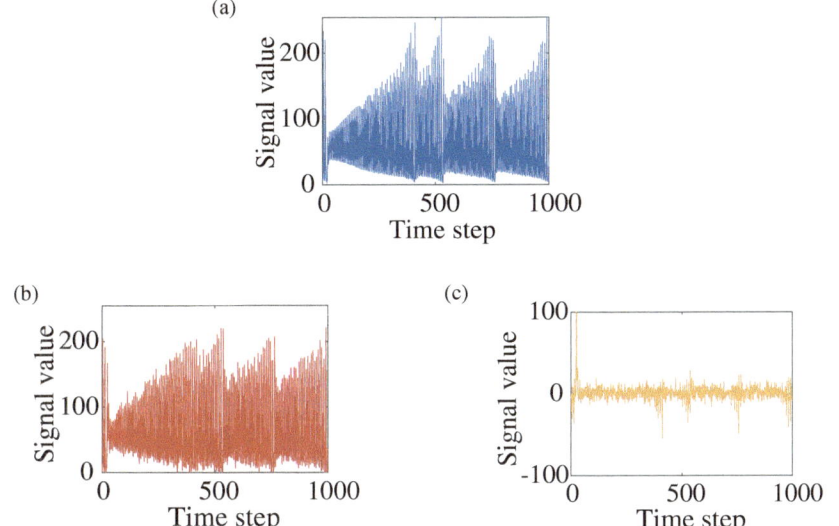

Fig. 8 Prediction result of Santa Fe time-series by optical setup. **a** Targeted data, **b** Predicted data, **c** difference between (**a**) and (**b**) [12]

21]. The results provide a promising perspective for IFS reservoir computing, which can tune the performance and flexibility in optical implementations.

6 Relationship Between Performance and Spectral Radius

To evaluate a property of reservoir computing, a spectral radius of the coupling matrix W_{res} in the reservoir layer is often used [22]. The spectral radius is the largest absolute value of eigenvalues of a matrix and is defined as follows:

$$\rho(W) = \max(|\lambda_i|, i = 1, 2, \ldots, n), \tag{8}$$

where $\lambda_1, \lambda_2, \ldots, \lambda_n$ are the eigenvalues of the matrix. The memory capacity increases as the spectral radius increases. In reservoir computing, a spectral radius less than one is preferred because the signal memory of the reservoir layer should be faded out [3].

In the IFS reservoir with leaking rate α, the coupling matrix was calculated as follows:

$$W = \alpha W_{res} + (1 - \alpha)I, \tag{9}$$

where I denotes a unit matrix. To investigate the characteristics of the IFS reservoir, the spectral radius and the NMSE of one-step-ahead prediction for the Santa Fe time-series were calculated. The individual parameters were set to the values listed in

Table 4 Combination of parameters in IFS reservoir computing [12]

Parameters	Value
Number of iterations	1, 3, 5, 10
Rotation angle θ (degree)	0, 30, 50, 80
Scaling ratio s	0.8, 1.0, 1.2
Horizontal shift t_x (pixel)	−10, 0, 10
Vertical shift t_y (pixel)	0

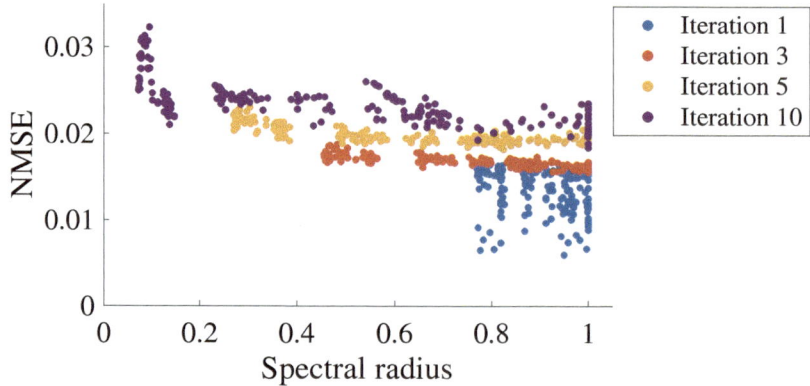

Fig. 9 Spectral radius and NMSE on one-step prediction for Santa Fe time-series [12]

Table 4, and the relationship between the spectral radius and NMSE were comprehensively verified. The size of the input image was 64×64, all the pixels were used for training, and the leaking rate was set to 1.0.

Figure 9 shows the relationship between the spectral radius and NMSE. Depending on the IFS parameters, the value of spectral radius is modulated, and the combination of scaling factors 0.8 and 1.0 generated a smaller spectral radius and improved the prediction performance. It was confirmed that the adjustment of IFS parameters provides modulation of coupling matrix W. In case of three and five iterations, the correlation coefficients were larger than 0.7, which indicates a relationship between the spectral radius and prediction performance. This result indicates that the one-step-ahead prediction of the Santa Fe time-series does not require rich memory capability for the task.

7 IFS Reservoir Computing with Optical Feedback Loop

It was demonstrated that the spectral radius changed with the number of iterations, and the prediction performance changed accordingly. However, in the experiment, it was necessary to repeat the electronic feedback process to change the number of

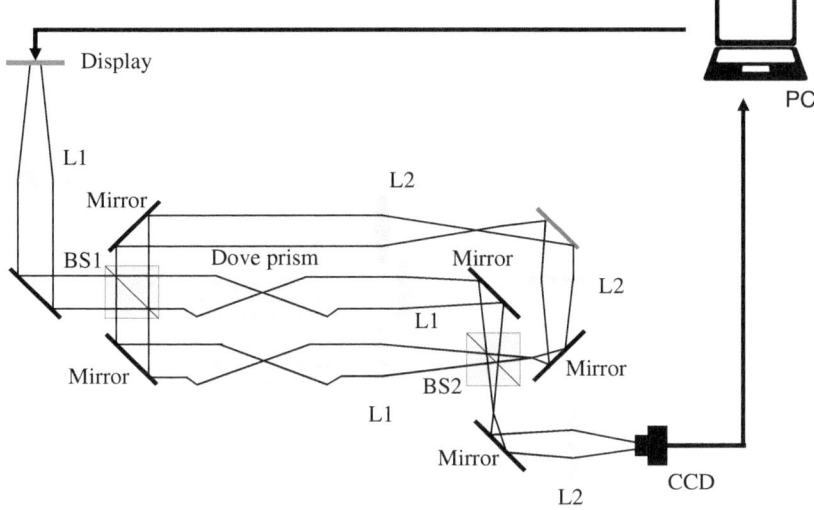

Fig. 10 Optical setup of IFS reservoir with an optical feedback loop

iterations. Therefore, optical feedback was introduced to realize optical control of the number of iterations. The experimental system is shown in Fig. 10. To facilitate optical feedback of the signals for multiple iterations, the combined image was transferred to BS1 through a relay lens. Consequently, a signal wherein the optical affine transformation is repeatedly executed can be generated. The combined matrix calculated in the experimental system is expressed as.

$$W_{\mathrm{res}} = (1 - \beta)A + \beta(1 - \beta)A^2 + \beta^2(1 - \beta)A^3 + \cdots \beta^{n-1}(1 - \beta)A^n, \quad (10)$$

where β denotes the feedback rate of the light signal branched by the beam splitter, and A is a coupling matrix when the number of iterations is one. By realizing multistage iterations, the range of combined matrix values was expanded. The signal after passing through the IFS processing was detected by an image sensor. Subsequently, the reservoir state was updated based on Eq. 3, and fed back to the display. The same procedure is repeated to develop the status of the IFS reservoir.

For evaluation, one-step-ahead prediction of the Santa Fe time-series data was performed. The feedback rate β, the number of training data, and the number of prediction data were set to 0.5, 3,000, and 1,000, respectively. The optical parameters are listed in Table 3. Figure 11a–c show the prediction results for the time-series data, label data, and their differences. The NMSE value obtained under these conditions was 0.098. When the feedback loop signal was removed under the same conditions, the NMSE was 0.105, demonstrating the potential for improved prediction accuracy with feedback.

The values presented in Eq. 10 demonstrate a decrease with an increase in the order. This is owing to $\beta < 1$. The sensitivity of the image sensor was used to adjust

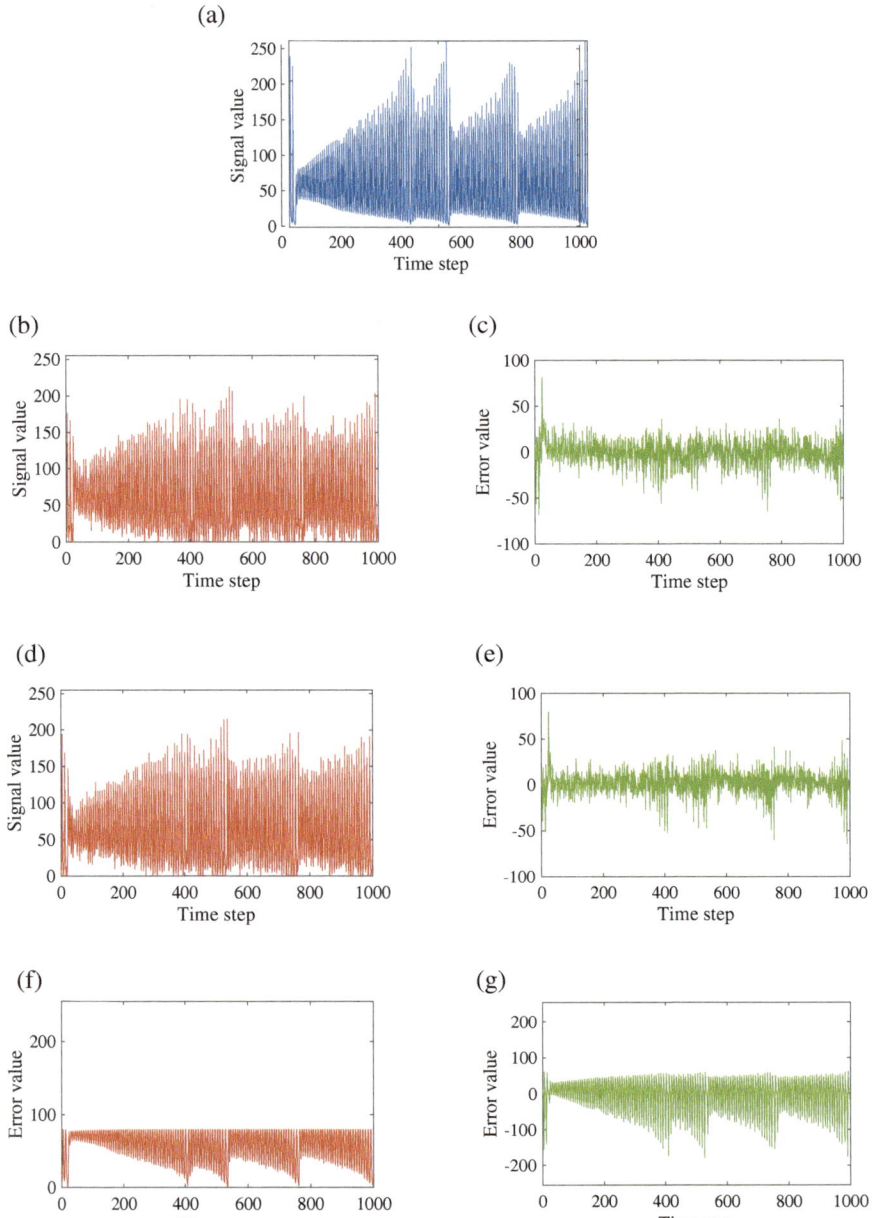

Fig. 11 **a** Label of Santa Fe time-series data. **b, d, f** One-step ahead prediction of the proposed system with an optical feedback loop and **c, e, g** the difference when the magnification of the gain in the image sensor was 1.0, 1.1, and 1.2, respectively

the signal acquisition of the calculated higher-order terms. Therefore, the change in the prediction ability due to the adjustment of the image sensor gain was verified.

Figure 11d, f show the prediction results when the gain was 1.1 and 1.2 times higher than that in case of Fig. 11b. In case of Fig. 11e The NMSE was 0.083 and the prediction accuracy improved. However, when the magnification of the gain was 1.2, the NMSE was 0.800, and the prediction accuracy decreased. When the gain is increased, a saturation of the light intensity occurs in the image sensor. . Consequently, a signal with effective prediction information cannot be obtained, and the prediction accuracy decreases. Therefore, it was demonstrated that it is necessary to adjust not only IFS parameters but also the sensitivity of the image sensor appropriately.

8 Discussion

The IFS reservoir computing allows for tuning of the parameters depending on the prediction task. Next step is to optimize individual parameters and maximize the performance. Various approaches have been suggested for optimizing hyperparameters in reservoir computing through computational processing [23–25]. In physical reservoir computing using FPGA, researchers have proposed methods for parameter tuning utilizing genetic algorithms [26]. Moreover, Bayesian estimation has been applied to realize more efficient parameters optimization compared to grid search methods [27]. Therefore, computational processing allows the optimization of hyperparameters efficiently, and the performance of reservoir computing can be optimized. In IFS reservoir computing, the number of IFS parameters is more than ten, which is twice as high as that in other studies. The number of parameters is corresponding to a degree of freedom in the tuning, and it is expected that higher prediction performance by the optimization can be realized. Furthermore, IFS reservoir computing provides the optimization of hyperparameters by adjusting the optical elements after construction of the system. By problem-specific parameter optimization, a reservoir computing predicting a wide range of time-series data can be built.

References

1. Y. Lecun, Y. Bengio, G. Hinton, Deep learning. Nature **521**(7553), 436–444 (2015). https://doi.org/10.1038/nature14539
2. T.W. Hughes, M. Minkov, Y. Shi, S. Fan, Training of photonic neural networks through in situ backpropagation. Optica **5**(7), 864–871 (2018)
3. H. Jaeger, The "echo state" approach to analysing and training recurrent neural networks-with an erratum note, Bonn, Germany: German National Research Center for Information Technology GMD Technical Report **148**(34), 13 (2001)
4. H. Jaeger, H. Haas, Harnessing nonlinearity: Predicting chaotic systems and saving energy in wireless communication. Science **304**(5667), 78–80 (2004)

5. W. Maass, H. Markram, On the computational power of circuits of spiking neurons. J. Comput. Syst. Sci. **69**(4), 593–616 (2004)
6. G. Tanaka, T. Yamane, J.B. Héroux, R. Nakane, N. Kanazawa, S. Takeda, H. Numata, D. Nakano, A. Hirose, Recent advances in physical reservoir computing: a review. Neural Netw. **115**, 100–123 (2019)
7. K. Nakajima, H. Hauser, T. Li, R. Pfeifer, Information processing via physical soft body. Sci. Rep. **5**(1), 10487 (2015)
8. R. Nakane, G. Tanaka, A. Hirose, Reservoir computing with spin waves excited in a garnet film. IEEE Access **6**, 4462–4469 (2018). https://doi.org/10.1109/ACCESS.2018.2794584
9. L. Appeltant, M. C. Soriano, G. Van der Sande, J. Danckaert, S. Massar, J. Dambre, B. Schrauwen, C. R. Mirasso, I. Fischer, Information processing using a single dynamical node as complex system. Nat. Commun. **2**(1), 468 (2011). https://doi.org/10.1038/ncomms1476
10. A.A. Ferreira, T.B. Ludermir, R.R. De Aquino, An approach to reservoir computing design and training. Expert. Syst. Appl. **40**(10), 4172–4182 (2013). https://doi.org/10.1016/j.eswa.2013.01.029
11. M. Rafayelyan, J. Dong, Y. Tan, F. Krzakala, S. Gigan, Large-scale optical reservoir computing for spatiotemporal chaotic systems prediction. Phys. Rev. X **10**, 041037 (2020). https://doi.org/10.1103/PhysRevX.10.041037
12. N. Segawa, S. Shimomura, Y. Ogura, J. Tanida, Tunable reservoir computing based on iterative function systems. Opt. Express **29**(26), 43164–43173 (2021). https://doi.org/10.1364/OE.441236
13. J. Tanida, A. Uemoto, Y. Ichioka, Optical fractal synthesizer: concept and experimental verification. Appl. Opt. **32**(5) (1993). https://doi.org/10.1364/AO.32.000653
14. T. Sasaki, H. Togo, J. Tanida, Y. Ichioka, Stream cipher based on pseudorandom number generation with optical affine transformation. Appl. Opt. **39**(14), 2340–2346 (2000). https://doi.org/10.1364/AO.39.002340
15. M.F. Barnsley, S. Demko, Iterated function systems and the global construction of fractals. Proc. R. Soc. London. A. Math. Phys. Sci. **399**(1817), 243–275 (1985)
16. L. Junges, J.A. Gallas, Intricate routes to chaos in the mackey-glass delayed feedback system. Phys. Lett. A **376**(30–31), 2109–2116 (2012)
17. A. Weigend, N. Gershenfeld, Results of the time series prediction competition at the santa fe institute, in *IEEE International Conference on Neural Networks*, vol. 3 (1993), pp. 1786–1793. https://doi.org/10.1109/ICNN.1993.298828
18. H. Hasegawa, K. Kanno, A. Uchida, Parallel and deep reservoir computing using semiconductor lasers with optical feedback. Nanophotonics **12**(5), 869–881 (2023) [cited 2023-04-20]. https://doi.org/10.1515/nanoph-2022-0440
19. J. Moon, Y. Wu, W. D. Lu, Hierarchical architectures in reservoir computing systems. Neuromorphic Comput. Eng. **1**(1), 014006 (2021). https://doi.org/10.1088/2634-4386/ac1b75
20. S. Sunada, K. Kanno, A. Uchida, Using multidimensional speckle dynamics for high-speed, large-scale, parallel photonic computing. Opt. Express **28**(21), 30349 (2020). arXiv:2104.00311, https://doi.org/10.1364/oe.399495
21. M. Freiberger, S. Sackesyn, C. Ma, A. Katumba, P. Bienstman, J. Dambre, Improving time series recognition and prediction with networks and ensembles of passive photonic reservoirs. IEEE J. Sel. Top. Quantum Electron. **26**(1), 1–11 (2019)
22. B. Schrauwen, D. Verstraeten, J. Van Campenhout, An overview of reservoir computing: theory, applications and implementations, in *Proceedings of the 15th European Symposium on Artificial Neural Networks* (2007), pp. 471–482
23. B. Ren, H. Ma, Global optimization of hyper-parameters in reservoir computing. Electron. Res. Arch. **30**(7), 2719–2729 (2022). https://doi.org/10.3934/era.2022139, https://www.aimspress.com/article/doi/10.3934/era.2022139
24. A.T. Sergio, T.B. Ludermir, Reservoir computing optimization with a hybrid method, in *2014 International Joint Conference on Neural Networks (IJCNN)* (IEEE, 2014), pp. 2653–2660
25. A.A. Ferreira, T.B. Ludermir, Genetic algorithm for reservoir computing optimization, in *2009 International Joint Conference on Neural Networks* (IEEE, 2009), pp. 811–815

26. B. Penkovsky, L. Larger, D. Brunner, Efficient design of hardware-enabled reservoir computing in FPGAs. J. Appl. Phys. **124**(16), 162101 (2018). https://doi.org/10.1063/1.5039826
27. P. Antonik, N. Marsal, D. Brunner, D. Rontani, Bayesian optimisation of large-scale photonic reservoir computers. Cogn. Comput. 1–9 (2021)

Bridging the Gap Between Reservoirs and Neural Networks

Masanori Hashimoto, Ángel López García-Arias, and Jaehoon Yu

Abstract While reservoir computing is drawing attention, its applications are limited to small tasks. This chapter proposes a solution to this issue by introducing the Hidden-Fold Network, a recursive model with fixed random weights that resembles reservoir computing. The model is constructed based on recent discoveries in neural networks, namely the Strong Lottery Ticket Hypothesis and folding. By pruning an overparameterized model that is randomly initialized, it is possible to find accurate neural networks without the need for weight optimization. It is conjectured that residual networks may contain better subnetwork candidates for inference time when transformed into recurrent architectures, as they may be approximating unrolled shallow recurrent neural networks. This hypothesis is tested in image classification tasks, where subnetworks within the recurrent models are found to be more accurate and parameter-efficient than those within feedforward models, as well as the full models with learned weights.

1 Introduction

Due to their complex structure, deep neural networks are capable of various types of learning. However, when it comes to time-series problems, recurrent neural networks (RNNs) are commonly used, but as their size increases, the cost of training becomes an issue. Reservoir computing (RC) is a special type of RNNs that offer a solution to this problem. In RC, the weights of the intermediate layer are fixed, and only the output weights from the intermediate layer to the output layer are trained using low-cost learners such as linear learners, which reduces the overall training cost. Physical reservoir computing takes this a step further by implementing the intermediate layer nodes of reservoir computing as physical phenomena on hardware. This approach reduces the need for design and can achieve even greater energy efficiency

M. Hashimoto (✉)
Department of Informatics, Kyoto University, Kyoto, Japan
e-mail: hashimoto@i.kyoto-u.ac.jp

Á. L. García-Arias · J. Yu
Tokyo Institute of Technology, Yokohama, Japan

© The Author(s) 2024   245
H. Suzuki et al. (eds.), *Photonic Neural Networks with Spatiotemporal Dynamics*,
https://doi.org/10.1007/978-981-99-5072-0_12

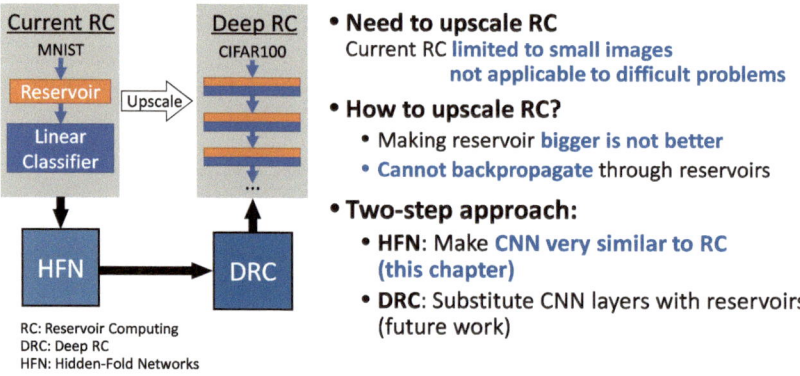

Fig. 1 Approach to deep reservoir computing

by selecting appropriate physical phenomena for the intermediate layer. There are various approaches for implementing physical reservoir computing [1].

Currently, all RC models are single-layered and limited to processing small datasets like MNIST. However, although there is a desire to scale up RC to larger models capable of handling bigger datasets like CIFAR100, there are no methods to achieve this. Increasing the size of the reservoir is not feasible since it makes the output too complex for the simple classifier, and backpropagation cannot be utilized. Figure 1 illustrates our approach toward achieving deep reservoir computing (DRC). We begin by developing a method to make convolutional neural networks (CNNs) similar to RC and replacing CNN layers with reservoirs to construct a DRC.

This chapter explores the compatibility between CNN and RC using two recent concepts: the Strong Lottery Ticket Hypothesis [2, 3] and folding [4]. According to the Strong Lottery Ticket Hypothesis, high-performing neural networks can be obtained by pruning overparameterized dense models since they contain already available high-performing subnetworks. These subnetworks are sparse, random, and tiny, and can achieve competitive performance in vision tasks, making them suitable for efficient hardware implementation [5]. However, finding optimal connectivity patterns using current training methods can be challenging [6]. Folding, on the other hand, is based on the observation that a specific type of shallow RNN is equivalent to a deep residual neural network (ResNet) with weight sharing among layers [4], where the original ResNet is proposed in [7] and remains the backbone of many SOTA models. Implementing a folded RNN with significantly fewer parameters than the corresponding ResNet leads to similar performance.

Figure 2 shows our baseline idea. We fold a ResNet to then find a strong lottery ticket within the network with random fixed weights. We have found that the obtained recursive model, which is called Hidden-Fold Network (HFN), is similar to reservoir computing.

We have observed that ResNets have an inherent ability to learn ensembles of all possible unrollings of a shallow recurrent neural network. This restriction on

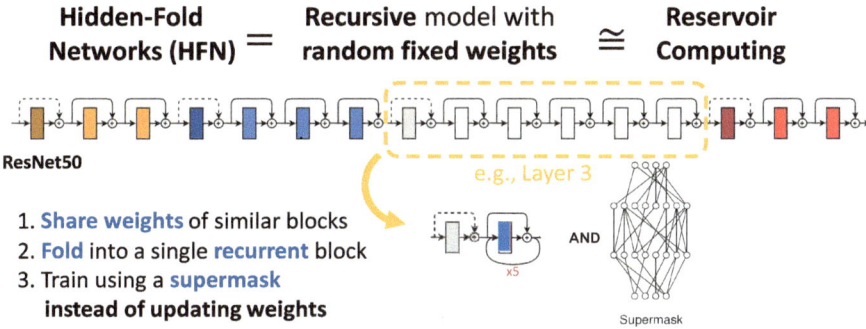

Fig. 2 A sketch of hidden-fold network

the hypothesis space actually increases the number of potential subnetworks available at initialization time [8]. These resulting subnetworks are efficient in terms of parameters and demonstrate competitive performance on image classification tasks. Additionally, these subnetworks can be compressed to small memory sizes and have a high degree of parameter reusability, making them ideal candidates for inference acceleration.

2   Background

Driven by the increasingly powerful computational power promised by Moore's Law, artificial neural networks have grown in size, leading to the field of deep learning [9]. This trend was initially led by image classification models [10] and has continued as researchers enhance neural networks at the cost of larger models. Natural language processing and generative models have also joined the trend, offering impressive capabilities at the expense of immense size [11, 12]. However, while this is a convenient approach for cutting-edge research, the high computational cost of these models makes them impractical for real-world applications. Therefore, researchers are also exploring small and efficient models as an alternative trend of research.

   Strong lottery tickets refer to a set of efficient neural networks obtained through a training process that combines learning, pruning, and weight quantization. The development of this approach, depicted in Fig. 3, is described in detail in Sect. 2.1. In this study, we introduce a method that converts a ResNet into a recurrent architecture to enhance the quality of the strong lottery tickets that can be extracted from it, where ResNet is introduced in Sect. 2.2.

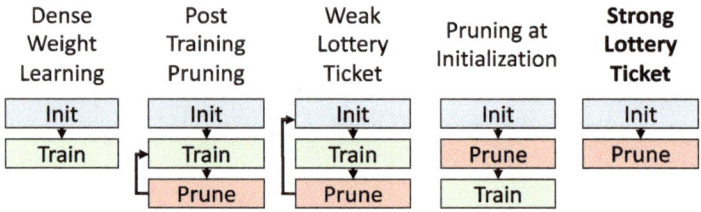

Fig. 3 Evolution of the training methods leading to the strong lottery ticket hypothesis

2.1 Lottery Ticket Hypotheses

Pruning is a widely used technique to compress trained neural networks into much smaller models by removing unnecessary weights [13–17]. By iteratively applying training and pruning, large portions of trained models can be removed without affecting accuracy, revealing that the original models are overparameterized. The sparsity of the resulting network can be leveraged for additional compression using entropy coding and for arithmetic optimization. This approach, combined with weight quantization, has resulted in highly efficient model compression schemes [18, 19] and specialized hardware neural accelerators [20].

In the past, it was observed that pruning did not separate the connectivity patterns from the pre-trained weights, making it impossible to reinitialize and train from scratch. However, a recent paper [21] introduced the concept of the Lottery Ticket Hypothesis (LTH), which states that overparameterized neural networks contain a subnetwork that can match the original model if trained in isolation. These subnetworks, known as winning tickets, are discovered by iteratively training, pruning, and resetting the remaining weights to their original value.

In an unexpected development during the analysis of the LTH, [2] discovered that learning weights is not essential: high-performing subnetworks exist within overparameterized neural networks at their randomly initialized state, which can be identified through pruning. Moreover, they proposed an algorithm for discovering these subnetworks by training a binary mask. Building upon this, [3] introduced a training algorithm and weight initialization method that produces sparse random subnetworks with competitive performance in image classification tasks.

Following a series of studies that explored the theoretical limits of necessary overparameterization to obtain these subnetworks [22–24], the notion of subnetworks obtained exclusively by pruning has been referred to as the *Strong Lottery Ticket Hypothesis* (SLTH). However, some researchers also refer to it as the multi-prize lottery ticket hypothesis [25], or simply as "hidden network."

Although the discovery of strong tickets is surprising, it has some related precedents. For example, extreme learning machines use fixed random weights in their hidden units, only learning the output layer. Reservoir computers also use random recurrent architectures in an analogous manner. Perturbative neural networks propose substituting convolutional layers with fixed additive random noise and a learned

Fig. 4 The residual network architecture. Specifically, ResNet-50, which has 3, 4, 6, and 3 blocks in each stage, from input to output

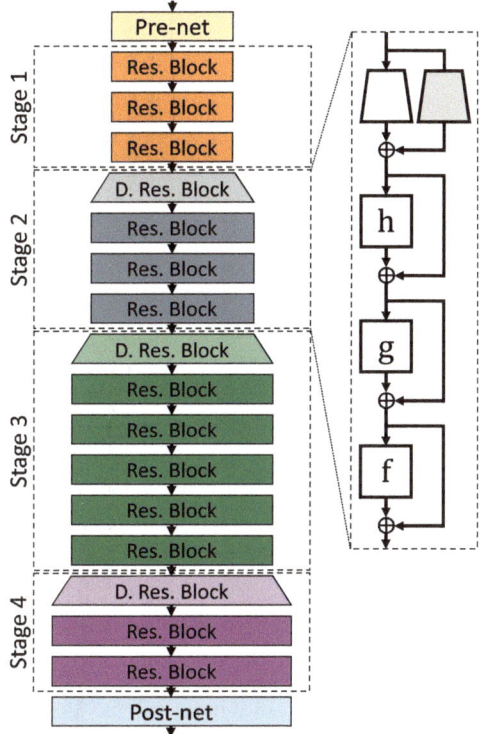

linear combination. Training the batch normalization parameters of a fixed random network can achieve non-trivial accuracy. The binary neural network training method has been adapted to learn binary masks that, when applied to a trained model, extract subnetworks that perform well on untrained tasks.

2.2 Residual Neural Networks

The residual neural network (ResNet) [7], as illustrated in Fig. 4, is a popular architecture among the continuously expanding variety of neural network architectures. It serves as the backbone of many state-of-the-art (SOTA) models. ResNetis a deep convolutional neural network that follows a pyramidal feedforward structure. It comprises a convolutional pre-net, four stages of residual blocks, and a fully-connected post-net classifier. The first residual block of each stage downsamples the feature map and doubles the number of channels, adjusting the representation space size. The remaining blocks in a stage have the same size and shape and maintain the same representation space size.

Residual blocks in the ResNet architecture consist of batch normalization, ReLU activation, and convolutional layers concatenated together. In this chapter, we focus on the bottleneck block variant [7], which consists of three convolutional layers applied in sequence with kernel sizes of 1×1, 3×3, and 1×1, respectively. Each residual block has a skip-connection, which is an identity function running parallel to the block that adds the block's input to its output. In order to accommodate different representation space sizes, the downsampling blocks have a learnable layer in the skip-connection called a projection shortcut.

The original motivation for the ResNet architecture was to provide a clear path for backpropagation to reach a layer directly, solving the vanishing gradient problem. However, studies have shown that shuffling or removing residual blocks does not severely impact its performance, but rather there is a gradual degradation proportional to the amount of corruption introduced [26]. This phenomenon is not observed in other feedforward models where similar lesions result in critical performance loss. As a result, two alternative views of ResNet have been proposed: one suggests that it is an ensemble of shallow networks, while the other argues that it is an approximation of an unrolled shallow recurrent neural network. This work aims to reconcile these two views into a single coherent explanation.

Each residual block in ResNet can be thought of as a path divergence point, which leads to the interpretation of ResNet as an ensemble of all possible paths within it [26], resulting in 2^n possible paths for a model with n residual blocks. Several improvements to ResNet have been proposed based on this perspective. For example, during training, removing random subsets of blocks makes the ensembled networks shallower and acts as regularization [27]. Furthermore, increasing the number of skip-connections per block increases the number of ensembled paths, thereby improving performance [28]. Another approach is to reduce network depth by increasing width, enabling the training of larger models for improved performance in less time [29].

An alternative explanation for the lesion and shuffling results is that all the blocks within a stage approximate the same function, as they have the same shape and partially receive the same inputs and gradients through the identity shortcuts. This implies that ResNet may naturally converge to the approximation of an unrolled shallow recurrent neural network, with each stage corresponding to a different hierarchical level of representation, including the downsampling block for feature map size adjustment. Meanwhile, the remaining blocks perform iterative refinement of features, according to proponents of this view [30].

The ensemble of unrollings in ResNet, which combines the two views mentioned above, the ensemble view and the unrolled iteration view, offers a vast search space to discover effective tickets within a model with limited parameters. Therefore, it is reasonable to assume that a ResNet with more recurrent approximations at its initial state may contain more potent tickets.

3 Hidden-Fold Networks

In this section, we present a technique for discovering a strong lottery ticket in a ResNet by first folding it. The approach, described in [8], yields Hidden-Fold Networks (HFNs), which outperform the strong tickets present in feedforward ResNets. Here, we provide details on the network's architecture and training. Refer to [8] for information on weight initialization and batch normalization.

3.1 *Folded ResNet Architecture*

In line with the unrolled iteration view, the chains of residual blocks with identical shapes in each stage approximate an iterative function. To achieve this, folding [4] is used, which converts these chains into recurrent blocks through weight sharing. In other words, $h \approx g \approx f$ in Fig. 4 is explicitly transformed into $h = g = f$. Since applying the same functions in succession is equivalent to repeatedly applying one of them, the feedforward chain can be transformed into a single recurrent block, in a process opposite to time unrolling.

The downsampling block in ResNet has a different shape than the rest of the blocks, and therefore cannot be folded with them. Strategies to address this issue have been explored in previous works, such as removing the block to create an isotropic architecture or substituting it with a simpler block, as discussed in [4, 31]. However, this work does not modify the downsampling blocks based on the view that different stages correspond to different hierarchical levels of features, which are composed of downsampling blocks [30]. Instead, the rest of the blocks within a stage are folded into a single recurrent residual block that is iterated the same number of times as the original number of blocks, as illustrated in Fig. 5.

Fig. 5 A ResNet stage folded into a recurrent residual stage through the opposite of time unrolling. The downsampling block is left untouched, whereas the rest of the stage is folded into a single recurrent block. That is, a stage of n blocks is folded into 2 blocks, the second of which is applied $n - 1$ iterations

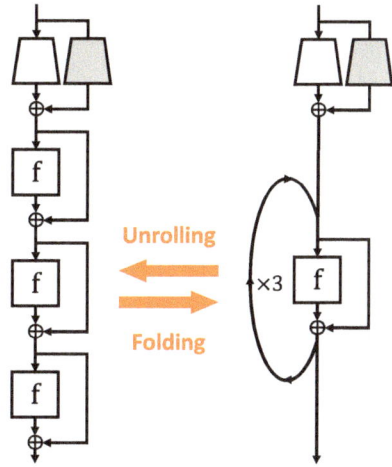

Folding has a dual effect on the search space. It restricts the hypothesis space of the model to iterative functions and reduces the number of parameters. Despite the exponential reduction in available subnetwork candidates, we argue that folded residual networks contain better performing strong tickets than their feedforward counterparts. If the weights of a ResNet naturally converge to approximations of unrolled iterative functions, then the strong tickets within it are likely approximations of recurrent networks. This restricts the number of relevant subnetworks to a small subset that includes consecutive blocks with similar random weights. By folding, all candidate subnetworks become recurrent, which increases the number of relevant subnetworks and their likelihood of containing stronger tickets.

Moreover, the parameter reduction occurs not only at inference time but also during training. By reducing the search space for strong tickets during training, folded tickets become easier to find. At inference time, the found subnetworks are even smaller and benefit from parameter reusability, making them ideal for efficient hardware implementation.

3.2 Supermask Training

Rather than optimizing the model's weights, the model is pruned to identify a high-performing subnetwork that is hidden within the randomly initialized folded model, which is referred to as an HFN. This connectivity pattern is learned by training a *supermask* [2], which is a pruning mask containing a binary element for each weight. During inference, the ticket is discovered by applying an element-wise product of the random weight tensor and the trained supermask.

This study adopts the `edge-popup` algorithm [3] for training the supermasks, as shown in Fig. 6a. The algorithm assigns a score to each weight, which is updated during backpropagation using straight-through estimation for the supermask (i.e., the supermask is not applied in the backward pass). The weights are sorted based on their scores, and the supermask is updated to include the weights with the highest top-k% scores and prune the rest. Although the value of top-k% is determined globally, the sparsity is enforced at the layer level. Folding does not impact this process; supermasks and scores are shared similarly to their corresponding weights, and backpropagation gradients are propagated through the unrolled model just like a feedforward model. Therefore, folded parameters receive distinct gradients from each iteration, as demonstrated in Fig. 6b.

4 Experiments and Results

In this section, we explore how to effectively integrate supermask training with a recurrent residual network and compare the outcomes with the baseline approaches outlined in Table 1.

Fig. 6 Training an HFN with a supermask. The supermask (H) includes the random weights (W) with the top-k% scores (S), updated via backpropagation. $\odot$ is the Hadamard product

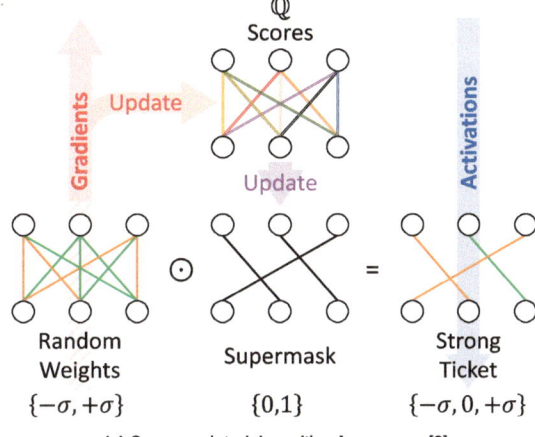

(a) Supermask training with edge-popup [3]

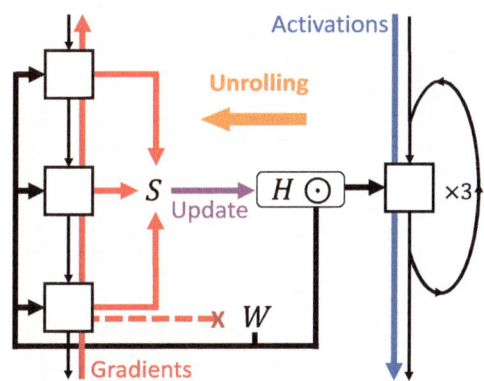

(b) Folded models are unrolled, so scores receive a gradient from each iteration.

Table 1 Summary of the four methods compared on ResNet in this work

Method	Architecture	Training
Standard ("Vanilla") [7]	Feedforward	Dense weight learning
Folding [4]	Recurrent	Dense weight learning
Hidden-Networks (HNN) [3]	Feedforward	Supermasks
Hidden-Fold Networks (HFN)	Recurrent	Supermasks

4.1 Experimental Settings

All experiments were implemented in PyTorch [32], using the original code of [3] available in their public repository [33]. The baseline model for all experiments was ResNet-50 [7] and its variations, as folding only applies to deep networks. The experiments were conducted on the CIFAR100 [34] dataset, which is relatively complex.

Unless stated otherwise, the experiments were conducted using the following methodology. We split the CIFAR100 dataset, consisting of $60,000$ images, into $45,000$ for training, $5,000$ for validation, and $10,000$ for the test set. Image pre-processing and augmentation were carried out as in [3]. We trained the models on CIFAR100 using stochastic gradient descent (SGD) with weight decay of 0.0005, momentum of 0.9, and batch size of 128 for 200 epochs. For models deeper than 100 layers or double width, we trained for an additional 100 epochs. The learning rate started at 0.1 and was reduced using cosine annealing with no warmup. We report the average of three runs of top-1 test accuracy scores measured at the highest scoring validation epoch. The standard deviation is shown with shaded areas on plots.

A vanilla ResNet-50 trained on CIFAR100 using an NVIDIA GeForce RTX 3090 requires 2.4 h. In comparison, the folded, HNN, and HFN versions of the same model require 2.2, 4.4, and 4.2 h, respectively.

4.2 Results

The comparison of accuracy and parameter count of various ResNet sizes trained with different methods on CIFAR100 is shown in Fig. 7a. HFN models are found to be the most parameter-efficient among the compared methods, with fewer parameters than equally performing models, and more accurate than models with similar parameter counts. Additionally, deeper and wider HFNs achieve the highest accuracies overall.

In addition, the superiority of HFNs is more pronounced when examining the model memory sizes under the compression scheme presented in [8], as shown in Fig. 7b. The memory size of HFNs is approximately half that of their feedforward counterparts, with ResNet-50 fitting into less than 2 MB. Moreover, the wider HFN models outperform the dense models that are more than $30\times$ larger in size.

5 Summary

We explored the similarity between the Hidden-Fold Network, a recursive model with fixed random weights, and reservoir computing, as depicted in Fig. 8. In this chapter, we presented a method for folding and training ResNet with supermasks, as a first step toward deep reservoir computing. We also tested the conjecture that

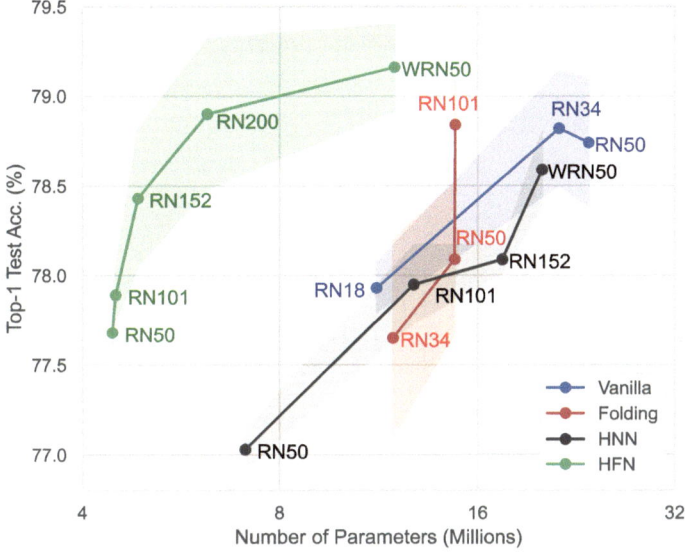

(a) Accuracy vs. number of parameters.

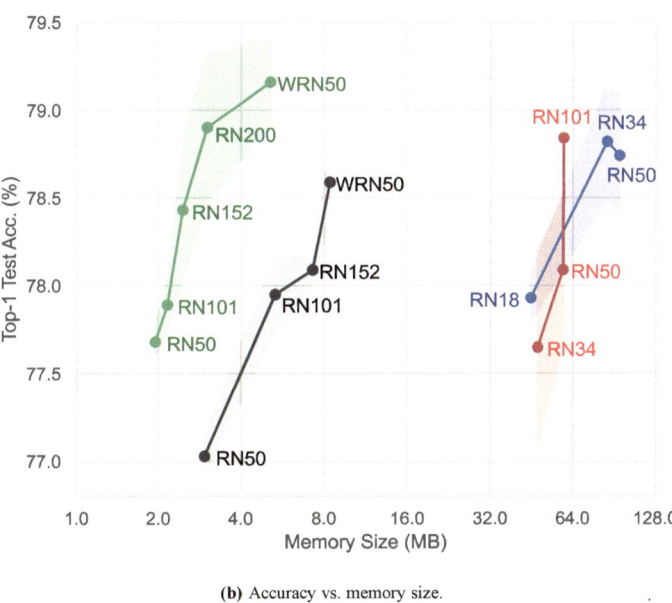

(b) Accuracy vs. memory size.

Fig. 7 Comparison of the four methods using different model sizes on CIFAR100. HFN is both the most parameter-efficient and the tiniest. RN and WRN are abbreviations of ResNet and Wide ResNet, respectively

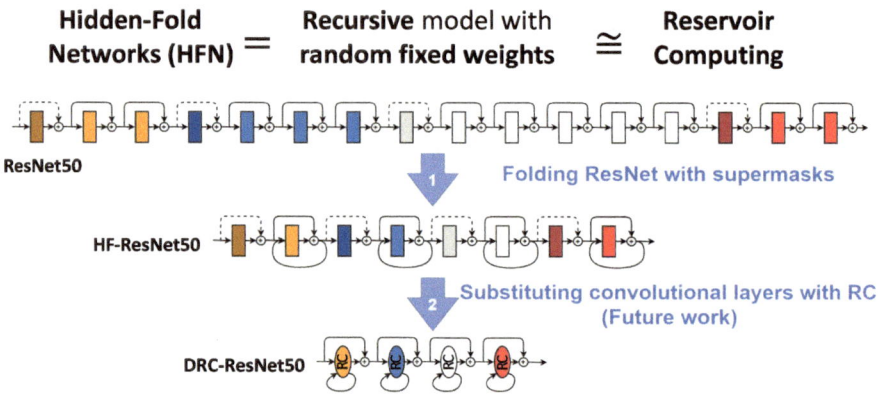

Fig. 8 The current status in this chapter and future work

recurrent residual networks have stronger tickets than their feedforward counterparts, which can be leveraged to significantly reduce the memory footprint while achieving comparable or superior accuracy to dense models. Since HFN's blocks are recurrent and unlearned, leaving all the learning load to a simple mask, the model bears strong resemblance to a deep reservoir computer. However, the second step of substituting convolutional layers with actual reservoir layers remains a future direction. Once this is achieved, we will bridge the gap between reservoirs and neural networks and approach deep reservoir computing that can solve complex tasks.

References

1. G. Tanaka, T. Yamane, J.B. Héroux, R. Nakane, N. Kanazawa, S. Takeda, H. Numata, D. Nakano, A. Hirose, Recent advances in physical reservoir computing: a review. Neural Netw. **115**, 100–123 (2019)
2. H. Zhou, J. Lan, R. Liu, J. Yosinski, Deconstructing lottery tickets: zeros, signs, and the supermask, in *Advances in Neural Information Processing Systems*, pp. 3597–3607
3. V. Ramanujan, M. Wortsman, A. Kembhavi, A. Farhadi, M. Rastegari, What's hidden in a randomly weighted neural network? in, *IEEE Computer Society Conference on Computer Vision and Pattern Recognition* (2020), pp. 11893–11902
4. Q. Liao, T. Poggio, Bridging the gaps between residual learning, recurrent neural networks and visual cortex. arXiv:1604.03640
5. K. Hirose, J. Yu, K. Ando, Y. Okoshi, Á. López García-Arias, J. Suzuki, T.V. Chu, K. Kawamura, M. Motomura, Hiddenite: 4K-PE hidden network inference 4D-tensor engine exploiting on-chip model construction achieving 34.8-to-16.0TOPS/W for CIFAR-100 and ImageNet, in *IEEE International Solid-State Circuits Conference*, vol. 65 (2022), pp. 1–3
6. J. Fischer, R. Burkholz, Plant'n'seek: Can you find the winning ticket? arXiv:2111.11153

7. K. He, X. Zhang, S. Ren, J. Sun, Deep residual learning for image recognition, in *IEEE Computer Society Conference on Computer Vision and Pattern Recognition* (2016), pp. 770–778

8. Á. López García-Arias, Y. Okoshi, M. Hashimoto, M. Motomura, J. Yu, Recurrent residual networks contain stronger lottery tickets. IEEE Access **11**, 16588–16604 (2023)

9. Y. LeCun, Y. Bengio, G. Hinton, Deep learning. Nature **521**(7553), 436–444 (2015)

10. A. Krizhevsky, I. Sutskever, G.E. Hinton, Imagenet classification with deep convolutional neural networks. Commun. ACM **60**(6), 84–90 (2017)

11. A. Ramesh, M. Pavlov, G. Goh, S. Gray, C. Voss, A. Radford, M. Chen, I. Sutskever, Zero-shot text-to-image generation, in *International Conference on Machine Learning* (2021), pp. 8821–8831

12. E. Strubell, A. Ganesh, A. McCallum, Energy and policy considerations for deep learning in NLP. arXiv:1906.02243

13. S. Han, J. Pool, J. Tran, W. Dally, Learning both weights and connections for efficient neural network, in *Advances in Neural Information Processing Systems* (2015), pp. 1135–1143

14. J. Liu, Z. Xu, R. Shi, R.C. Cheung, H.K. So, Dynamic sparse training: Find efficient sparse network from scratch with trainable masked layers. arXiv:2005.06870

15. M. Zhu, S. Gupta, To prune, or not to prune: exploring the efficacy of pruning for model compression. arXiv:1710.01878

16. T. Gale, E. Elsen, S. Hooker, The state of sparsity in deep neural networks. arXiv:1902.09574

17. D. Blalock, J.J.G. Ortiz, J. Frankle, J. Guttag, What is the state of neural network pruning? arXiv:2003.03033

18. S. Han, H. Mao, W.J. Dally, Deep compression: Compressing deep neural networks with pruning, trained quantization and Huffman coding, in *The International Conference on Learning Representations*

19. A. Dubey, M. Chatterjee, N. Ahuja, Coreset-based neural network compression, in *European Conference on Computer Vision* (2018), pp. 454–470

20. S. Han, X. Liu, H. Mao, J. Pu, A. Pedram, M.A. Horowitz, W.J. Dally, EIE: Efficient inference engine on compressed deep neural network, in *International Symposium on Computer Architecture*

21. J. Frankle, M. Carbin, The lottery ticket hypothesis: Finding sparse, trainable neural networks, in *The International Conference on Learning Representations* (2019)

22. E. Malach, G. Yehudai, S. Shalev-Schwartz, O. Shamir, Proving the lottery ticket hypothesis: Pruning is all you need, in *International Conference on Machine Learning* (2020), pp. 6682–6691

23. A. Pensia, S. Rajput, A. Nagle, H. Vishwakarma, D.S. Papailiopoulos, Optimal lottery tickets via subset sum: Logarithmic over-parameterization is sufficient, in *Advances in Neural Information Processing Systems* (2020)

24. L. Orseau, M. Hutter, O. Rivasplata, Logarithmic pruning is all you need, in *Advances in Neural Information Processing Systems* (2020), pp. 2925–2934

25. J. Diffenderfer, B. Kailkhura, Multi-prize lottery ticket hypothesis: Finding accurate binary neural networks by pruning a randomly weighted network, in *The International Conference on Learning Representations* (2021)

26. A. Veit, M.J. Wilber, S. Belongie, Residual networks behave like ensembles of relatively shallow networks, in *Advances in Neural Information Processing Systems*, p. 29

27. G. Huang, Y. Sun, Z. Liu, D. Sedra, K.Q. Weinberger, Deep networks with stochastic depth, in *European Conference on Computer Vision* (Springer, 2016), pp. 646–661

28. M. Abdi, S. Nahavandi, Multi-residual networks: Improving the speed and accuracy of residual networks. arXiv:1609.05672

29. S. Zagoruyko, N. Komodakis, Wide residual networks, in British Machine Vision Conference (2016)

30. K. Greff, R.K. Srivastava, J. Schmidhuber, Highway and residual networks learn unrolled iterative estimation. arXiv:1612.07771

31. S. Jastrzębski, D. Arpit, N. Ballas, V. Verma, T. Che, Y. Bengio, Residual connections encourage iterative inference. arXiv:1710.04773
32. A. Paszke, S. Gross, F. Massa, A. Lerer, J. Bradbury, G. Chanan, T. Killeen, Z. Lin, N. Gimelshein, L. Antiga, A. Desmaison, A. Kopf, E. Yang, Z. DeVito, M. Raison, A. Tejani, S. Chilamkurthy, B. Steiner, L. Fang, J. Bai, S. Chintala, PyTorch: An imperative style, high-performance deep learning library, in *Advances in Neural Information Processing Systems* (2019), pp. 8024–8035
33. V. Ramanujan, M. Wortsman, A. Kembhavi, A. Farhadi, M. Rastegari, What's hidden in a randomly weighted neural network? (2020) https://github.com/allenai/hidden-networks
34. A. Krizhevsky, Learning multiple layers of features from tiny images, Master's thesis, Department of Computer Science, University of Toronto, Toronto (2009)

Brain-Inspired Reservoir Computing Models

Yuichi Katori

Abstract This chapter presents an overview of brain-inspired reservoir computing models for sensory-motor information processing in the brain. These models are based on the idea that the brain processes information using a large population of interconnected neurons, where the dynamics of the system can amplify, transform, and integrate incoming signals. We discuss the reservoir predictive coding model, which uses predictive coding to explain how the brain generates expectations regarding sensory input and processes incoming signals. This model incorporates a reservoir of randomly connected neurons that can amplify and transform sensory inputs. Moreover, we describe the reservoir reinforcement learning model, which explains how the brain learns to make decisions based on rewards or punishments received after performing a certain action. This model uses a reservoir of randomly connected neurons to represent various possible actions and their associated rewards. The reservoir dynamics allow the brain to learn which actions lead to the highest reward. We then present an integrated model that combines these two reservoir computing models based on predictive coding and reinforcement learning. This model demonstrates how the brain integrates sensory information with reward signals to learn the most effective actions for a given situation. It also explains how the brain uses predictive coding to generate expectations about future sensory inputs and accordingly adjusts its actions. Overall, brain-inspired reservoir computing models provide a theoretical framework for understanding how the brain processes information and learns to make decisions. These models have the potential to revolutionize fields such as artificial intelligence and neuroscience, by advancing our understanding of the brain and inspiring new technologies.

Y. Katori (✉)
The School of Systems Information Science, Future University Hakodate, 116-2
Kamedanakano-cho, Hakodate City 041-8655, Japan
e-mail: katori@fun.ac.jp

© The Author(s) 2024
H. Suzuki et al. (eds.), *Photonic Neural Networks with Spatiotemporal Dynamics*,
https://doi.org/10.1007/978-981-99-5072-0_13

1 Introduction

The brain's capacity to process sensory information and make decisions can be attributed to the intricate neural dynamics within a highly interconnected network of nonlinear elements known as neurons. However, the specific mechanisms underlying this framework are not yet fully understood. The primary objective of engineering and computer science researchers is to develop models that replicate the brain's information processing capabilities. A promising approach in this regard is the brain-inspired reservoir computing model, which has demonstrated effectiveness in diverse applications.

Reservoir computing (RC) is a framework for constructing recurrent neural networks that can model time-varying complex sensory signals [1, 2]. In the RC framework, recurrent connections are randomly and sparsely configured and do not require training. The readout connections from the reservoir are trained to reproduce a given target time series, reducing the network's computational cost. An important feature of RC is that it requires extremely low computational cost for learning because only the connections in the readout part are acquired through training. In addition, RC has many applications, including time-series generation and prediction, pattern recognition in time series, and robot control. The key requirement for RC is the presence of high-dimensional, that is, a large number of nodes or neurons that give rise to complex dynamics. Another important feature of the RC framework is the several possible physical implementations [3, 4], including electrical and optical systems, among other numerous possibilities. Provided that it has high dimensionality, non-linearity, and echo state property, it can serve as a reservoir for computing. The framework of reservoir computing is being actively researched as an approach for modeling brain regions, such as the prefrontal cortex and cerebellum [5, 6].

Predictive coding is a theory of brain function, in which the brain processes sensory information by generating and updating internal models of the external world [7–9]. These models enable the brain to predict future sensory inputs. When the actual input deviates from the predicted input, prediction error signals are generated and sent back through the neural network to update the models. The iterative process of generating and updating predictions improves the accuracy of the model and reduces prediction error over time. Predictive coding is a widely accepted framework for understanding perception, attention, and learning in the brain and has been applied to various sensory modalities, including vision, audition, and touch. The predictive coding model is a key component of many neural network models of the brain and has been used to explain different neural phenomena, such as adaptation, attentional modulation, and perceptual illusions. Despite its success, however, the predictive coding model remains an active area of research with ongoing debates over the specific mechanisms and neural substrates underlying predictive coding in the brain [10].

The brain's reward system and the process of reinforcement learning are essential components of decision-making and learning. The reward system is a collection of neural circuits that processes information related to motivation, pleasure, and rewards. It plays a critical role in shaping behavior, such as learning and motivation, by providing feedback on the outcomes of an action. The neurotransmitter dopamine, which is released in response to a reward or the anticipation of a reward, is a key mediator of the activity of the reward system. In reinforcement learning, an agent learns to select actions based on rewards or punishments [12]. This involves learning to maximize long-term cumulative rewards by taking actions in an environment. Reinforcement learning algorithms often use trial and error to learn the optimal behavior, explore the environment, and observe the consequences of different actions. The brain's reward system and the process of reinforcement learning are closely related. The reward system provides feedback that drives the learning process, and reinforcement learning provides a framework for understanding how the brain learns to make decisions based on rewards. Computational models of reinforcement learning have been successful in explaining different behaviors, including goal-directed behavior, habit formation, and addiction [12].

In this chapter, we discuss brain-inspired reservoir computing models for sensory-motor information processing in the brain [13, 14]. First, we introduce the reservoir predictive coding model that corresponds to sensory information processing in the cerebral cortex [11]. Subsequently, we discuss the reservoir reinforcement learning model that corresponds to action learning based on rewards in the basal ganglia [14]. Finally, we present an integrated model that combines these two RC models based on predictive coding and reinforcement learning [14]. This integrated model has the potential to provide a more comprehensive understanding of the brain's information processing mechanisms.

2 Reservoir-Based Predictive Coding Model

Predictive coding is a neuroscience theory explaining how the brain processes sensory information by constantly making predictions about the world and updating them based on incoming data [7–9]. This concept is particularly relevant to the hierarchical organization of the visual system in the brain, which consists of multiple processing stages, each of which is responsible for detecting specific features of the visual input. For example, lower-level neurons may detect simple features such as edges, whereas higher-level neurons may identify more complex patterns or objects. The same principle applies to the architecture of CNNs, which have multiple layers that learn to extract increasingly complex features from input images.

Predictive coding posits that the brain actively generates predictions regarding sensory input at each level of the hierarchy. These predictions are based on information gathered from higher hierarchical levels and on previously learned internal models. These internal models, also known as generative models, make predictions and propagate them to lower levels via a top-down pathway. The difference between the actual input and the prediction, known as the prediction error, then propagates up the hierarchy in a bottom-up manner. This error signal helps the brain update its internal models and refine future predictions.

In the field of neural networks, the predictive coding with reservoir computing (PCRC) model proposed by Katori et al. [11] is a novel approach for processing time-varying sensory signals. The PCRC model employs a reservoir as the generative model for predictive coding, wherein the reservoir generates multidimensional, time-varying sensory signals. The prediction error is subsequently transmitted back to the reservoir, allowing for the rectification of the network's internal state. This model demonstrates the capability of reconstructing and predicting time-varying sensory signals.

The network architecture within each module comprises four key components: the prediction layer, input layer, prediction error layer, and reservoir (Fig. 1a). Within the module, the input signal located in the input layer is replicated in the prediction layer, which is facilitated by the complex motion of the reservoir. This prediction error is then fed back into the reservoir to minimize errors. During the training phase, the connection between the reservoir and the prediction layer is modulated using the first-order reduced and controlled error (FORCE) algorithm [15].

In the testing phase, the model operates in two distinct modes: error-driven and free-running. The error-driven mode involves feedback on the prediction error to the reservoir to further reduce the error. In contrast, the free-running mode does not involve the transmission of the prediction error to the reservoir, allowing for the autonomous operation of the reservoir. This dual-mode functionality highlights the versatility and adaptability of the PCRC model for processing time-varying sensory signals.

The PCRC module consists of a reservoir, prediction layer, input layer, and prediction error layer, which are mathematically described as follows: The membrane potential, or internal state, and the neuron activities within the reservoir are represented by $\boldsymbol{m} \in \mathbb{R}^{N_x}$ and $\boldsymbol{r} \in \mathbb{R}^{N_x}$, respectively, where N_x denotes the size of the reservoir. The states of the reservoir are updated according to the following equations:

$$\boldsymbol{m}(n+1) = \boldsymbol{m}(n) + \frac{1}{\tau}\{-\boldsymbol{m}(n) + W_{\text{rec}}\boldsymbol{r}(n) + W_{\text{back}}\boldsymbol{y}(n) + \alpha_e W_e \boldsymbol{e}(n) - \boldsymbol{b}(n)\}, \quad (1)$$

$$\boldsymbol{r}(n+1) = \tanh(\beta_m \boldsymbol{m}(n+1)), \quad (2)$$

where $W_{\text{rec}} \in \mathbb{R}^{N_x \times N_x}$ represents the matrix for recurrent connections, and τ is the time constant. The parameter β_m scales the neuron activities. The reservoir receives inputs from the prediction layer $\boldsymbol{y}^{(i)} \in \mathbb{R}^{N_y}$ through the feedback connection $W_{\text{back}} \in \mathbb{R}^{N_x \times N_y}$, the prediction error layer $\boldsymbol{e} \in \mathbb{R}^{N_y}$ with a coefficient α_e that determines the error feedback strength and model operation mode, and the top-down input from the higher area network $\boldsymbol{b}^{(i)}(n)$. The states of the prediction and prediction error layers are given by

$$\boldsymbol{y}(n) = \max(0, W_{\text{out}}\boldsymbol{r}(n)), \quad (3)$$

$$\boldsymbol{e}(n) = \boldsymbol{d}(n) - \boldsymbol{y}(n). \quad (4)$$

(a)

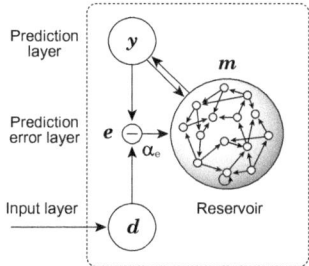

(b)

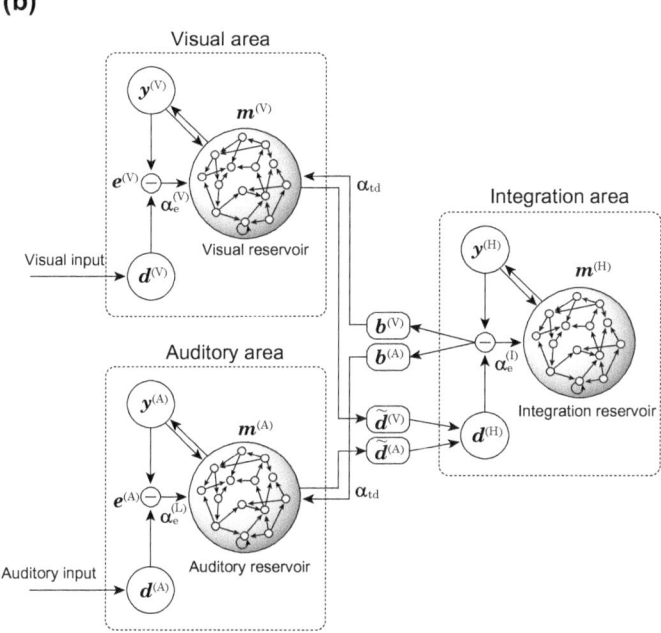

Fig. 1 Network structure of the PCRC models. **a** Module of the predictive coding based on reservoir computing. **b** PCRC-based hierarchical model for the multimodal processing of the visual and auditory processing

In the error-driven mode ($\alpha_e = 1$), the reservoir is updated using the prediction error, and the state of the prediction layer follows the state of the input layer. In the free-running mode ($\alpha_e = 0$), the reservoir states are updated based on the internal dynamics, independent of the sensory input.

The network's configuration and learning process involve the following steps. The recurrent connections within the reservoir and the feedback connections from the prediction layer to the reservoir are configured in a random and sparse manner, with no need for training their connectivity. During the training phase, the network operates in error-driven mode, and the connections from the reservoir to the prediction layer are trained using the FORCE learning algorithm with a given time-series dataset. Recurrent connections W_{rec} are set up using the following procedure. First, create a matrix W_0 filled with zeros. Then, assign the non-zero values of either -1 or 1 to randomly chosen $\beta_r \times N_x \times N_x$ elements. Then, compute the spectral radius of W_0: $|\rho_0|$. Define $W_{\text{rec}} = \alpha_r W_0/|\rho_0|$, where α_r indicates the strength of recurrent connections. Feedback connections W_{back} and W_e are set up using the following procedure. Similar to W_{rec}, generate a zero matrix W_0 and assign non-zero values of -1 or 1 to the randomly selected $\beta_b \times N_x \times N_y$ elements, where β_b specifies connectivity of the recurrent connection. Define $W_{\text{back}} = \alpha_b W_0$, with the strength of the feedback connections given by α_b. Use the same procedure to generate W_e with the coefficient α_e.

The readout connections from the reservoir W_{out} are updated using the FORCE learning algorithm [15] as follows:

$$v(n) = P(n)r(n) , \tag{5}$$

$$P(n + 1) = P(n) - \frac{v(n)v^T(n)}{1 + v^T(n)r(n)} , \tag{6}$$

$$W_{\text{out}}(n + 1) = W_{\text{out}}(n) - \frac{e(n)v^T(n)}{1 + v^T(n)r(n)} . \tag{7}$$

The initial value of $P(n) \in \mathbb{R}^{N_x \times N_x}$ is $P(0) = \frac{I}{\alpha_f}$, where matrix I is an identity matrix and α_f is a scaling parameter. Once the readout connection training is complete, the module can reconstruct the given input in error-driven mode and predict the input in free-running mode.

The PCRC-based hierarchical model for multimodal processing comprises three modules (Fig. 2b). Each module in the hierarchical model is distinguished by superscript (i), where $i \in \{V, A, I\}$ denotes the visual, auditory, and integration modules, respectively.

The configuration and learning of this model were performed using the following steps: The recurrent and feedback connections were established in accordance with the previously described procedure. The connection matrices between the lower and higher levels, $U_{(A)}$ and $U_{(V)}$, are defined using the method below; their inverse matrices are used for dimensionality reduction.

Firstly, operate the lower area network (visual and auditory areas) in error-driven mode ($\alpha_e^{(V)} = 1$ and $\alpha_e^{(A)} = 1$) without top-down signals ($\alpha_{td} = 0$), and gather the time course of the reservoirs $r^{(V)}$ and $r^{(A)}$ in the state collecting matrices $R^{(V)}$ and $R^{(A)}$, respectively. Next, compute the dimension reduction matrices $U_{(V)}$ and $U_{(A)}$. Assuming that T timesteps of reservoir states are collected in $R^{(i)}$ ($i \in \{V, A\}$), $R^{(i)}$ can be decomposed by principal component analysis (PCA) as $R^{(i)} = S^{(i)} U_{(i)}^T$. Here, $S^{(i)}$ is a $T \times 20$ matrix, and $U_{(i)}$ is an $N_x^{(i)} \times 20$ matrix. The dimension reduction matrix $U_{(i)}^{-1}$ can be obtained as the pseudo-inverse matrix of $U_{(i)}$. Finally, connect the sensory modules (visual and auditory modules) and the integration using the obtained $U_{(i)}$, and operate the entire network in error-driven mode ($\alpha_e^{(V)} = \alpha_e^{(A)} = 1$) with $\alpha_{td} > 0$. The matrices $W_{\text{out}}^{(V)}$, $W_{\text{out}}^{(A)}$, and $W_{\text{out}}^{(I)}$ are acquired using FORCE learning.

Within the hierarchical PCRC model for visual and auditory processing, the integration reservoir is responsible for reconstructing and predicting the compressed and concatenated states of the sensory reservoirs. Consequently, the integration reservoir is expected to reconstruct information from one modality using information from another modality. Both the visual and auditory reservoirs are driven by the prediction error on each sensory module and the integration reservoir.

The multimodal model is assessed using time-series data, consisting of pairs of hand-written digit images and their corresponding spoken number utterances. Three hand-written digit images ("2," "5," and "9") from the MNIST dataset are employed as visual signals [16]. Each image comprises 28×28 (784) grayscale pixels. These images undergo preprocessing via non-negative matrix factorization (NMF) and are converted into a 20-dimensional signal. Assuming V is an $L \times 784$ matrix with each row representing an individual image, and V is a collection of L images, NMF decomposes V into two matrices: $V = HW$, where H is an $L \times 20$ coefficient matrix and W is a 20×784 feature matrix. The transformed 20-dimensional vector serves as the input to the visual area network. The coefficient vector reconstructed by the PCRC module can be converted back into images using W.

In addition, linguistic data containing spoken number utterances from the Ti46 dataset are utilized as auditory signals [17]. This dataset comprises uncompressed audio data. Each dataset is preprocessed using a cochlear filter model [18], an auditory model that simulates sound propagation within the inner ear, and the conversion of acoustic energy into neural representations. The auditory signals are transformed into 55-dimensional signals. Figure 2 displays samples of the dataset. In the auditory signal, the initiation of spoken number utterances exhibits jitter, starting anywhere from 60 to 90 timesteps. The corresponding visual signals are presented from 80 to 160 timesteps without jitter.

After training, the network is expected to reconstruct sensory information from one modality based on input signals originating from the other modality. In the subsequent analysis, the focus is on reconstructing visual information in the presence of corresponding auditory signals. In this case, the auditory and integration reservoirs operate in error-driven mode, whereas the visual reservoir functions in free-running mode.

In the association process, a given auditory signal is initially presented to the input layer of the auditory area. At this time, the reservoir maintains a silent state; as no signal is formed in the prediction layer, a significant prediction error occurs. This prediction error serves as a trigger, inducing the motion of the auditory reservoir and generating the auditory signal in the prediction layer. Subsequently, the prediction error gradually decreases. A spatial pattern reflecting the temporal pattern in the auditory signal is represented within the reservoir. This information is spatially compressed and conveyed to the integration reservoir, where only the auditory information is input. The prediction layer in the integration area initially remains

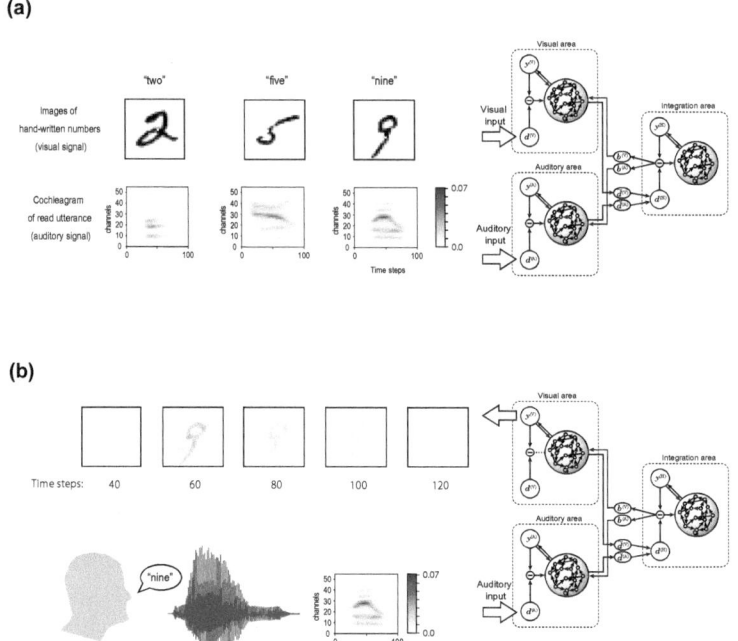

Fig. 2 Evaluation of the PCRC-based multimodal model with auditory and visual signals. **a** Training phase: All modules are operated in error-driven mode. The model is trained with the datasets comprising time-series pairs of visual signals (hand-written digits) and auditory signals (spoken utterances of corresponding digits). Each signal is displayed for 100 timesteps. Visual signals undergo preprocessing using NMF, resulting in a 20-dimensional signal that serves as sensory input for the visual area network. Auditory signals are preprocessed with a cochlear filter and converted into 55-dimensional signals. These signals are then provided as sensory input to the auditory area network. **b** Cross-modal association from the auditory signal to the visual signal after the training phase: Auditory and the integration modules are operated in error-driven mode, whereas the auditory module is operated in free-running mode. The model is driven by an auditory signal, and the corresponding visual image appears in the prediction layer of the visual area

silent, and the prediction error triggers the activity of the integration reservoir. As the integration reservoir begins to move, predictions are generated to compensate for the prediction error. At this time, both the auditory and corresponding visual signals are generated. Because there is no signal coming from the lower visual layer, the prediction error regarding the visual information is larger. This prediction error is then transmitted to the visual area, inducing activity in the visual reservoir. Based on the fluctuations in the visual reservoir, visual signal prediction is performed. In summary, a visual signal corresponding to the input auditory signal is generated in the prediction layer of the visual area.

During the processing of multidimensional complex time courses, the proposed hierarchical model combines the mechanisms of temporal structure accumulation and spatial pattern compression. The input signal is reconstructed using a reservoir that captures the temporal structure of the signal within its high-dimensional nonlinear dynamics. Subsequently, the high-dimensional state vector in the reservoir, which encompasses a short history of the signal, undergoes spatial compression and is transferred to the integration area network. This combination of accumulation and compression results in a higher-order abstraction of the intricate time course. In cross-modal association, the processes of compression and abstraction are reversed, allowing the generation of sensory information through expansion and instantiation.

3 Reservoir-Based Reinforcement Learning Model

Reinforcement learning is a type of learning in which an agent learns to choose actions based on rewards or punishments, with the goal of maximizing long-term cumulative rewards by taking actions in an environment. Reinforcement learning algorithms often employ trial and error to learn optimal behavior, explore the environment, and observe the consequences of various actions. The learning process is fueled by feedback from the environment, providing information on the outcomes of an agent's actions. Among the various reinforcement learning approaches, TD-learning is a model-free technique that merges ideas from dynamic programming and Monte Carlo methods [12, 19]. It estimates the value function (expected future reward) by learning from the difference between consecutive predictions, which is known as the temporal difference error. This method enables agents to learn online and update their value estimates incrementally as new experiences are acquired, making it particularly suitable for learning in dynamic environments.

In recent years, RL has been combined with deep learning to create deep reinforcement learning (DRL), which has achieved remarkable success in solving complex control tasks with high-dimensional sensory inputs, such as images and sounds [20]. DRL algorithms, such as Deep Q-Networks (DQN) [20], proximal policy optimization (PPO) [21], and actor-critic methods [22, 23], have been successfully applied to various applications, such as video game playing, robotics, and autonomous driving.

The two important frameworks within RL are Markov decision processes (MDPs) [24] and partially observable Markov decision processes (POMDPs) [25]. MDPs are a mathematical framework used to model decision-making problems in reinforcement learning, where the environment's state transitions and rewards are assumed to be Markovian; that is, the future state depends only on the current state and action taken and not on previous states or actions. Although MDPs have been successfully applied to various problems, they exhibit certain limitations, particularly in partially observable environments.

In real-world situations, an agent may not have full access to the environment's state owing to noisy sensors, occlusions, or other factors. This lack of complete information about the environment's state can lead to suboptimal decision-making, as the agent cannot accurately estimate the value of different actions. This is where POMDPs play a significant role, extending the MDP framework to handle environments with partial observability.

POMDPs are a generalization of MDPs that consider uncertainty in perceiving the environment's state. Instead of using the environment's true state, the agent maintains a belief state, which is a probability distribution over the possible environment states. The belief state is updated as the agent takes actions and receives observations, allowing it to make better-informed decisions even with incomplete information. However, solving POMDPs is generally more computationally demanding than MDPs owing to the increased complexity associated with maintaining and updating belief states.

One approach to address POMDP is the use of reservoirs. Reservoirs, which store not only the current state but also the history of sensory inputs reflecting the environmental state in high-dimensional state vectors, can be expected to function effectively in POMDP environments by compensating for information that cannot be directly observed. In the following, we introduce how the reservoir reinforcement learning model, which is a model that reads action values from the reservoir states where the history of sensory information is accumulated, can effectively function in POMDP environments.

The proposed model consists of a sensory layer, a dynamical reservoir, and an output layer (Fig. 3a). The reservoir receives sensory input from the environment through the sensory layer and generates action values in the output layer, which are then converted to action commands. The reservoir state, comprising N_x neurons, is denoted by $x(t) \in \mathbb{R}^{N_x}$. The dynamical reservoir state $x(t)$ evolves as follows:

$$x(t + \Delta t) = x(t) + \frac{\Delta t}{\tau_x} \left(-x(t) + W^{\text{in}} u(t + \Delta t) + W^{\text{rec}} r(t) + W^{\text{back}} q(t) \right) , \quad (8)$$

where τ_x is the time constant; W^{in} is the $N_x \times N_u$ sensory matrix from the sensory layer to the dynamical reservoir, W^{rec} is the $N_x \times N_x$ recurrent weight matrix in the dynamical reservoir, and W^{back} is the $N_x \times N_y$ feedback matrix from the output layer to the dynamical reservoir. These weight matrices W^{in}, W^{rec}, and W^{back} are randomly and sparsely generated and remain fixed. The neuron firing rate in the dynamical reservoir, $r(t)$, is defined as $r(t) = f_r(\beta x(t))$, where β specifies the firing rate responses and $f_r(x) = \tanh(x)$. The output layer state, denoted by $q(t) \in \mathbb{R}^{N_y}$, represents the N_y action values and is specified according to:

$$q(t) = W^{\text{out}} r(t) . \quad (9)$$

W^{out} is the $N_y \times N_x$ output weight matrix from the dynamical reservoir to the output layer. Reservoir-based TD-learning is performed on the matrix W^{out} to minimize the temporal difference and approximate the action quality (Q-value). Exploration noise, $s(t) \in \mathbb{R}^{N_y}$, is added to the output. The Q-value for the exploration noise is denoted by $\tilde{q}(t)$. The connection from the reservoir to the output layer, W^{out}, is trained using the following equations in an online learning manner:

$$W^{\text{out}}_{a,j}(t + \Delta t) = W^{\text{out}}_{a,j}(t) + \eta(t) f_q \left(R(t) + \gamma \tilde{q}_a(t + \Delta t) - q_a(t) \right) r_j(t + \Delta t),$$
$$(10)$$

(a)

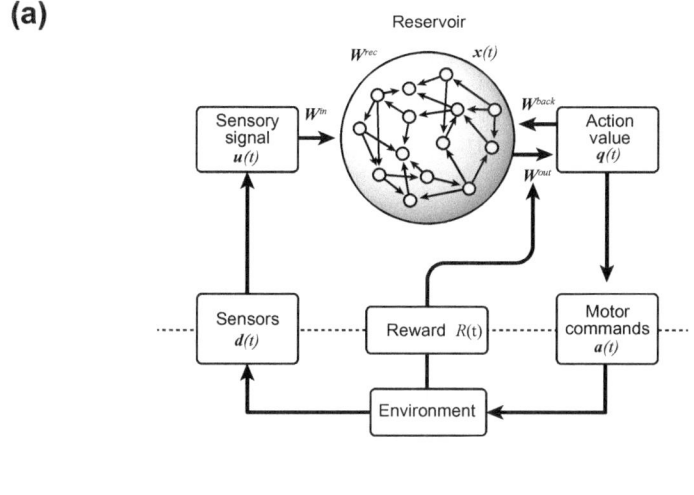

(b)   **(c)**

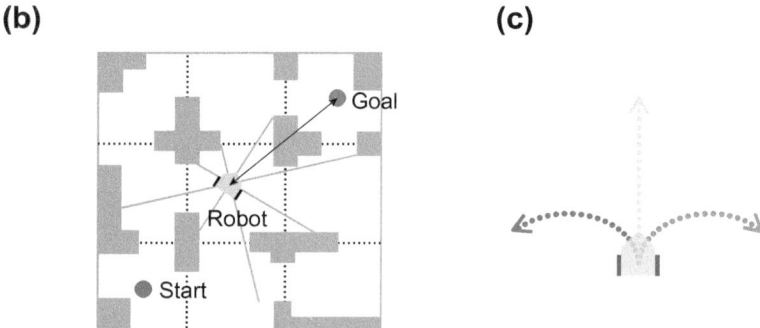

Fig. 3 Reservoir-based reinforcement learning and its evaluation. **a** Network structure of the reservoir-based reinforcement learning. **b** Environment of the autonomous robot: The robot (agent) is required to move from the start to the goal position. The agent receives a positive reward depending on the distance between the goal position and the robot and receives a negative reward (punishment) if the robot crashes into the wall. **c** The robot sequentially chooses from one of the possible three actions (move left, right, or forward)

where a is the index of the action commands, and the action command at time t is given by $a(t) = \arg\max_i(q_i(t))$. $R(t)$ represents the reward received from the environment, and $f_q(x) = \tanh(x)$. γ is the discount factor, and $\eta(t)$ is the learning rate. The exploration noise is temporally correlated and changes according to the following equation:

$$s(t + \Delta t) = \left(1 - \frac{\Delta t}{\tau_s}\right) s(t) + \sigma_s N(0, 1) , \tag{11}$$

$$\tilde{q}(t + \Delta t) = q(t + \Delta t) + s(t + \Delta t) , \tag{12}$$

where τ_s is a time constant, and σ_s represents the noise strength. $N(0, 1)$ is a random variable following a normal distribution with a mean of 0 and a standard deviation of 1.

The proposed reservoir reinforcement learning model was assessed within a simulation environment, in which the model was tasked with navigating a robot to a designated goal (Fig. 3b, c). The information received by the agent from the environment is the distance to the obstacles in eight directions around the robot. The sensory layer state is given by $u(t) = \exp\left(-\frac{d(t)}{d_0}\right) \in \mathbb{R}^{N_u}$, where $d(t)$ is the sensory signal from $N_u = 8$ distance sensors. Note that the position and direction information of the robot are not provided to the agent. The agent is required to continuously choose one of three possible actions (move left, right, or forward). The agent receives a positive reward depending on the distance between the robot and the given goal position and a negative reward if the robot crashes on the obstacle. In the simulations, we use the following parameter values: $N_x = 500$, $N_u = 8$, $N_y = 3$, $\Delta t = 1$, $\beta = 1$, $\tau_x = 2$, $\tau_s = 20$, $\gamma = 0.9$, and $d_0 = 100$.

Figure 4a illustrates the typical robot trajectories during training. Initially, the robot quickly collided with obstacles and failed to reach the goal. However, as training continued, the robot learned to avoid obstacles; after 300 episodes, it successfully circumvented most obstacles and reached the goal.

Figure 4b shows a typical time course of the network state after 300 training episodes. The reservoir state fluctuated in response to the sensory signals, which varied based on the distance between the robot and the obstacles. In the output layer, the Q-values for the three potential actions were determined, reflecting reservoir fluctuations. The action corresponding to the highest Q-value was selected as the motor command. The Q-value for moving forward was lower than those for turning right or left. The maximum Q-value alternated between turning right and left, thereby restricting the robot's motion to either turning right or left.

The proposed reservoir reinforcement learning model effectively learned the action sequence required to reach a given goal within the environment. The sequence of sensory signals induced substantial fluctuations in the reservoir state, and reward-based training resulted in an appropriate action sequence. Future studies should focus on refining the network model in several ways. From a neuroscience perspective, the function demonstrated in this study, specifically the transformation of sensory information into motor information, underlies the prefrontal cortex. Additional neuroscience-inspired functions should be incorporated, such as the amygdala, which offers a gating mechanism for sensory signals based on their importance, or hippocampal grid and place cells, which enable flexible representation of the agent's position.

(a)

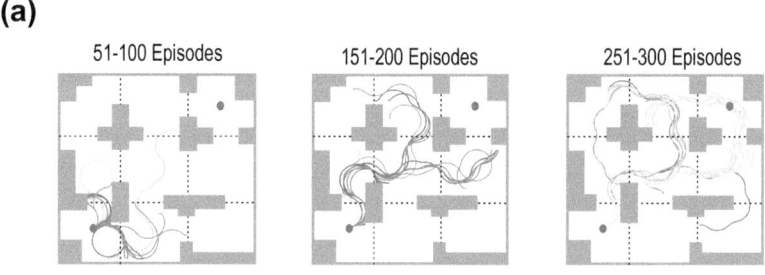

(b)

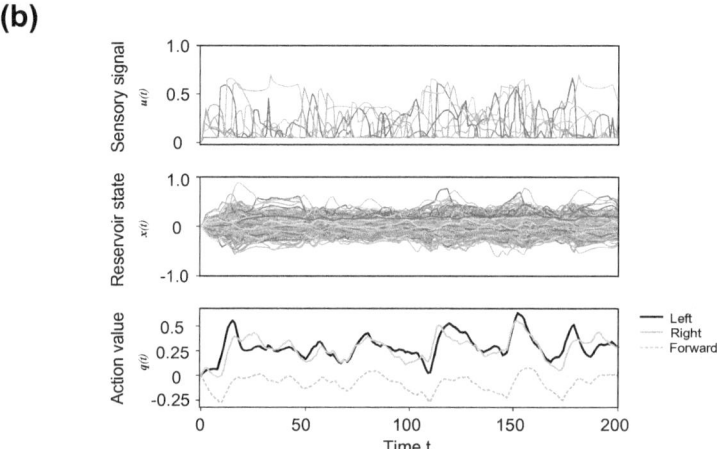

Fig. 4 Robot navigation task **a** Trajectory of the robot during the learning process. At the beginning of the learning stage (episodes 51–100), the robot collides with obstacles shortly after starting and does not reach the goal. However, in the middle of the learning stage (episodes 151–200), the robot learns to avoid collisions with obstacles. At the end of the learning stage (episodes 251–300), the robot learns to reach the goal while avoiding obstacles. **b** Temporal changes of sensory input, reservoir, and action value after learning. (Top) Sensory signal reflecting changes in the distance to obstacles as the robot moves. (Middle) Internal state of the reservoir that fluctuates according to the sensory signal. (Bottom) Action-value functions corresponding to the three actions

4   Integrated Model and Mental simulation

Mental simulation is a cognitive process in which an individual mentally enacts or imagines a scenario or action without physically performing it [29]. This mental rehearsal can be used for various purposes, such as problem-solving, planning, decision-making, and skill development [28]. Mental simulation allows individuals to predict the outcomes of various actions, assess risks, and evaluate potential solutions without committing to a specific course of action in the real world.

In the context of artificial intelligence and robotics, mental simulation refers to an agent's ability to internally model and predict the consequences of its actions in a given environment [30]. One approach for implementing mental simulation in AI systems is to use world models, which is an internal representation of the agent's environment, capturing relevant information regarding the relationships, objects, and dynamics within that environment. By simulating potential actions and their consequences within the world model, the agent can make better-informed decisions, adapt to new situations, and learn from hypothetical scenarios without requiring actual interaction or trial-and-error experiences. This approach can improve the learning efficiency and reduce the time and resources required for training.

A reservoir-based mental simulation model combining the predictive coding and reinforcement learning models described above (Fig. 5) has been proposed [27]. In this model, the reservoir generates predictions of the sensory input and action values as readouts. After a pretraining phase, the model operates in two distinct modes: execution and mental simulation. In execution mode, the agent and environment are connected, allowing the reservoir to receive information from the environment and output actions that influence the environment. In contrast, the mental simulation mode involves decoupling the agent's actions from the environment, with the predictive error feedback disconnected. In this mode, the reservoir functions as a world model, simulating environmental changes within the agent's internal network.

The process of action planning using mental simulation consists of two phases: pretraining and test. The pretraining phase involves collecting fundamental information about the environment and constructing a world model within the reservoir. In the test phase, the reservoir is detached from the environment, and action planning is conducted by simulating the constructed world model. This enables the optimization of action sequences required to achieve the desired state.

The overall network structure is illustrated in Fig. 5a. The agent receives sensory signals d from the environment and generates sensory information predictions y from the reservoir. The agent generates action values q based on sensory information predictions and the state of the reservoir. This action value is modulated by the bias input b, which is determined by optimization. Actions a are determined based on the action values, and these actions are sent to the environment while simultaneously being fed back into the reservoir.

In the pretraining phase, the agent is connected to the environment and updates the connections from the reservoir to the layers representing sensory information predictions and action values (Fig. 5b left). In the planning mode of the test phase, the agent is disconnected from the environment, and environmental changes and action selection are simulated through the reservoir's internal dynamics (Fig. 5b center). In the execution mode of the test phase, the agent reconnects to the environment and performs actions in the real environment using the bias determined in the planning mode (Fig. 5b right). During the planning mode of the test phase, the bias terms are optimized such that the state of the simulated environment is close to the desired state.

The model is evaluated in the context of a mobile robot environment. The robot receives the following sensory signals: the distance to obstacles in eight directions

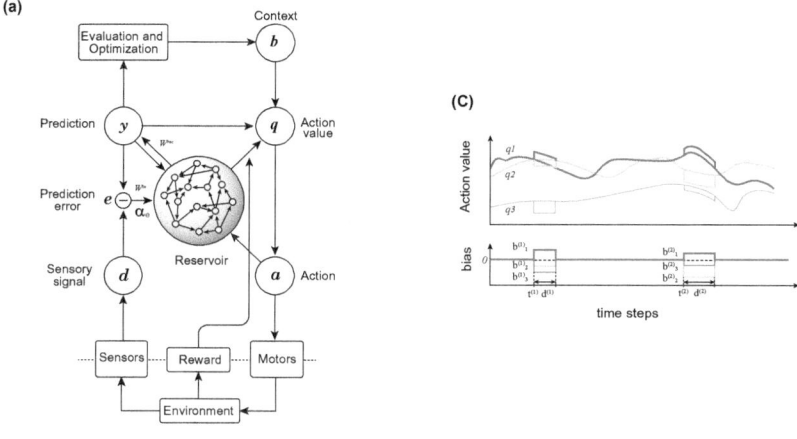

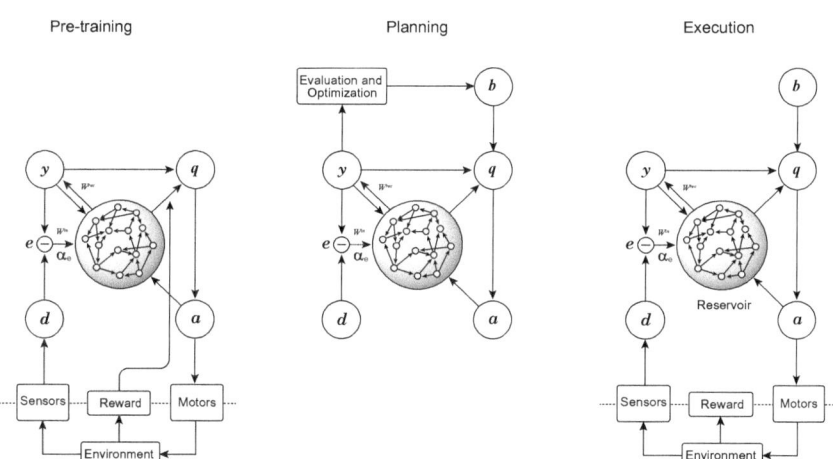

Fig. 5 Integrated reservoir model of predictive coding and reinforcement learning. **a** Overall network structure. **b** Network components operating in each phase and mode: pretraining phase (left), planning model in the test phase, and the execution mode of the test phase. **c** Optimization of the bias terms of the action value. Optimization of the bias terms is performed so that the state of the simulated environment is close to the desired state. The sequences of the action values generated by the reservoir (upper panel) and the bias term (lower panel). The case of the three possible actions is shown

around the robot, the position of the robot with a place-cell representation, and the direction of the robot [31].

During the pretraining phase, the robot learns to move through the environment while avoiding collisions with the walls. This pretraining involves generating outputs for both the predictive layer and action value readouts, following predictive coding

Fig. 6 Action planning task. The goal location is set at the position marked by a star. Robot trajectories (solid curves) are shown when operating in the environment using the action value bias obtained through optimizing the action sequence by mental simulation. Trajectories without action planning (no bias) are represented by dashed curves. When using the action value bias obtained through action planning, the robot reaches the vicinity of the goal location. The starting orientation of the robot is facing right (upper panel) and upward (lower panel)

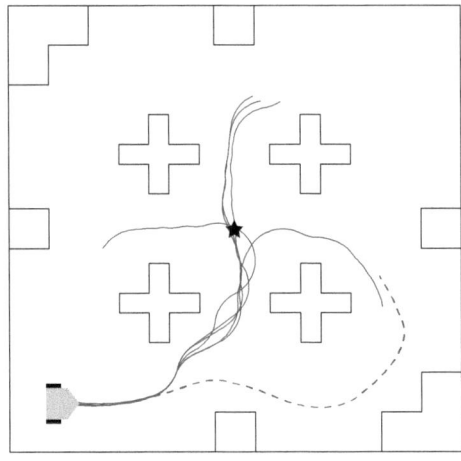

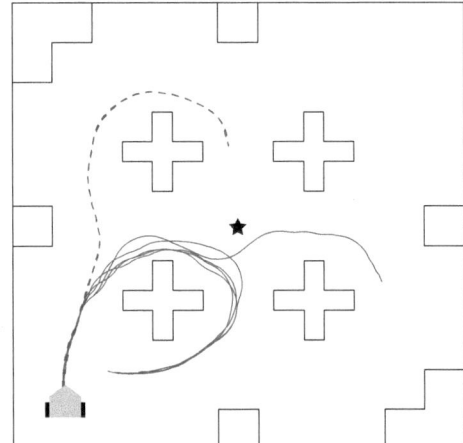

and temporal difference learning models. In this phase, no specific goal location is defined; however, a negative reward (penalty) is given upon collision with a wall.

During the test phase, action planning is conducted, with the task requiring the robot to navigate to a specific location within the environment. In the mental simulation of the planning mode, the robot generates and optimizes action sequences from the starting point to the desired position in the environment. Action values are augmented with the bias term to modify the actions.

The bias terms consist of three parameters corresponding to the possible number of actions: $N_y = 3$. In addition to these three parameters, the start time and duration of the bias application must be optimized. In the case of $N_y = 3$, there are a total of five parameters to be optimized. Depending on the task, these bias terms can be combined into multiple sets of modifications to optimize actions. These parameters are optimized by minimizing the distance between the robot's current position and target location. In the mental simulation, the robot's position can be estimated from the states generated in the predictive layer, and the distance to the desired position can be measured. The Nelder-Mead method [32] is utilized for parameter optimization.

Figure 6 demonstrates that the action sequences planned through mental simulation can be successfully applied in a real environment, allowing the robot to effectively reach the target location. The solid line represents the trajectory of the robot's actions. Without context-vector-based action modification, the robot cannot reach its destination (dashed line).

This example illustrates how the dynamic characteristics of reservoir computing can be effectively employed in action planning. Although the robot task presented herein involves only three possible actions and relatively simple planning, more complex environments require further evaluation in the future. In addition to action planning, mental simulation may help accelerate learning processes. Reinforcement learning typically requires trial-and-error learning, involving numerous interactions between the agent and the environment. However, this process can be replaced by mental simulation through internal dynamics.

5 Summary

The brain is an important organ that allows us to perceive the world around us, learn from our experiences, and make decisions based on this learning process. However, understanding the brain's information processing mechanisms is challenging because of its complexity and dynamism. Brain-inspired reservoir computing models are one approach that seeks to elucidate these mechanisms. These models are based on the idea that the brain processes information using a large population of interconnected neurons, where the dynamics of the system can amplify, transform, and integrate incoming signals.

In this chapter, we discussed brain-inspired reservoir computing models for sensory-motor information processing in the brain. We began by introducing the reservoir predictive coding model based on the theory of predictive coding. Predictive coding posits that the brain constantly generates expectations regarding the sensory input it receives and uses these expectations to interpret and process incoming signals. The reservoir predictive coding model incorporates a reservoir of randomly connected neurons that amplify and transform sensory inputs and generates predictions regarding future sensory inputs. This model also highlights the role of feedback connections between different levels of processing in the brain, which can refine and update these predictions.

Subsequently, we discussed the reservoir reinforcement learning model, which corresponds to action learning based on rewards in the basal ganglia. This model explains how the brain learns to make decisions based on rewards or punishments received after performing a certain action. This model uses a reservoir of randomly connected neurons to represent various possible actions and their associated rewards. The reservoir dynamics allow the brain to learn which actions lead to the most rewards. The reservoir reinforcement learning model also highlights the role of neuromodulators, such as dopamine, in shaping the learning and decision-making processes of the brain.

Finally, we presented an integrated model that combines these two reservoir computing models based on predictive coding and reinforcement learning. This integrated model has the potential to provide a more comprehensive understanding of the brain's information processing mechanisms. This model demonstrates how the brain integrates sensory information with reward signals to learn the most effective actions for a given situation. It also explains how the brain uses predictive coding to generate expectations about future sensory inputs and accordingly adjusts its actions.

Overall, these brain-inspired reservoir computing models offer a new perspective on the workings of the brain. They provide a theoretical framework for understanding how the brain processes information and learns to make decisions. By incorporating principles from both predictive coding and reinforcement learning, these models offer a more complete picture of the brain's information processing mechanisms. This could have important implications in fields such as artificial intelligence and robotics, where researchers are trying to build machines that can learn and adapt similar to the human brain.

There are several directions for future research on brain-inspired reservoir computing models. First, it is important to understand the computational and neural mechanisms underlying these models. This could involve conducting simulations and experiments to validate the models and test their predictions. Second, it is interesting to explore how these models can be applied to real-world problems, such as robotic control or natural language processing. Finally, it is important to consider the ethical and societal implications of developing more intelligent machines based on these models.

In conclusion, brain-inspired reservoir computing models offer a promising approach for understanding the brain's information processing mechanisms. They provide a theoretical framework for guiding future research and motivating new technologies. By advancing our understanding of the brain, these models have the potential to revolutionize fields such as artificial intelligence and neuroscience.

Acknowledgements This work was supported by JST CREST(JPMJCR18K2) and JSPS KAKENHI (21H05163, 20H04258, 20H00596, 21H03512) and was based on the results obtained from a project, JPNP16007, commissioned by the New Energy and Industrial Technology Development Organization (NEDO).

References

1. H. Jaeger, Tutorial on Training Recurrent Neural Networks, Covering BPPT, RTRL, EKF and the "Echo State Network" Approach, GMD Report, vol. 5 (2002)
2. W. Maass, T. Natschläger, H. Markram, Real-time computing without stable states: a new framework for neural computation based on perturbations. Neural Comput. **14**(11), 2531–2560 (2002). https://doi.org/10.1162/089976602760407955. (Nov.)
3. G. Tanaka, T. Yamane, J.B. Héroux, R. Nakane, N. Kanazawa, S. Takeda, H. Numata, D. Nakano, A. Hirose, Recent advances in physical reservoir computing: A review. Neural Netw. **115**, 100–123 (2019). https://doi.org/10.1016/j.neunet.2019.03.005. (Jul.)
4. K. Nakajima, Physical reservoir computing—An introductory perspective, nlin.AO, 2005.00992 (2020). https://doi.org/10.35848/1347-4065/ab8d4f
5. T. Yamazaki, S. Tanaka, Computational models of timing mechanisms in the cerebellar granular layer. Cerebellum **8**(4), 423–432 (2009). https://doi.org/10.1007/s12311-009-0115-7
6. K. Tokuda, N. Fujiwara, A. Sudo, Y. Katori, Chaos may enhance expressivity in cerebellar granular layer (2020). arXiv:2006.11532v1 [q-bio.NC]
7. R.L. Gregory, Perceptions as hypotheses. Philos. Trans. R. Soc. B Biol. Sci. **290**(1038), 181–197 (1980). https://doi.org/10.2307/2395424
8. R. Rao, D. Ballard, Predictive coding in the visual cortex: a functional interpretation of some extra-classical receptive-field effects. Nat. Neurosci. **2**, 79–87 (1999). https://doi.org/10.1038/4580
9. K. Friston, Hierarchical models in the brain. PLoS Comput. Biol. **4**(11), e1000211 (2008). https://doi.org/10.1371/journal.pcbi.1000211
10. S. Shipp, Neural elements for predictive coding. Front. Psychol. **7**, 1792 (2016). https://doi.org/10.3389/fpsyg.2016.01792
11. Y. Katori, Network model for dynamics of perception with reservoir computing and predictive coding, in *Advances in Cognitive Neurodynamics (VI)*, eds. by J.M. Delgado-Garcia, X. Pan, R. Sanchez-Campusano, R. Wang (Springer Nature, Singapore, 2017), pp. pp. 89–95. https://doi.org/10.1007/978-981-10-8854-4_11
12. R.S. Sutton, A.G. Barto, *Reinforcement Learning: An Introduction*, 2nd edn. (MIT Press, 2018)
13. E.A. Antonelo, D. Stefan, S. Benjamin, Learning navigation attractors for mobile robots with reinforcement learning and reservoir computing, in *Proceedings of the X Brazilian Congress on Computational Intelligence (CBIC)* (Fortaleza, Brazil, 2011)
14. M. Inada, Y. Tanaka, H. Tamukoh, K. Tateno, T. Morie, Y. Katori, Prediction of sensory information and generation of motor commands for autonomous mobile robots using reservoir computing, in *Proceedings 2019 International Symposium on Nonlinear Theory and its Applications (NOLTA2019)* (2019), p. 333
15. D. Sussillo, L.F. Abbott, generating coherent patterns of activity from chaotic neural networks. Neuron **63**(4), 544–557 (2009)
16. Y. LeCun, C. Corinna, C. Burges, *MNIST Handwritten Digit Database* (Florham Park, NJ, USA, 2010)
17. Texas Instruments Inc, The TI-46 Word Speech Corpus (1990). Visit https://catalog.ldc.upenn.edu/LDC93S9 Linguistic data consortium, TI 46-Word
18. R. Lyon, A computational model of filtering, detection, and compression in the cochlea. Proc. IEEE **86**(11), 2278–2324 (1998); *ICASSP '82. IEEE International Conference on Acoustics, Speech, and Signal Processing*, vol. 7 (1982), pp. 1282–1285. https://doi.org/10.1109/ICASSP.1982.1171644
19. R.S. Sutton, Learning to predict by the methods of temporal differences. Mach. Learn. **3**(1), 9–44 (1988)
20. V. Mnih, K. Kavukcuoglu, D. Silver, A.A. Rusu, J. Veness, M.G. Bellemare, A. Graves et al., Human-level control through deep reinforcement learning. Nature **518**(7540), 529–33 (2015)
21. J. Schulman, F. Wolski, P. Dhariwal, A. Radford, O. Klimov, Proximal Policy Optimization Algorithms, arXiv [cs.LG] (2017). arXiv. http://arxiv.org/abs/1707.06347

22. V. Konda, J. Tsitsiklis, Actor-critic algorithms, in *Advances in Neural Information Processing Systems* (2000), pp. 1008–1014
23. R.S. Sutton, D. McAllester, S. Singh, Y. Mansour. Policy gradient methods for reinforcement learning with function approximation, in *Advances in Neural Information Processing Systems* (2000), pp. 1057–1063
24. M.L. Puterman, *Markov Decision Processes: Discrete Stochastic Dynamic Programming* (Wiley, New York, 1994)
25. L.P. Kaelbling, M.L. Littman, A.R. Cassandra, Planning and acting in partially observable stochastic domains. Artif. Intell. **101**(1–2), 99–134 (1998)
26. I. Szita, G. Viktor, L. András, Reinforcement learning with echo state networks, in *International Conference on Artificial Neural Networks* (Springer, Berlin, Heidelberg, 2006), pp.830–839
27. Y. Yonemura, Y. Katori, Mental simulation on reservoir computing as an efficient planning method for mobile robot navigation, in *2020 International Symposium on Nonlinear Theory and Its Applications (NOLTA2022)* (2022), pp.83–86
28. X. Xiao, B. Liu, G. Warnell, P. Stone, Motion Planning and Control for Mobile Robot Navigation Using Machine Learning: a Survey. Auton. Robot. 1–29 (2022)
29. S.E. Taylor, L.B. Pham, I.D. Rivkin, D.A. Armor, Harnessing the imagination: mental simulation, selfregulation, and coping. Am. Psychol. **53**(4), 429–439 (1998). (April)
30. J.B. Hamrick, Analogues of mental simulation and imagination in deep learning. Curr. Opin. Behav. Sci. **29**, 8–16 (2019)
31. K. Zhang, I. Ginzburg, B.L. McNaughton, T.J. Sejnowski, Interpreting neuronal population activity by reconstruction: unified framework with application to hippocampal place cells. J. Neurophysiol. **79**(2), 1017–1044 (1998)
32. F. Gao, L. Han, Implementing the Nelder-Mead simplex algorithm with adaptive parameters. Comput. Optim. Appl. **51**, 259–277 (2012)